P9-EDY-061

Designing the Networked Enterprise

The Artech House Technology Management and Professional Development Library

Bruce Elbert, *Series Editor*

Applying Total Quality Management to Systems Engineering, Joe Kasser

Designing the Networked Enterprise, Igor Hawryszkiewycz

Engineer's and Manager's Guide to Winning Proposals, Donald V. Helgeson

Evaluation of R&D Processes: Effectiveness Through Measurements, Lynn W. Ellis

Global High-Tech Marketing: An Introduction for Technical Managers and Engineers, Jules E. Kadish

Introduction to Innovation and Technology Transfer, Ian Cooke, Paul Mayes

Managing Engineers and Technical Employees: How to Attract, Motivate, and Retain Excellent People, Douglas M. Soat

Preparing and Delivering Effective Technical Presentations, David L. Adamy

Successful Marketing Startegy for High-Tech Firms, Eric Viardot

Survival in the Software Jungle, Mark Norris

The New High-Tech Manager: Six Rules for Success in Changing Times, Kenneth Durham and Bruce Kennedy

For further information on these and other Artech House titles, contact:

Artech House
685 Canton Street
Norwood, MA 02062
617-769-9750
Fax: 617-769-6334
Telex: 951-659
email: artech@artech-house.com

Artech House
Portland House, Stag Place
London SW1E 5XA England
+44 (0) 171-973-8077
Fax: +44 (0) 171-630-0166
Telex: 951-659
email: artech-uk@artech-house.com

R

Designing the Networked Enterprise

Igor Hawryszkiewycz

Artech House, Inc.
Boston • London

blication Data

Designing the networked enterprise/ Igor Hawryszkiewycz.
p. cm.
Includes bibliographical references and index.
ISBN 0-89006-920-4 (alk. paper)
1. Business enterprises—Computer networks. 2. Groupware (Computer software). 3. Teams in the workplace. 4. Reengineering (Management)
I. Title.
HD30.35.H386 1997
658'.0546—dc21 97-4214
CIP

British Library Cataloguing in Publication Data
Hawryszkiewycz, I. T. (Igor Titus)
Designing the networked enterprise
1. Work groups—Computer network resources 2. Business enterprises—Computer network resources
I. Title
658'.05'4678

ISBN 0-89006-920-4

Cover design by Marcelle Lapow Toor

685 Canton Street
Norwood, MA 02062

International Standard Book Number: 0-89006-920-4
Library of Congress Catalog Card Number: 97-4214

10 9 8 7 6 5 4 3 2 1

Contents

Preface

The combination of computers and communications has opened up many new opportunities for getting new commercial and social benefits for business, organizations, and communities. These opportunities range from better communication in the office to the ability to distribute work globally to get better expertise to work on a problem. These improved communication methods have opened up the way for networked enterprises made up of organizations, businesses, and individuals, all working together to some common goal. Such networked enterprises are increasingly going to rely both on public networks, in particular the Internet, and on their own intranets to support their work. This book describes how computer communication systems can make such collaboration across distance work, how to choose the best services for a particular need, and how to integrate them into effective networks. It particularly focuses on designs, and differs from other books in this area by describing a design method for designing cooperative systems specifically for enterprise intranets.

The book is made up of three parts. The first part (Chapters 1 to 6) concerns the way that people work in networked environments. The next part (Chapters 7 to 13) describes the wide range of technologies, sometimes known as groupware, that are available to support teams in such environments. The third part (Chapter 14 to 17) covers a design method that can be used to analyze requirements and design new systems, while Chapter 18 is a summary chapter that looks to the future.

The first part begins by describing typical systems that can benefit from using groupware. In Chapter 1, the book includes real-life examples that cover the broad spectrum of systems concentrating on networked organizations, ranging from large organizations to connected individuals. It also presents a broad view of the support provided by computer networks, ranging from information exchange, through interpersonal communication, to support for work processes. Chapter 2 covers work practices across distance, including telework and mobile work. It also defines the broad requirements of such work practices, stressing the need to provide connectivity through intranets that support seamless integration of tools to support work processes. Chapter 3 defines group dynamics and processes that are followed by groups and stresses the need to pay more and more attention to these processes. It sees maintaining awareness for people working in networked enterprises as an important issue, and describes how this can be done. The importance of process is stressed in Chapter 4, which covers processes and their definition, again in the context of examples. Chapter 5 then describes some effects on organizations by the new technologies, the trend to electronic commerce, and how processes are re-engineered to support cooperation. Chapter 6 covers the importance of interface issues.

The second part considers the technology itself. It starts with Chapter 7, which defines the idea of network services, and then continues with three core chapters (8, 9, and 10) that deal with individual support services and how they are put together into platforms useful in business. It compares the approach of using middleware or a core technology to provide such platforms, identifying the World Wide Web and LOTUS Notes as major candidates for such technologies, and shows how they can be used as the basis for constructing intranets. This section also describes the usefulness of electronic mail and the impact of the World Wide Web in electronic commerce. Chapters 11, 12, and 13 then go into the details of joint authorship, supporting meetings, workflows, and videoconferencing.

The third part describes a design method that provides a systematic way for gathering requirements and putting together systems that satisfy these requirements. System services are chosen in a way that meets both technical and social requirements within an organizational context. The book sees that the whole idea of the networked enterprise is to bring together people across distance. This bringing together must maintain the social interpersonal relationships that are necessary in any endeavor and also support these relationships across distances. The chapters describe a systematic way of approaching this goal, first by developing an understanding of what is needed and then choosing the best network services in a systematic way.

As such, the book is general, but especially useful for both the student and the professional to help them understand the issues that must be faced in the ever-growing penetration of computers into everyday life. It is also useful to the different application areas to see the way that computers can contribute to systems in these areas. There is a wide cross-section of cases to cover major industry areas such as health, construction, and manufacturing.

Readers may only wish to read parts of the book, according to their background and needs. The road map that follows illustrates alternate approaches (see Figure P.1). Thus, for example:

- A designer already familiar with the field may wish to follow path A and just concentrate on Part 3, the design issues.

- A business analyst may also do this, but follow path B and also cover the core technologies while skipping the detailed part of the technologies. However, someone who just needs a business overview need only cover Part 1, following path M.
- A computer specialist interested in cooperative network design may follow path D to get a background of the technologies and design method.
- A computer specialist with a background in groupware technologies may, on the other hand, wish to follow path C, reading the background on social and business issues in Part 1, skip Part 2, and then go on to Part 3.

The book can be used in courses specifically covering cooperative systems as well as courses on office support or group support. It can also be used in courses on electronic commerce or system design. It includes discussion questions as well as some exercises for use by students. The parts of the book used may depend on the type of course. A course on cooperative systems can cover the whole book, while electronic commerce courses can concentrate on Parts 1 and 2. Information systems design courses will find Parts 1 and 2 useful, especially where they wish to introduce the benefits of intranets to enterprises. Courses that emphasize social issues may also find the book useful in describing the impact of computer communications systems on work in enterprises.

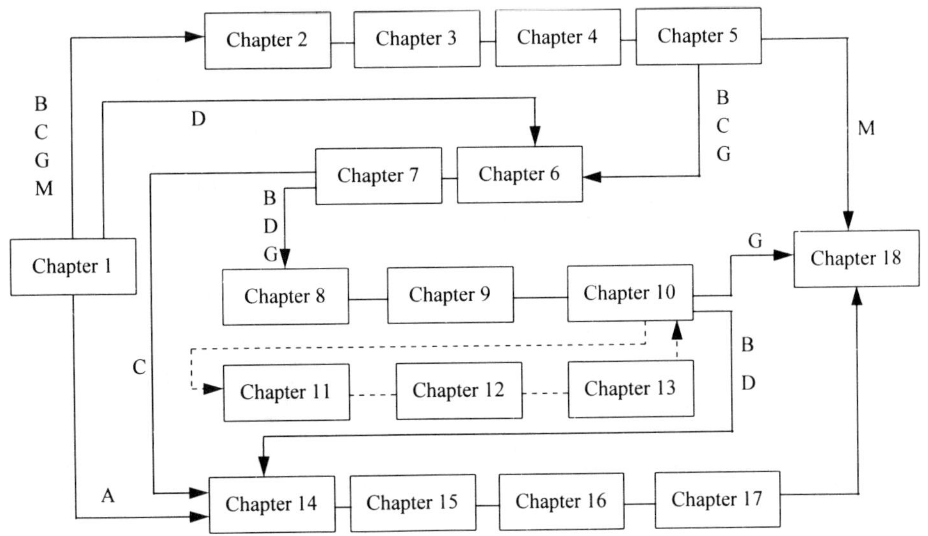

Figure P.1 Road map.

Part A

The Environment

Chapter 1

The New Environment

LEARNING OBJECTIVES

- ❑ *The opportunities from combining computers and communication*
- ❑ *Information highway and the networked enterprise*
- ❑ *The importance of standards*
- ❑ *New cultures*
- ❑ *Expanding work contexts*
- ❑ *Quality*
- ❑ *Services*
- ❑ *Design criteria for cooperating systems*
- ❑ *The importance of interface*
- ❑ *The virtual world*
- ❑ *Groupware*

1.1 INTRODUCTION

There are now many text books about the analysis and design of computer systems, including one by the author of this book. Many if not most such textbooks have stressed the more structured aspects of computer use- where the computer is simply used by someone to carry out a well-defined task or to monitor the progress of some artifact, usually a form, through a sequence of tasks. Systems that manage the progress of artifacts are often known as transaction-based systems. Such systems do things like producing an invoice for a service or placing an order for a part. Such systems are often made up of structured and repetitive processes, and many if not most of them have now been implemented as computer-based transaction systems.

However, there are now opportunities for using computers in ways other than for structured tasks or managing transactions. These opportunities have come about from many factors. Perhaps the most important is the combination of computers and communications. This combination has made it possible for any person to get information stored on any computer through communication links. Furthermore, people can now also use computers to exchange information between themselves and to communicate as they work together in their everyday work. Such communication through computers has made a significant impact on organizations and society because of the possibility of closer collaboration between people separated by time and distance. This book begins by describing the impact of the combination of computers and communications, and elaborates the kinds of work made possible by this combination in later chapters.

1.2 COMPUTERS AND COMMUNICATIONS

Over the last few years there has been a phenomenal growth of distributed computing, made possible by combining computers and communications. Whereas earlier most computing was carried out using mainframes, now there is a substantive component of work carried out in workgroups using local area networks (LANs) and people using personal computers. Figure 1.1 shows some of the statistics and changes experienced in that time. For example, there are about 40 million new personal computers installed every year; the number of mobile phones is also growing rapidly. It is conceivable that many of these mobile phones will become the remote terminals of the future. Many of these personal and mobile systems are still independent, but in time they too will become part of computer networks, connecting even more people to computing networks.

This has all been possible because of dramatic changes in computer and communication costs. As shown in Figure 1.2, both computer and communications costs have fallen dramatically. For example, it is now possible to make calls between distant countries like Canada and Australia for as little as 89 cents a minute whereas a few years ago such costs were around four or five dollars. Similarly, the costs of computer systems has dropped considerably over the last few years. Most people are no doubt aware of the fact that the power of many of today's small personal computers could previously only be obtained using large mainframe systems occupying large air-conditioned rooms. At the same time, there has been a corresponding growth in the number of end users and the amount of data available on computers. As well as increasing numbers of businesses, more and more individuals as well as organizations such as schools or senior citizen groups will be connected to networks in the future, thus dramatically increasing this end-user population. All of these users generate information that becomes available to other connected users.

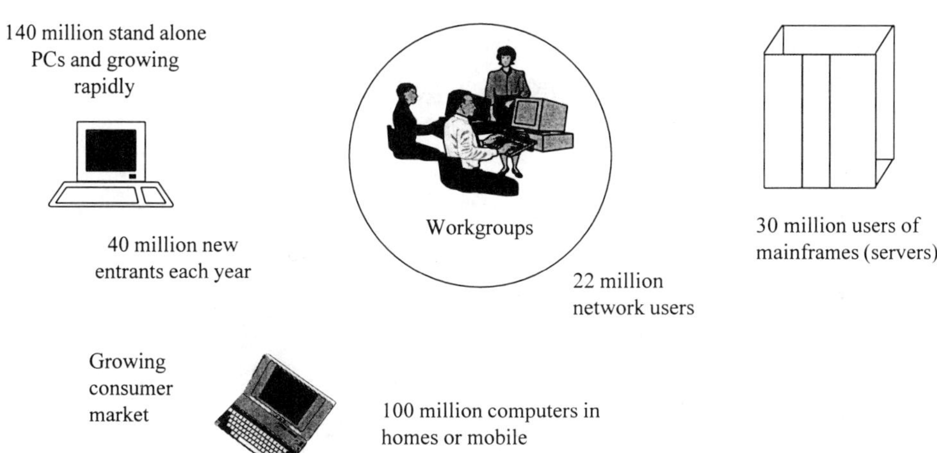

Figure 1.1 Changes in computer use.

The combination of computers and communications means that it is easy now to link people across distances as well as making much more knowledge available to these people. Before such combinations became possible, most people simply saw the computer as a tool on their desk to do things like word processing, prepare a spreadsheet, or perhaps enter data into a database. They would then prepare a report, send a letter or fax about their work, and generally discuss their activities and results with other people. The connection of computers to a network gives people new opportunities to organize their work. Now this combination means that people can directly communicate with each other, so why send mail or write letters when you can do all this conveniently using the computer on your desk. Thus it becomes possible to maintain your business and personal contacts using computers, arrange events such as conferences, or even trade. Use of computer communication technologies has raised issues such as why have meetings when you can perhaps discuss things over the network, especially so if the meeting participants are widely distributed because it is now easier to communicate over distances. So, just like a mobile telephone allows you to contact anyone on the telephone network, soon a mobile computer will allow one to contact any resource or person on the computer network. The result is that now we can extend the idea of social groups across distances to also include *electronic groups*, which can meet and discuss topics of interest to them electronically rather than physically.

The term *information highway* [1] has in the past been used to express the idea of the collection of technologies that make this communication possible. This highway is not some finite object that is opened by cutting a ribbon, but a network that evolves like some natural organism connecting an ever-growing number of

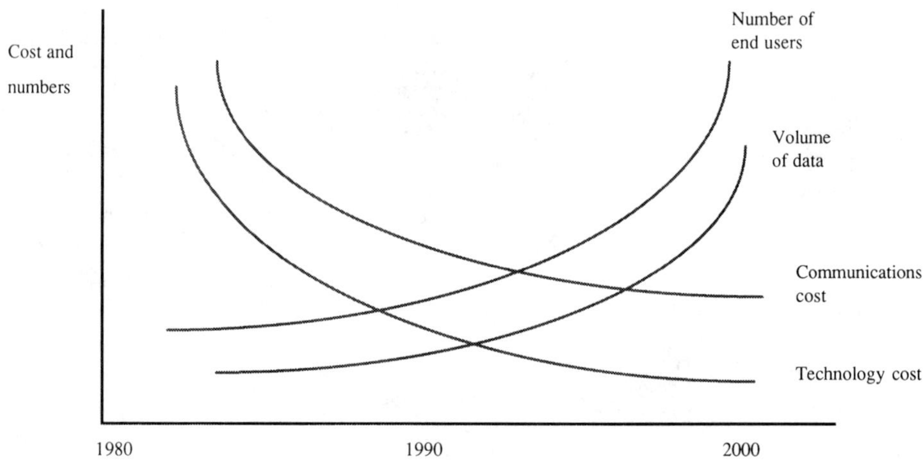

Figure 1.2 The changing patterns.

users and allowing information to be transmitted between any locations. The information can use media such as voice, video, diagrams, and other forms, all mixed together in some way that is determined by the user. Thus you can send any information to anywhere using the highway facilities. At the same time, it now becomes easier to become aware of what is happening in one's field of work by using the network.

What is more important about the idea of the information highway is shown in Figure 1.3. Here people in institutions such as factories and businesses can all be interconnected, making it possible to work together. Such connected institutions can be provided with software to trade over a network, place orders, or manage a joint venture. A person in their home can also participate in such arrangements and in fact work from their home. The term *connected house* is now also being increasingly used to imply the ability of all services to be made available in the home. Thus in addition to the home office, entertainment can be provided through easy access to a wide variety of video services. Rapid growth of such services is expected once costs for obtaining a video replay directly into a home drop to about $3 or so, making it unnecessary to use video stores. Another possibility is to deliver health services through connection to health delivery systems for patients to get advice or monitor their progress.

1.2.1 The Networked Enterprise

The growth of networking within and between organizations is gradually leading to what may be called the networked enterprise. The networked enterprise is a generic term that means the interconnection of a variety of entities to construct yet a

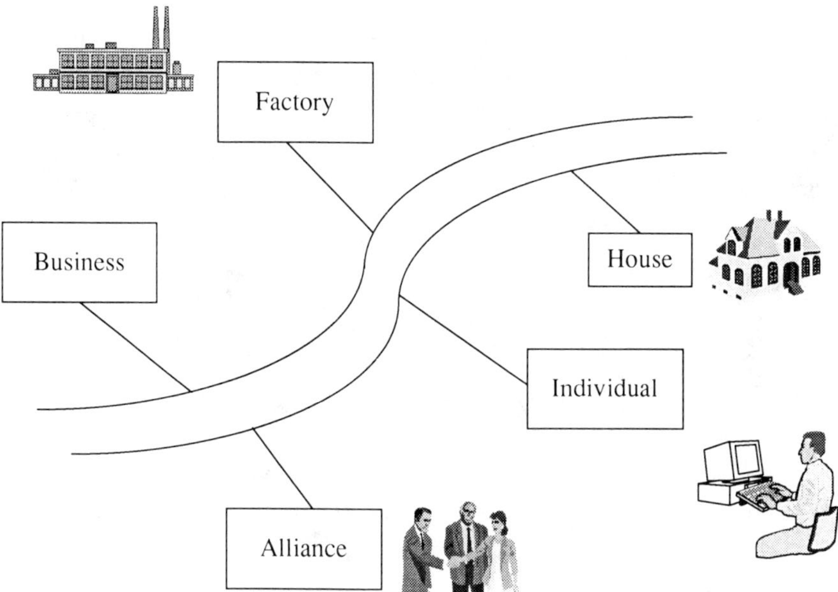

Figure 1.3 The information highway.

larger entity conducting some form of business. There is no such thing as a standard networked enterprise. It can be an entity that only includes people in a single organization. It may be a number of independent individuals cooperating in some venture. Or, what is more likely, it will be a combination of all of these. In the enterprise shown in Figure 1.4 there are three networks, each catering for a different organization and connecting people in that organization. Then, there are also links between the organizations through different individuals in each organization. The network also has connections with clients of each organization and with external consultants. Such networked enterprises are now increasingly supported by a computer communication network.

The networked enterprise must support a large number of different ways of communication. This book makes a distinction between three types of communication, although obviously gray areas exist between these three types. They are the following:

- *Information exchange* where information in the form of documents is exchanged between people in relatively short exchanges;
- *Interpersonal relationships* where specific individuals, through several related lengthier exchanges, discuss, and perhaps decide on, actions to be taken (for

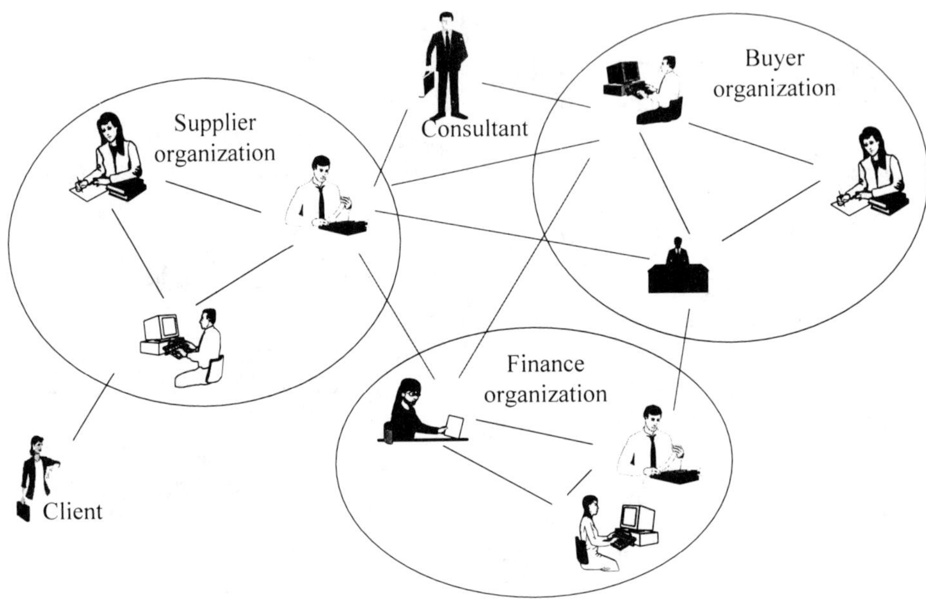

Figure 1.4 The networked enterprise.

example, in help desk situations or where wide expert advice must be sought to make a decision);

- *Work processes* that define the flow of work through the enterprise.

Each of these classes of communication often require different kinds of computer support.

1.2.2 Computer Communications for the Networked Enterprise

There is a large variety of network types that can be chosen for the networked enterprise. The choice in itself presents a challenge to provide a network support systems most suitable for the enterprise. The kinds of systems now commonly used are

- The Internet and other public information networks (for example, CompuServe) that provide the ability to interchange information between organizations and individuals. Perhaps the most well-known network to support such flexibility for networked enterprises is the Internet. The Internet connects any number of computers or other networks into the one

network, supporting cooperation across organizations or between many enterprises.
- Electronic data interchange (EDI) has now been used for many years to exchange transaction data between organizations. However, EDI only applies to predefined arrangements; that is, where organizations make arrangements for specific connections that will be stable over a period of time.
- Intranets, which are increasingly used to provide the same services within the organization as the Internet provides between organizations.

All of these networks are based on services, often known as *value-added services,* that are added on top of the communications hardware and software to provide support for work practices used in enterprises. Networks can then be seen as combinations of services chosen to meet the goals of the networked enterprise.

1.2.2.1 Services in the Networked Enterprise

An ever-growing number of such value-added computer communication services are becoming available to build computer networks. Each kind of service supports a different activity and designers of networked enterprises must choose the appropriate services for the type of work process to be supported in the enterprise. In general, people need to use a variety of services and it is common in design to classify services into the different communication methods. The term service can be used in two ways here. One way covers the network services provided by the network vendor or any value-added services, and the other covers user services provided for use by the users. Users often use the network services directly, but often network services are *customized* to meet special user needs. This book classifies network services into the following:

- Document services, which provide access and the ability to work on documents by groups;
- Meeting services that enable people to discuss various issues;
- Workflows to support the organized flow of documents;
- Messaging that allows people to exchange information.

These kinds of network services have existed for some time, but what is important now is that they are being integrated into people's everyday work. Thus databases have existed for many years to store data for organizations. However, now databases are being combined with meeting services and workflows to give people easy access to the data and also to be able to discuss and change it while working across distance. Figure 1.5 illustrates the way that such services can be combined. Here, there is a workflow service that support the flow of forms (in electronic form) needed to obtain a part. Messaging services are provided for informal

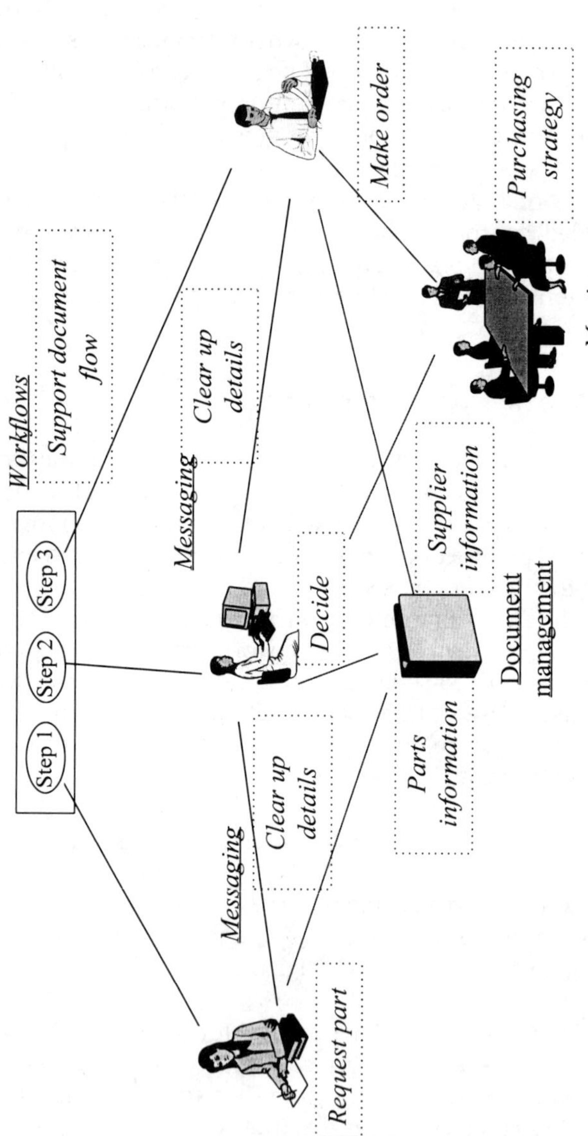

Figure 1.5 Combining collaborative services.

exchange such as, for example, clearing up some detail about a part or its availability. Document services are needed to store information about suppliers and parts. Meeting services are needed to discuss alternative strategies for purchasing parts. This is typical of many of the applications in the networked enterprise, with different services combined to provide new ways of working using the services to support networked enterprises.

These services are described in detail in later chapters. They include the following:

- *Messaging Services for the following:*
  - Sending mail messages;
  - Sending files or documents between individuals with annotations and comments;
  - Distributing news messages, tips, important events, or policy updates.

- *Meeting Services including:*
  - Videoconferencing, where two or more people at different locations can talk to each other while seeing each other on their computer screens;
  - Brainstorming, where ideas are quickly generated by a group of people, and the computer system collects the ideas and the responses to them;
  - Calendaring systems that support making group appointments;
  - Discussion databases, where groups can make contributions to particular topics, with the discussion sorted by statements and responses to them;
  - Forums, for exchanging ideas about particular issues; here an issue can be raised by one person and other people comment on this issue.

- *Document Services for the following:*
  - Storage and management of multimedia documents;
  - Joint editing, where two or more people at different locations can work on the same document and see each others actions;
  - Keeping track of document changes.

- *Workflow Services for the following:*
  - Document flow system to support documents and forms flowing through a set of predefined steps in the organization;
  - To-do lists, which present group members with lists of things that they must do, sorted in some priority order.

The Internet, which is described in Chapter 9, is itself made up of a variety of such services. Intranets are often made up of similar services, although it can be fairly said that the Internet is primarily concerned with information exchange, whereas intranets must place more emphasis on interpersonal relationships and business processes.

1.2.2.2 The Importance of Standards

Combining services into networks requires standards to enable different organizations to easily connect their computers and interfaces that make these computers easy to use, and to connect different services to support work processes. Thus the Internet and intranets are collections of such services-the difference is that the Internet combines services needed for the more open exchange between organizations, whereas intranets combine services needed to support work processes with an organization.

The most important network standard for connecting computer systems is known as TCP/IP. Any computer user that purchases software that adheres to this standard can connect their computer to the Internet. The Internet provides a number of services that include sending electronic mail messages, transmission of files, newsgroups, as well as various browsers that can be used to search for information on the net. The addition of the World Wide Web as an Internet service has further extended the Internet. The WWW makes the Internet easy to use by enabling multimedia information to be stored on the Internet and providing special ways of browsing through this information. It is becoming the standard service for information exchange.

Standards for connecting other value-added services are not as universal. Later, in Chapter 7, we will describe two approaches for combining such services-using *middleware* to bring together a variety of services or using a *core technology* that provides some selected services but has interfaces to others.

1.2.2.3 Designing to Support Work Processes

It is also interesting that services are not only used by individuals, but are now very often combined to support specific work practices. In fact, design closely parallels the idea shown in Figure 1.6. The basis of any design is to identify the work practices to be supported and the relationships between people needed to support the practices. Then, the tools and services to support the relationships are chosen. Furthermore, most work situations have a variety of communication needs, which in turn need a variety of services. These services must be integrated into a single platform so that any user can easily move from one service to the next.

1.3 THE MAJOR IMPACTS

The fact that people can communicate and access more and more information will have major impacts on work practices and the entire work environment. Global networks such as the Internet open up new frontiers where people can see themselves interconnected with many individuals or organizations, with

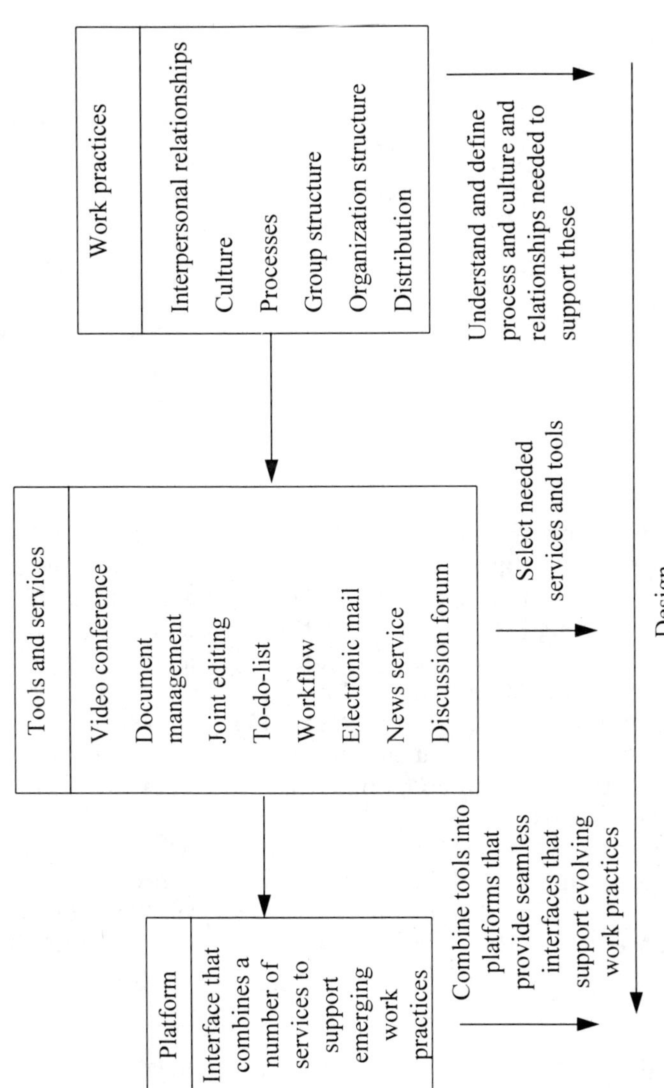

Figure 1.6 The fundamental design approach.

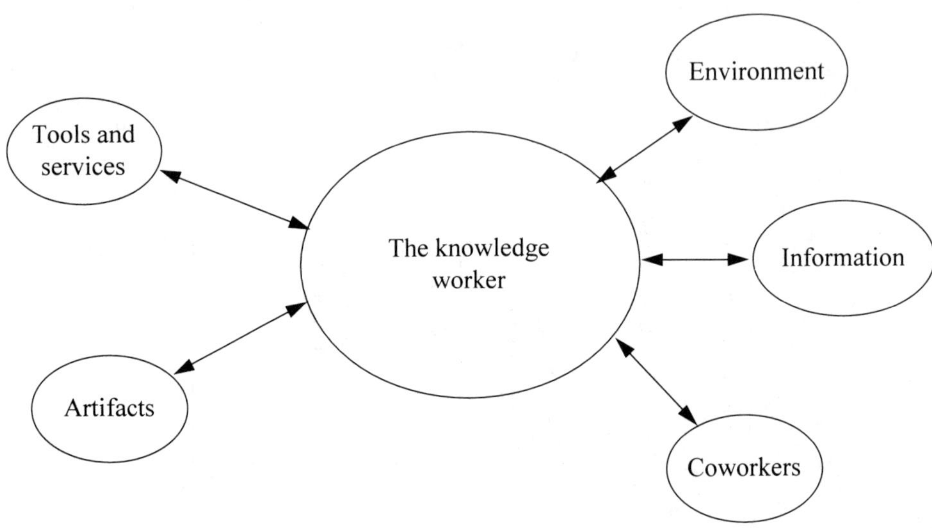

Figure 1.7 The widening context.

access to almost unlimited information. If you have a look at the Internet, for example, the opportunity for sharing information is almost boundless. You can now get tourist information, sport scores, contact your friends, or distribute research reports. Although the Internet has been predominantly used more for personal communication, its ideas can be easily extended to most organizations and businesses. Why should an individual not have easy access to what is going on in their organization. It will only make individuals into more effective workers because of better knowledge about the organization. Often the first steps here are to distribute news items about important developments, policy changes, or personal achievements.

Sometimes the word *context* is used to describe everything that one sees or is aware of when taking an action. People's context will now become broader. Just as television has let us see what is going on in the world, we will be able to use communication services to see what is going on in our work environment. A person's perspective will now change to that shown in Figure 1.7. In this wider context, they will

- See artifacts, such as reports, with which they are directly concerned;
- Be easily able to access other people who may also be interested in these documents;
- Know what other people in the organization are doing;
- Access information (or knowledge) from many sources and needed services to accomplish their task.

Again it only remains to ensure that people can make good use of this broader context. Simply providing a group with technology will not mean that advantages will automatically be gained. Organizations will thus have to make sure that they go beyond simply connecting their people and also provide them with effective tools that fit into their ways of working. Important goals here are to provide:

- Interfaces that allow people access to the context;
- Search directories to browse through large volumes of information;
- Services that enable people to maintain good relationships.

The problem of excessive browsing has gained considerable attention. Excessive browsing leads to the possibility of forever looking through information but never putting it to good use. There have, for example, been many reports about the difficulty of finding relevant information on networks such as the Internet. What is needed are better tools and services to quickly search through this information. Such tools are now rapidly appearing. The World Wide Web is one example of a general tool for linking information to simplify searches.

The *broadening of context* and ability to communicate easily will have many impacts on the workplace and the individual. It is often said that it will impact on the work *culture* itself. The word *culture* is sometimes used to specify accepted ways of doing things in an environment or organization. Culture in this sense has nothing to do with national boundaries or the appreciation of art. It is the culture of the workplace-what is accepted as good and normal and what should not be done. Culture includes such things as innovation, or even more precisely how, if at all, innovation happens in an organization. It includes the way decisions are made, by individuals or committees. It includes the way we work together—it's the culture to work more as individuals or to work as teams. Impacts, however, are often a two-way process. The new structures of work create new opportunities; these in turn lead to new requirements on the system, and so on.

1.3.1 Quality

In this new environment, with easier communication, there is the expectation that organizations must become adaptable, more responsive and flexible, and capable of making decisions quicker [2]. They are also now expected to turn out their products quicker because of better communication methods. This in turn leads to new structures to meet new demands while preserving *quality*. Quality in itself is a term that is hard to define. It means doing things well, such as delivering products on time, producing products that do not fail, and so on. Often it is a term that is defined within organizations and communities. Often, though, it is not actually defined but we know what it is. However, the trend now is to define quality explicitly and to work to achieve it. But what is more important is that it is necessary to preserve quality while improving product delivery times.

1.3.2 Providing Service

The idea of services identifies another new and important cultural shift. People in an organization now often have a specific goal or mission and require services to achieve it. Such services may be an accounting service, an analysis service, or a cleaning service. Any knowledge or expertise possessed by a person can be a service provided by that person. A computer department (or perhaps its external equivalent) is thus seen as providing a quality service to the organization rather than dictating how the organization should work. The gradual trend to a service culture can mean that organizations no longer have workers, but simply individuals who provide a service to an organization.

1.3.3 Continuous Innovation

Another important goal is to be better and better to stay ahead of the competition. This calls for rapid improvement of services. Most organizations can no longer afford to carry out research independently of production, then test new products resulting from the research, then install these products. This process just takes too long before a new product or service emerges. Research and production must now all happen at the same time-there must be continuous innovation. Again it is necessary to look at new forms of research within the organization to facilitate new products and question existing approaches. The existing concept of a laboratory-based research leading to some result, which is then transferred to industry is thus losing its attraction. This occurs for a number of reasons. One is the time taken to get an idea into a product. First of all, there is the research that takes time, then the transfer that takes more time, and then the training to use the system. Another reason is that laboratory-based research will not always address business problems. Hence more time is lost adapting the results to industry. The idea of continuous innovation overcomes this-the research now takes place within the work context. Research must now be done in industry and research goals must be continually redefined.

1.3.4 Concurrent Processes

Many businesses are also now looking at ways of expediting their work by doing things concurrently. In this case, the different tasks tend to overlap. For example, why should the diagramming not be happening for Chapter 1 while Chapter 2 is being edited. But what if the work is to overlap? But at the same time, it is important to ensure that there is not too much work discarded because of mismatch between the chapters. It is also important in manufacturing, where agility in rearranging production is needed to quickly adapt to changes in demand.

In the wider context, the term concurrent engineering is now becoming common. This has come to mean that tasks, which have previously been carried out in sequence, are now carried out at the same time. The issues addressed here are how to help people working concurrently on strongly related tasks coordinate their activities in a way that reduces wasted effort while improving delivery times.

1.3.5 Intellectual Property Rights, Copyright, and Security

The value of information is now receiving considerable attention. Information on a publicly shared network is usually intended for the general public (for example, marketing information whose goal is to attract clients to the provider). But there is also information that may have taken considerable time and expense to produce. A computer program may have required considerable effort to produce, and the producer wants to recover these costs. This means that the producer wishes to retain the copyright to the program and restrict others from making multiple copies and reselling the same product. However, complete access would mean that anyone connected to the network can get access to this information at minimal cost, copy it, and sell it. There is also information that is confidential to a small group of people (for example, financial information). The network must provide adequate security to prevent unauthorized access to the data.

Clearly, if networked enterprises are to grow, then information on the network will have to be protected and some ways set up to trade this information. Furthermore, there must be some assurance that sold products, such as a market survey, are then not resold or simply passed on by the buyer, but the right to sell remains with the seller. This right is sometimes known as intellectual property, and its protection is both a legal matter as well as one that must be technically enforced. It is a problem that is assuming vast proportions, especially where networks cross national boundaries with different legal systems.

1.4 PEOPLE WORKING TOGETHER

The changing environment is leading to new work practices that support greater distribution of work while at the same time giving access to better expertise. New work processes must evolve to coordinate such dispersed groups while at the same time making it possible for more and more people to work at their chosen location. Why, for example, should organizations provide offices when its workers can work with others through computers in their home? There is also wide agreement that such new ways of work will lead to new social structures. An important question is how the new technologies will effect individual work. Issues that are important here include the following:

- The way that individuals will work. The general consensus here is that people will be more likely to work in teams rather than as individuals. The computer can let us concentrate on what we do best. We can be part of a

network and simply share our knowledge with others and together with them produce services that satisfy the market.
- The locations at which people will work. Again the consensus is that it will be possible to work together at different locations, or even from home.
- The volatility of work. There are prognoses here that see stable jobs disappearing and people moving from one place to another in rapid succession. Here, people will sell their services via networks to form alliances that may only last a short time. This has been happening with consultants for a long time, but now more and more people may work in this way, with perhaps a reduced need to travel as more and more contacts are maintained using computer networks.
- On the negative side there are concerns that this way of working may lead to overspecialization and isolation. People may no longer feel part of the organization. Thus any new structures here must consider both technical and social requirements, and reach a balance between social satisfaction and technical efficiencies.

There are advantages to working together in teams when compared to working as individuals. It leads to the possibility of resolving many issues through discussion rather than using criteria imposed external to the business process. Discussion can also take place within wider contexts, leading to more informed decisions. For example, consider a situation where two order clerks receive requests for the same item, say one for 40 parts and another for 60 parts, but there are only 30 parts in the store. The computer solution is one on a first come, first served basis, which may not always be the best from a business point of view. Thus one customer will probably be served but another not. The latter could, of course, be the firm's most important customer. So why not instead have the two order clerks discuss the matter first, decide priorities and perhaps in discussion with customers satisfy both, but in a progressive manner. Thus perhaps each customer will get 15 parts next week and the rest later. In that case, there are still two happy customers. The problem here is how to organize teams to get such beneficial outcomes. This book will address team structures in more detail in Chapter 3.

The remainder of this chapter describes some of the new structures that are arising as a result of networking. It covers large organizations, small businesses, and individuals.

1.4.1 Representing Networked Enterprises

Before we go on describing such systems, we touch on one important issue-how to represent networked enterprises. Such enterprises are complex in their nature, involving people, artifacts, computers, and so on. Furthermore, they are not structured to some formal rules as found, for example, in systems such as transaction systems. And yet some unambiguous ways must be found to describe them. In this book we develop some such ways, but gradually. In this chapter, we have introduced the idea of service as a central element of such systems. But it is also important to introduce concepts such as people, the roles that they take, the artifacts that they use, and how they interact. We use the idea of rich pictures to do this initially. These ideas are expressed in detail later in the book, but are introduced in this chapter in the following examples.

1.4.2 Working Together in Large Organizations

Working together can extend to individuals groups, organizations, or governments. With large organizations there is the need to connect many people together and at the same time provide them with the necessary services to keep them aware of developments in the organization and to have them contribute to these developments.

1.4.2.1 The German Government

Perhaps one of the major examples is that of the German Government. Following unification of Germany in 1990, a decision was made to move the capital from Bonn to Berlin, with the resultant need to move the majority of government offices to Berlin. This is not a minor task and is expected to take a number of years. Because departments will be moved individually, for a considerable period of this time government administration will be split between the two cities. However, these departments must still work together. Computer and communication technology will play a major role in bringing the areas of administration together and a number of significant projects are currently in progress to make such joint work possible. Such projects will range across large activities such as the distribution of information throughout distributed departments or support for small groups working on report production, planning, or design work.

Apart from providing the technical systems, designers must ensure that these systems provide a united government face and also maintain the procedures and processes that have been part of the national culture. The way that technology is adapted to these processes has been described in the 1995 Conference on Computer Supported Collaborative work by a team from the German National Research Center [3]. They have described the importance of carefully analyzing existing practices and gradually adapting technology to these practices. Analysis required implementation of the idea of circulation folders that carry a history of actions on a particular task. At the same time, the system is being supplemented by notification schemes to keep people aware of document progress.

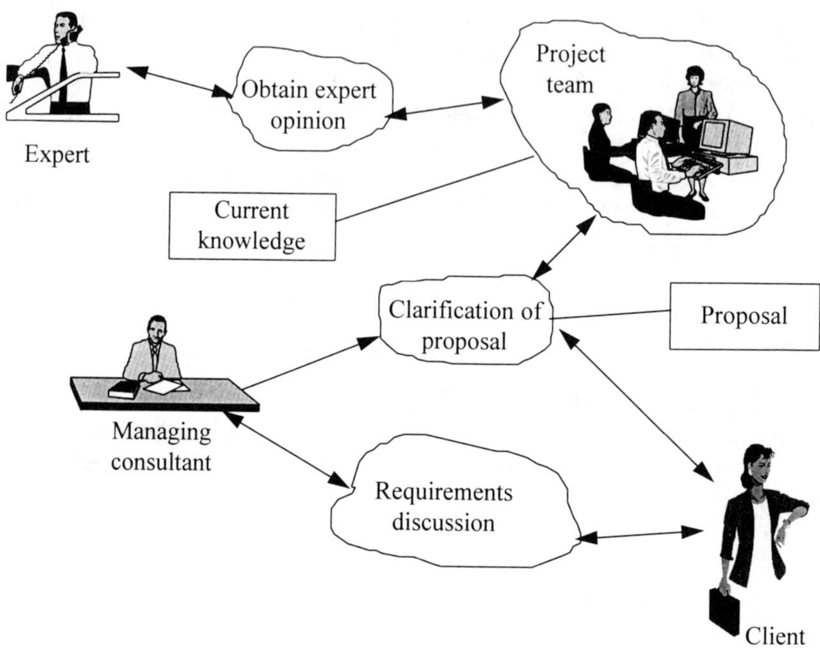

Figure 1.8 Preparing proposals.

1.4.2.2 Networks for Large Organizations

Just as is the case with the German Government, many large distributed organizations are now connecting many of their computers into networks. Organizations in this group include businesses or groups loosely connected with a common goal. This presents a major challenge on how to use such networks effectively. Often, the first step is to distribute notices or important events. One of the first things that people connected in this way also do is prepare reports.

Case Study: Worldlink Consultants

Worldlink consultants are required to provide advice and support that requires a wide range of expertise while at the same calling for quick response to client requests. Such responses must include technical data, costing, and pricing information as well as legal requirements. To do this they need to contact their experts, who may be at various locations throughout the world, to get quick inputs, and use these to prepare responses to clients. At the same time, they are looking at better ways of providing their widely distributed consultants and other employees with easy access to their company's current position and knowledge on current topics. Figure 1.8 illustrates the people involved in this process, the artifacts used (as rectangles), and the kind of interpersonal interactions that take place. It shows that a managing consultant interacts with a client to obtain client requirements and develop a proposal. A project team is usually charged with responsibility for the proposal, and this team seeks advice from experts as well as records of information on various topics held within the organization. Members of this team may themselves often be at different (and changing) locations, and the managing consultant must coordinate their activities.

Proposals are divided into a number of parts, with the preparation of each part assigned to one member of the project team. Drafts of the parts are often sent to people for comment. Once comments are obtained, the managing consultant must check the consistency of the parts, request further changes if needed, and assemble them for the client. Problems arise where there are many copies of the same document with different comments, or with delays in getting the comments. This often makes it difficult for everyone to keep track of changes and ensure that a consistent document is produced in the required time.

Worldlink consultants are looking at ways of shortening proposal preparation time while still allowing all experts to contribute their expertise to the proposals and improve proposal quality. Any supporting system must let each of the team members have considerable freedom in developing their part of the proposal, but in the end ensure (often through personal interaction) that all parts are consistent with each other. Ways of doing this using the services of a newly installed computer network are being examined. These services must improve information exchange while supporting the interpersonal relationships needed to resolve the more difficult issues. Services are needed to enable work to be subdivided so each team member can work on their proposal parts while still keeping each person aware of what the others are doing. Simply distributing proposals electronically can result in the same problem as occurred in the paper form case-many electronic copies with different comments. Another approach is to store the proposal centrally, but make it accessible to all team members so that they can electronically annotate the one copy of the document. Still another solution is to look for a document management system that keeps track of versions of documents as they evolve. However, the system must also support some way of reaching agreement where people propose different alternatives. Videoconferencing with the latest document versions displayed to all members is being considered for this purpose.

The previous example has described how collaborative technologies can be used in a relatively well-defined task of producing a document. Such technologies can also be used to support more unstructured work such as strategic planning. A survey by Stephanie Teufel [5] (see Sauter, et. al, 1995) of a large number of large organizations in Switzerland has shown that most managers now use the Internet.

Some interesting outcomes were that a majority of managers already use electronic mail and also consider workflows in planning important. Planning can extend beyond individual organizations and involve large groups, as illustrated in the following case.

Case Study: Global Planning Support

Computer communication is making it increasingly possible for people to communicate and share knowledge across international boundaries. Thus, as an example, the goal of the WHO (World Health Organization) is to improve health throughout the world by progressively reaching agreed upon targets for the various world regions. Research is often required to reach these targets. To do this, the Advisory Committee for Health Research (ACHR) is looking at ways to provide member nations with a facility to develop collective views on research needs without the need for excessive face to face meetings. Such a facility must support both the gathering of views from these regions as well as agreement on priorities for addressing different issues and on how to define better health. At the FAW, a research institute in applied knowledge processing at the University of Ulm in Germany, Work is in progress to develop, test, and implement such a planning network for health research, now known as Planet HERES [4]. The broad goal here, as illustrated in Figure 1.9, is to integrate the work of policy makers, research groups, and project teams engaged in improving the health status of member nations.

Figure 1.9 shows a loop where on the one hand research must address and solve health problems identified in the field. The loop must then be closed by

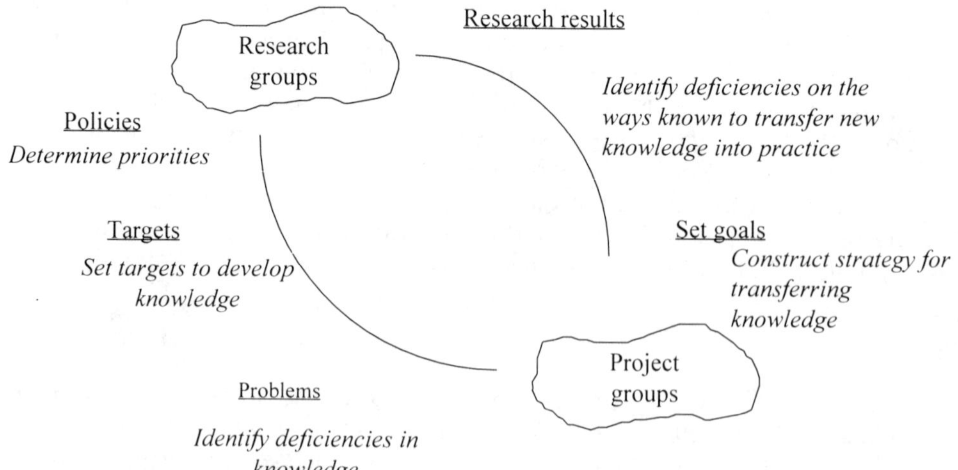

Figure 1.9 Global planning.

transferring research results into practice. Both of these not only require the integration of the work of the two groups, but also setting goals for each group. The goal is to facilitate information exchange so that researchers or policy makers are made aware of the health status of world regions and identify knowledge deficits needed to improve the health status. On the other side is knowledge needed on how to transfer research results into the field. This can be quite difficult given the spread of researchers, policy makers and related experts, and the project teams.

One approach being considered is to provide retrieval facilities through networks such as the Internet for researchers to exchange their results. It is also proposed to use these networks to also support interpersonal relationships by providing communication services such as discussion databases or bulletin boards for defining targets to be reached and setting priorities for these targets. These services will enable people separated by distance to

- Make anytime contributions to a specific target;
- Make comments on other people;
- Participate in decision making;
- Be kept actively informed of significant new developments.

Apart from the ability to communicate in structured ways, another goal will be to arrange contacts between parties that can work together, and secondly to provide facilities through which they make arrangements to set up relationships to work towards a target. Discussion forums are being widely considered here to allow people to activate issues and seek comments from others.

1.4.3 Business Networks

The previous case has described the use of technology to bring together people from various countries to work towards a common goal. Such strategic alliances are now beginning to cross national boundaries. One reason for doing this is to allow organizations to grow by increasing their marketplace. For example, a telephone company may have found it difficult to grow within their national boundaries as more and more people obtain phone services. There comes a limit to growth where everyone has a service. The communication company must then look at products that cross national boundaries, mostly concentrating on supporting communication between businesses forming global alliances. But not all alliances need concern only large organizations. Another important use of technology is to support small businesses to form alliances. There can be many goals of such alliances, some of which are described later in Chapter 5. Most, however, concern expanding both product and market range.

1.4.3.1 Case Study: Business Network Processes

Many countries have recognized the potential of small businesses as innovative organizations, but have seen that their potential can often not be fully realized because of their small size. As a consequence, full national potential is not being realized by businesses because of limitations imposed by their size. Consequently, they have looked at the possibility of encouraging alliances between small businesses to enable them to extend both their product range and marketing strength. A number of countries and governments now actively and financially support formation of such networks, including Norway, New Zealand, Denmark, and Australia as well as Quebec and Oregon state.

Figure 1.10 shows the kinds of communication processes found in a typical networking program. The major interactions between the different roles are enclosed here in clouded shapes, with links to the roles in the interaction. One of the critical roles here is that of the broker. Brokers are often specially accredited people who help businesses to identify opportunities for networking. Such brokers can identify existing small businesses that can combine into networks to take up these opportunities. They also play an extremely important role in the maintaining of trust and confidentiality between partners. Often, potential networkers do not want to expose themselves publicly as seeking network support. Instead, they express their need to the broker, who then searches for partners. The broker initially identifies the partners anonymously and then helps them set up the necessary business arrangements; the network then continues to operate without further broker assistance.

Networking can impose considerable overheads on individual businesses in the time required for their senior staff to ensure coordination both in setting up the network and its later operation. The coordination leads to considerable communication between people, who are often at different locations and busy running their own businesses, calling for a significant investment of their time. This especially impacts with brokers who must find matching partners, assist in developing proposals, and maintain contact with program management. Computer communication services are being proposed to facilitate communication and reduce the investment of time needed to set up and operate networks. Improved communication services will facilitate network formation by both shortening the time to find business partners and by minimizing the time needed by business partners to operate the network. Leads, tips, and opportunities are seen as one of the first steps in providing such support. Suggestions for such networks have included bulletin boards, World Wide Web sites, as well as just simple electronic mail to broadcast information. One important issue, however, has been raised-how to preserve confidentiality where information only concerns a few businesses or brokers. We will follow up some of these suggestions in later sections.

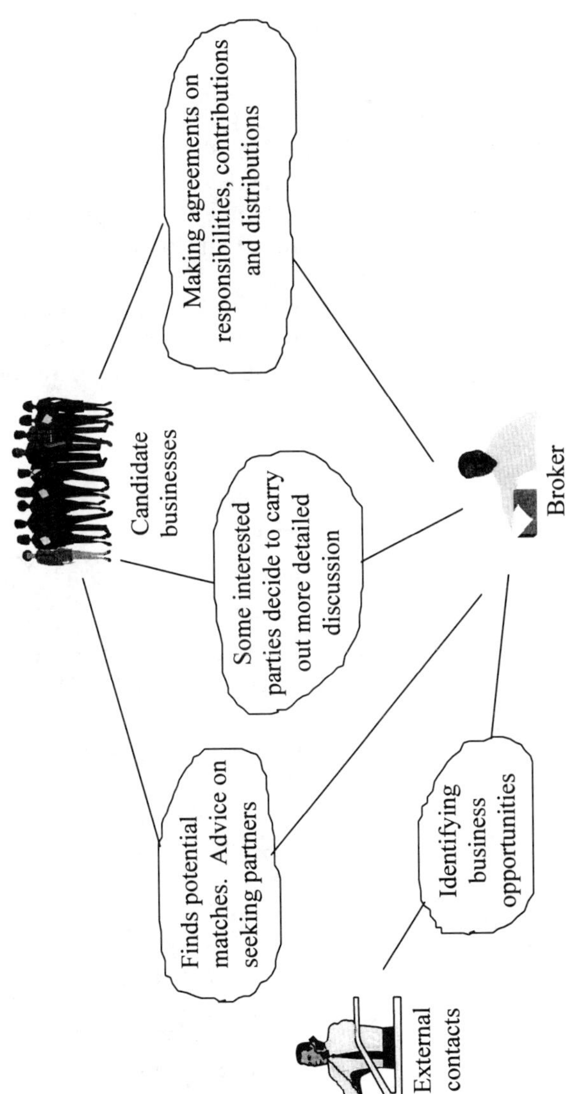

Figure 1.10 Business network formation.

1.4.4 Small Traders

There is also the increasing trend to exchange services over the network. A trader can advertise her service on a network or, alternatively, people may advertise their requirements on the same network.

1.4.4.1 Case Study: Specialist Designer

A small firm, made up of one individual designer with an assistant, produces interior designs for its clients as shown in Figure 1.11. To remain competitive they need to produce innovative and customized designs in minimum time. The process usually begins with a preproduction stage, where the designer obtains a contract through negotiation with a client, and then follows this up with the production stage by producing a design, leaving some of the detailed drawings to the assistant. Design production often requires frequent, but short, discussions with the client on various design aspects such as, for example, cost of alternative furniture items or furniture layouts. Often, because of the highly descriptive nature

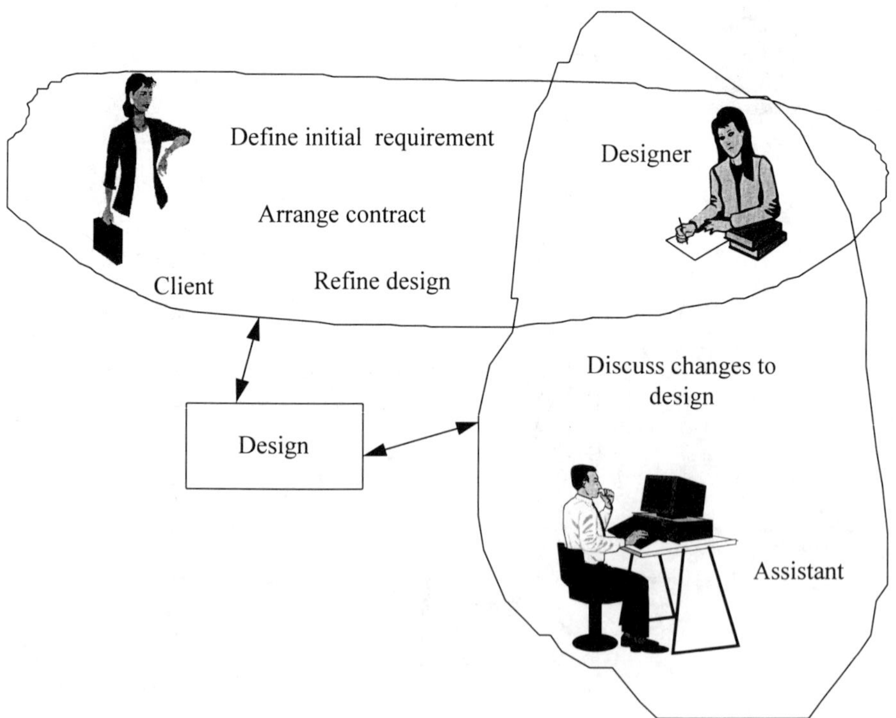

Figure 1.11 The small trader environment.

of the products, such discussion cannot be made by a phone as it must be accompanied by design layouts and pictures of the furniture items. Discussion thus often requires face to face contact, requiring travel by the designer. Where a designer has a large number of clients, such discussion can lead to an inordinate amount of traveling time for meetings with the client, putting pressure on the amount of work that can be done in design.

There is then a bind. Travel restricts the trader to the number of jobs that can be processed at the same time. Hiring additional assistants can solve this problem, but now most of the work will be carried out by assistants-either requiring additional supervisory time by the designer or leaving assistants to make important decisions, thus losing the designer's expertise. The designer is thus looking for an alternative approach to reduce travel time while still maintaining contacts.

The designer must then examine the possibilities offered by communication systems. One of course, is to use faxes to transmit diagrams and then discuss them over the phone-but this is sometimes difficult given the complexity of some figures and the fact that detailed explanations needed about them are difficult over the phone. There is then some advantage in using computer communication services for this purpose. Facilities such as videoconferencing with simultaneous display of diagrams at either end are exciting, but also expensive. The question then becomes whether and how to use these facilities in the most effective way. Using a videoconferencing facility with joint editing facilities looks ideal here, but it is expensive. The question is whether the expense is worth it.

To answer this question needs a better definition of how design, and discussion about it, takes place. How often and what do we need to discuss face to face, and what can be done through the exchange of notes? These issues are now being considered by the trader as ways of improving the way business is done.

1.5 THE IMPACT ON PROFESSIONS

A report in the October issue of the *Financial Times* described trends in the kinds of skills needed in the computer profession. It described a drop in the need for analysts of about 54% from 1988 to 1994. At the same time, the number of analysts went up about 17%. What, however, is significant is the substantive growth of user support staff, exceeding 200%. Much of this latter growth was attributed to the growth of networking, requiring more and more people to both set up the networks and assist users in choosing the best software services for their needs. The growth in the use of computer communication will continue to have further impacts on the professions, which will have to address issues that arise from the ever-growing network complexity. Most issues will center around the fact that the users of these systems do not want to spend their time learning the technologies, but instead may prefer to use the services of a facilitator to construct an environment for them to suit their work. Such facilitators will assist noncomputer users to set up services and provide advice on how to adapt the services to emerging processes.

Reducing information overload is another important issue. The amount of information available through networks can lead to a person spending the major portion of his time sifting through this information and not carrying out his task. Some means will have to be provided to manage such volumes of information. This task will most probably be associated with what up till now has been the library profession, but is now assuming a wider role in information management. One can also foresee a greater role for information scientists, who will advise not only on the retrieval of information, as is often the case with librarians, but also on information organization and use. These professionals must understand the services needed by new networks and provide these services in the most effective way. Thus the library profession will become more concerned with provision of information services other than text and work closer with users to set their context. Perhaps the idea of this impact is illustrated in Figure 1.12. This illustrates the effect in process terms. Currently, information begins with the author, who constructs an artifact, which is published by a publisher. The artifact can then be distributed by a distributor. The librarian then purchases the artifact for the library and makes it available to a user.

The network can substantially change this process, where many more channels will be provided between the author and the user. The claim is that the role of the librarian is likely to become one of providing the necessary interfaces and training rather than in the distribution process.

At the same time, some professions whose creative skills are considered as paramount may begin to disappear as their processes begin to be better understood and slowly automated. Such automation has been a historical paradigm. Gradually, many professions who were judged to be artisans found their processes better understood and eventually automated, with their professional

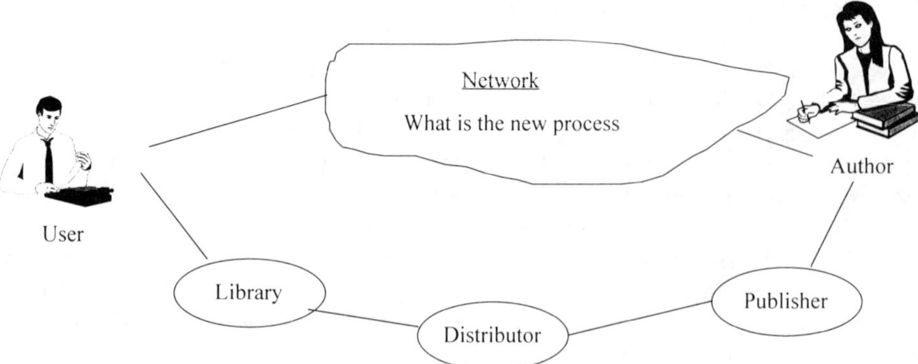

Figure 1.12 Changing processes.

services no longer needed. For example, if one looks at candidate professions for such automation, one cannot overlook the profession of teaching, especially at the higher levels. As the teaching process is proceduralized (as is happening in many institutions), and more structured ways and tools for teaching develop and are provided through networks, there will be less and less need for actual face to face contact in teaching.

1.6 WHAT ARE THE NEW PARADIGMS?

The new network environment requires organizations to develop different ways of working, and in effect produces a new view of the world. But people, as a rule, do not like rapid changes in their work environment. They still want to think in familiar ways. One approach to introducing new ways of working is to present people with work in familiar surroundings, but now depicted on a computer screen. Thus they still see a familiar paradigm (be it a document), but are in fact working differently.

There are many new paradigms now coming into this area. This will allow people to view the situation in some natural way so that they can work better. Furthermore, it is no longer necessary for all working forms to have a physical realization. Thus we can do the same things as we do in an office, without having an office. In an office, we pass documents to each other, discuss matters, and so on. But this can be done using communications. We can do the same things as we do in an office, but without having an actual physical office-in fact, we then have a virtual office. We still think in terms of an office, but the office is no longer there. We thus enter a *virtual world*. We can have a virtual classroom, a virtual shop, a virtual organization, and so on and so on.

1.6.1 The Virtual World

The virtual world allows us to see things as we have expected them in the past. However, we see them in their logical rather than physical form. Thus, for example, we have often seen our work as being in an office, so why not set up an office group on a screen, although the office need not exist physically. Thus we still feel that we are in an office but it is not actually there. We can call a meeting by selecting a collaboration service. We can have a discussion by (electronically) entering an office.

A similar effect can be made to create a virtual shop. We can show shelves on the screen and a cursor or some other device can be used to allow us to browse through the shelves, even picking up (virtually) items and rotating them. Some such paradigms will be described in Chapter 2, especially on how to reproduce our feelings of belonging to an organization while working remotely.

1.7 TERMINOLOGY

New paradigms often introduce new terminologies, and the computer-communication combination is no exception. The range of terminologies used in this work is shown in Figure 1.13. Furthermore, the terminology is still evolving and changes can be expected in the future. Terms often found in workgroup computing are illustrated in Figure 1.13.

One important part of the terminology is the idea of collaboration. This means people working together and agreeing on the kinds of actions taken. This differs from the idea of sharing, where we use the same information but do not have to agree on what information goes into the system. For example, entering a new sale into a database can be done without seeking agreement. However, adding a new paragraph to a joint report often requires agreement from other coauthors. Workgroups are predominantly people who collaborate, and workgroup computing is often the term used to indicate the technology needed to support workgroups. Another term often used to indicate such support is *group support* or *group support systems*. Other terms here are groupware, which is often used to describe workgroups, and computer-supported cooperative work (CSCW), which usually includes both the technology and the processes and methods used.

Coordination often comes up in workgroup computing. This has a narrower meaning than collaboration, as it only requires people to follow particular rules, but not necessarily devise them. For example, shared database access is a form of collaboration. Coordination is a more generic term for cooperation and is often used to imply the theoretical aspects of cooperating objects.

Workflow is also a common term used in this area and defines the flow of documents through an organization. It refers to the software used to define document workflow processes in an organization. Another common term that is perhaps subsidiary to workgroup computing is the term *client-server*. Although initially used in the system context of providing computing services, it is now finding its way into workgroup computing to describe a particular way of collaboration where one person supplies agreed-upon services to another.

1.8 SUMMARY

This chapter described the many new ways of working now arising from the combination of computers and communications using networks such as the Internet and the variety of services it provides. In summary, this chapter outlined some requirements that system designers must meet to construct a good system. One important requirement is for standards that make it easy to move between computers and between services. Another important requirements is to provide people with access to information without excessive browsing. This calls for tools that enable people to quickly access the information that they need and to work together with other people on this information.

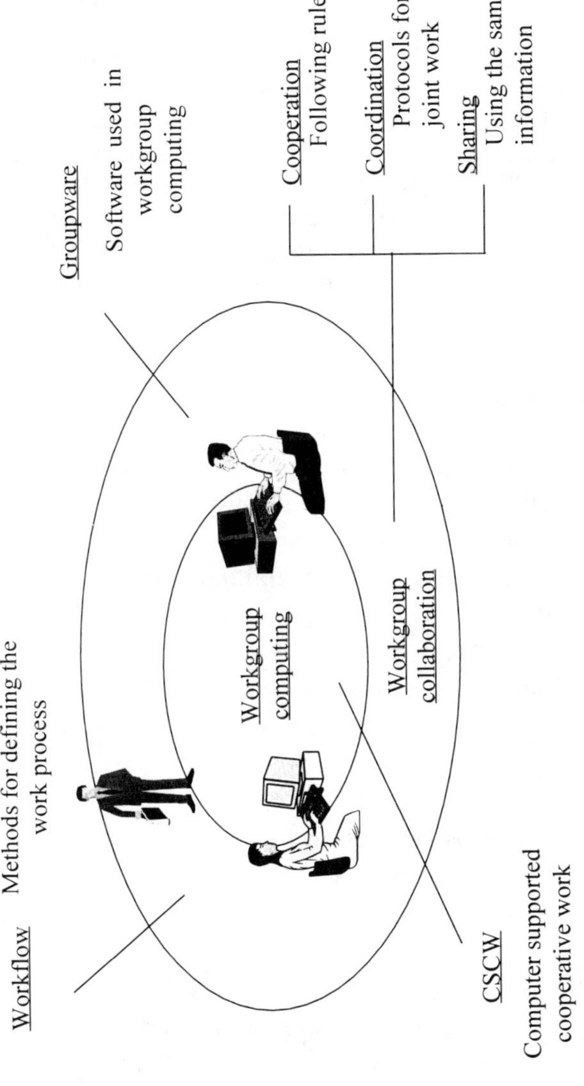

Figure 1.13 The terminology spectrum.

The ability to provide focus, especially with large groups, is also important. It means that people in such groups must be kept fully *aware* of what is happening around them-with a liberal interpretation on the word "around." It can mean everything within the context of a person's responsibilities. The person must be informed of important changes in their context, without necessarily searching through a number of databases for such changes.

Another requirement is for methods to design systems for collaboration. These methods will center on choosing the tools and services that fit in with the way that people work. They will require careful analysis of everyday work practices to gain a better understanding of how alliances are formed, or meetings organized, or ideas developed. The analysis will include the complex social interactions within these processes, which must be understood to build the best support systems for people. The book will describe group interaction in Chapter 3 and methodologies for designing systems for collaboration in Chapters 13 to 16.

1.9 DISCUSSION QUESTIONS

1. What further effects on distribution can you see coming from continuing reduction of computer and communication costs?
2. What do you understand by the term information highway?
3. Describe situations of computers in communications environments with which you may be familiar.
4. Describe some different ways of communication.
5. What are the major classes of network communication services?
6. Can you think of other examples of processes based on coordination of skilled specialists?
7. What do you understand by the term context?
8. What is the difference between research and innovation?
9. What are the advantages of forming alliances?
10. Why is the idea of service becoming important?
11. Examine the process shown in Figure 1.9. Can you see the loop in this figure applying in other situations.
12. Suggest some professions that may become redundant or radically changed because of the computer communication combination.
13. Why is it necessary to determine the ratio of communication that must be face to face and that which can take place by the exchange of notes?
14. What do you understand by the virtual world?
15. Why is interface design becoming more important?

1.10 EXERCISES

1. What kind of information do you think is needed to maintain a useful context for the specialist designer.
2. Perhaps you should begin to think here about building a virtual university. Do you really need to come to classes if is there some other way that you could get the skill required? Again, do not think here about simply automating current activities, but what is it that you actually do when you learn and how would you see a virtual university? Identify broadly the kinds of information exchange, interpersonal relationships, and work processes that must be supported and suggest some network services that can support them.

References

[1] Bell, G., and J. Gemmell, "On-ramp Prospects for the Information Superhighway Dream" *Communications of the ACM*, Vol. 39, No. 4, 1996, pp. 55–61.

[2] Busch, E., et al., "Issues and Obstacles in the Development of Team Support Systems," *Journal of Organizational Computing*, Vol. 2, No. 1, 1991, pp. 161–186.

[3] Klockner, K., et al., "POLITeam Bridging the Gap between Bonn and Berlin for and with the Users," in *Swiss Enterprises: An Empirical Study*, Marmolin, H., Y. Sundblad, and K. Schmidt, (Eds.), *Proc. of the Fourth European Conference on CSCW*, Sept. 1995, pp. 17–32.

[4] Greiner, C., I. T. Hawryszkiewycz, T. Rose, and T. M. Fliedner, "Supporting Health Research Strategy Planning Processes of WHO With Information Technology," Proceedings of the GI Workshop on Teleo Operation Systems in Decentralized Organizations, Berlin, March 1996, pp. 121–145, ISSN-0943-1624.

[5] Sauter, C., et al., "CSCW for Strategic Management" in Swiss Enterprises: An Empirical Study," Marmolin, H., Y Sundblad, and K. Schmidt, (Eds.), *Proc. of the Fourth European Conference on CSCW*, Sept. 1995, pp. 117–132.

Selected Bibliography

Coleman, D., and D. Khanna, *Groupware: Technology and Applications*, Upper Saddle River, NJ: Prentice-Hall, 1996.

Baecker, R. M. (Ed.), *Readings in Groupware and Computer Supported Cooperative Work - Assisting Human-Human Collaboration*, San Mateo, CA: Morgan Kaufman Publishers.

Conklin, E. J., *Groupware '92*, Morgan Kaufmann Publishers, 1992, pp. 133–137.

Drucker, P. F., "The Age of Social Transformation," *The Atlantic Monthly*, November 1994, pp. 53–80.

Greenbaum, J. *Windows on the Workplace*, New York: Cornerstone Books, 1996.

Grudin, J., "Computer Supported Cooperative Work: History and Focus," *IEEE Computer*, Vol. 37, No. 5., 1994, pp. 19–26.

Kline, P., and B. Saunders, "Ten Steps to a Learning Organization," Arlington, Virginia: Great Oceans Publications, 1993.

Krant, R., guest editor, "The Internet@Home," special issue of *Communications of the ACM*, Vol. 39, No. 12, December 1996.

Malone, T. W. and K. Crowston, "The Interdisciplinary Study of Coordination," *ACM Computing Surveys*, Vol. 26, No. 1, March 1994, pp. 87–119.

Wainwright, E., "The Big Picture: Reflection on the Future of Libraries and Librarians" *Australian Academic and Research Libraries*, Vol. 27, No. 1, March 1996, pp. 1–14.

Chapter 2

Working in the Networked Enterprise

LEARNING OBJECTIVES

- ❑ *Work environments*
- ❑ *Synchronous and asynchronous work*
- ❑ *Kinds of cooperative work*
- ❑ *Changes to work practices*
- ❑ *Support for work practices*
- ❑ *Tasks and missions*
- ❑ *Workspace*
- ❑ *Metaphors for distant work*
- ❑ *Benefits of distributing work*

2.1 INTRODUCTION

The challenge of supporting collaborative teamwork within enterprises is to enable people who may be at different places to work together on the same task at different times. Before going on, perhaps we should take some time to explain what is meant by different times and places. *Different place* simply means that people are at different locations, even countries, and cannot regularly meet face to face but have something important to do together. This may be preparing a report where different people write different parts of the report; it may be completing a technical design, where the designer is at one location but the reviewer or customer at another. From time to time the designer will have to communicate with the customer and the reviewer using communication facilities. Then we come to the idea of working together, at the same time or at a different time. Working at the same time means that people communicate directly by talking to each other-either face to face or over the phone-and discuss ideas spontaneously. If they work at different times, then they communicate through messages or document exchange rather than direct discussion. Thus the designer may send a design for comment and react to the comments in her own time. Often, work involves a mix of same-time and different-time communication, and good support systems will provide the most effective mix.

Collaboration between people at different places can lead to many benefits. Such benefits include things like getting access to the best expertise on a problem or getting the latest information on some project. Or, it may be simply a reduced amount of travel. The challenge is to obtain these benefits by using computers. The usual temptation is to automate the way we work together at the same location and time, and then try and use the automated system across distance and time. However, this simple approach ignores many social issues. Things like finding out what is going on through informal comments or chats in the corridor. We often find this out by discussion between ourselves, even at tea breaks. Our actions often depend on the information we obtain by these informal means. But how can we do this when we are working at separate locations? Even when we have facilities that support spontaneous discussion? This is perhaps one of the challenges facing designers today. How to provide distant workers with facilities that enable them to visualize their work environment from a distance and feel part of it? It is also one of the major problems confronting intranet designers-how to go beyond simply connecting their computers, and to provide support for particular work practices.

Distributed teamwork will become attractive once its benefits are understood and technology can be adapted to realize these benefits. It is clear that to obtain such benefits one should not simply reproduce the process that is followed at the same location. It requires people to adopt new practices as well as making changes to organizations to support these practices. This chapter also describes practices being adopted for working on the same task at different locations and what must be done to design systems that effectively support such practices.

2.2 NEW WORK ENVIRONMENTS

A variety of new working environments have been made possible by the connection of computers and communications. It is now common to use a matrix like that shown in Figure 2.1, which was initially put forward by Johansen [1], to classify these new ways of working. Figure 2.1 also shows some of the technologies that are used to support these new ways.

The two major dimensions in Figure 2.1 are time and place. The common terms in the time dimension are *synchronous* to indicate people who are communi-

Place

Time

	Colocated (same place)	Distant (different place)
Synchronous (same time)	Electronic meeting rooms Electronic boards Shared screen	Video conference Cooperative design Group editor
Asynchronous (different time)	Shared files Design tools Team rooms	Structured workflow Electronic mail Voice mail Bulletin board Conversation support

Figure 2.1 Working environments

cating at the same time and *asynchronous* to indicate people who may work together but at different times. The two major terms in the place dimension, *collocated* or *distant*, are used to describe work at the same or different places. Often, work proceeds in a *semisynchronous* way where some of the work is carried out synchronously by a group and then parts of the work are allocated to particular members to work on in their own time. It is also possible to have a combination of synchronous and asynchronous work where, for example, joint work by some people is forwarded to others for comment.

The time and space dimensions have significant impact on the technology needed to support a group. It is often found that synchronous work, especially over distance using videoconferencing, can be expensive because it requires high-density transmission links to transmit video and images almost instantaneously. There are therefore advantages in identifying which parts of a collaboration can be done synchronously and asynchronously and devising a process that uses a combination of both of these approaches, and support them with matching services. Figure 2.1 also shows how some of the services introduced in Chapter 1 can be used for different time-space combinations.

2.2.1 Size of the Group

Most early cooperative work concentrated on small groups and consequently on synchronous communication. However, synchronous work is not always possible with large groups. It is often difficult for such large groups to work together on the same artifact at the same time, or even to meet. Videoconferencing does not help much here, as it would be difficult to imagine groups of 20 or more working on the same artifact on the screen concurrently. Consequently, large groups as a rule mostly operate asynchronously and are often not collocated. The situation is different with smaller groups. Here, synchronous work is easier as a few people can often concurrently discuss some issue and work on the same artifact.

Systems, such as intranets, that go beyond a group and cover major parts of an enterprise are also now being built. Such systems connect groups together so that each group is aware of other group activities. Most such systems are asynchronous and give groups mutual access to each other's information so that people can be kept aware of activities within the organization and better coordinate their own work within the organization.

It is of course common in many work processes to have a mix of synchronous and asynchronous work. Groups may regularly meet to subdivide work on a large project. They may then carry out their parts individually, meeting occasionally on a small group basis to discuss common problems and then having a larger group meeting to smooth out major problems.

2.2.2 The Type of Applications

The new working environment leads to a number of new ways for using computers. Figure 2.2 illustrates a range of group applications by a two-dimensional spectrum. One dimension is the degree of problem structure, the other is whether the application supports a group of people or one individual. Structure defines whether there is a standard way of doing something. Thus, for example, arranging the purchase of a part is usually quite standard. The process followed is well-defined, and everyone concerned knows what is to be done given when some condition arises. For example, consider a purchasing process. A request is made and a purchase approved. Once a purchase is approved, the purchase order is prepared and sent to a chosen supplier. However, the preparation of a proposal for a project is often less structured. New ideas may arise as the plan evolves and new ways of doing things come up. The action chosen will depend on the ideas, and often cannot be predicted. User applications can fall anywhere in the spectrum shown in Figure 2.2. Time and space are the other two dimensions in this spectrum.

Figure 2.2 thus covers the entire spectrum of user applications, and not only those made possible by the new technologies. It should, however, be noted that most computer applications to date have been in the top left-hand corner, people working to defined procedures, such as arranging a purchase or trip. These structured procedures can be predefined, allowing people to simply follow predefined work procedures. Similarly, there are many applications at the bottom left-hand corner of the spectrum made possible by the use of personal computers by individuals for storing structured data. Thus local budgets, addresses, and contacts are often maintained on personal computers.

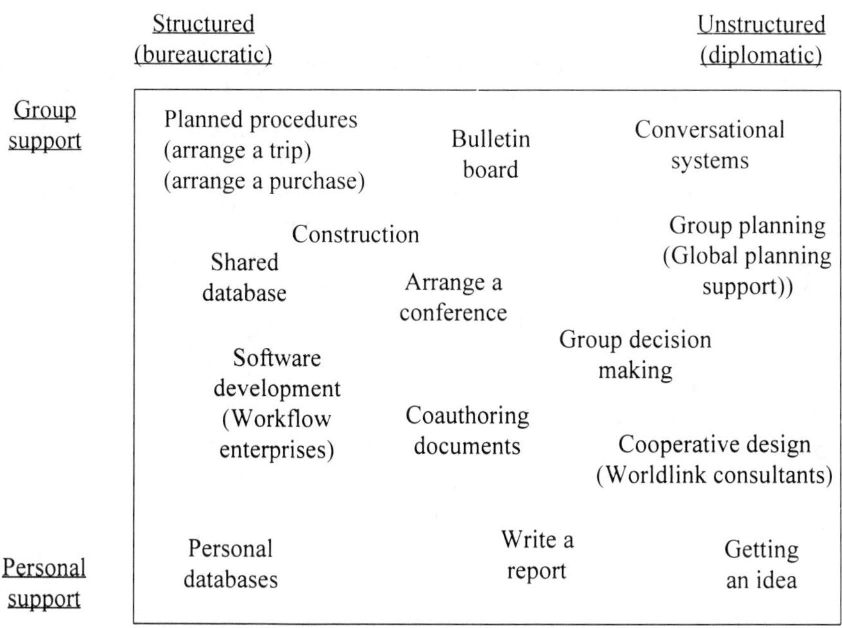

Figure 2.2 Application spectrum.

However, computers can be used to improve communication in the other parts of the spectrum. For example, getting an idea is often an unstructured personal activity. However, even here computers can assist people to share ideas, comment on them, and ultimately improve them. Discussion databases can be used for purpose and they are described in Chapter 11. Many applications need a mix of structured and unstructured work. Decision support is one area where information is developed in a structured way-but decisions are made using a less structured process involving considerable negotiation. So is strategic planning. Many of the systems described in Chapter 1 fall at different positions of this spectrum. The planning of health strategy across nations is not very structured, but requires getting agreement from many people. Furthermore, it involves a large group of geographically separated people and as a result is asynchronous in nature, but often requires a defined process to ensure that a plan is produced by a required time. Design carried out by the *expert designer* involves a smaller number of people and can be more structured.

Computer networks also make it easier for people in different places to share information, thus leading to better organizational decisions or services. Thus, for example, a marketing group can collaborate with a production group to provide customer feedback to product development.

2.2.2.1 Managing Distributed Work

The previous chapter described some examples of people working across distances. Often, such distributed work requires coordination support to be provided to groups, particularly when they are closely working on related documents with tight delivery schedules. These usually fall towards the group and structured axis of Figure 2.2. One example here is software development, where people must contribute individually (for example, writing a program module), but at the same time must coordinate their work with others (for example, the system designer). It could, however, apply in any work process that requires coordination between a number of people, such as, for example, preparing business proposals that require people with different expertise to work on different parts of the proposal but coordinate their work so that all the parts fit together. What distinguishes software development or development of business proposals from other kinds of projects is that the product (software or a document) can flow electronically with the task. Furthermore much of this work now takes a global nature, where many people in the one group may also work in different time zones.

Case Study: Workflow Enterprises

Workflow enterprises is developing a project support system for teams that may be geographically distributed but work on related documents. Its first pilot project is to support large computer development project, where clients specify requirements that are converted to working systems following the process made up of the activities shown in Figure 2.3. The next activity, once requirements are clearly identified, is to produce a broad architectural design. This is followed by systems designers producing a detailed system specification, and then by programmers developing modules. These modules are assembled into systems, which are then tested and put to use. Each process step produces a document that is used in the next step. Each of the people involved in each of these steps may be at a different location, but must exchange information with each other. Thus the designer must get requirements documents, the programmer must be aware of the system specification, and so on. It should also be noted that in practice there is considerable informal exchange of information between people in different groups. Furthermore, it should be noted that there are well-defined roles, such as designer or programmer, in the system. Any support system must identify these roles and provide them with the necessary communication paths.

One activity that is particular to software development is testing, especially testing large systems. The goal of testing is to assure the quality of the developed system. Modules are first tested individually in what is known as module testing. Groups of related modules are then tested to see if they correctly interface to each other, a process known as integration testing. System testing is the final step, where the system is tested by the client in a simulated working environment. Testing is quite time-consuming. It requires test data to be created as well as devising tests that represent all process alternatives. It is important that the testing be carried out by people other than the developers, with tests developed against specifications rather than against standards set by the developers. Such tests are developed in parallel with development, and testers must be kept aware both of the progress of development and any changes to requirements. To manage this testing, a test plan is developed during architectural design, followed by the preparation of test cases, which are then used in integration and system testing. Furthermore, there is considerable iteration between testing and development. This calls for special support to keep track of test reports and subsequent program changes. More details on software testing can be found in [2].

The process as described here could be supported by a workflow system that initiates tasks on completion of preceding tasks. This workflow system must be integrated with a document management system to pass documents between people who are carrying out these tasks. However, there are a number of further characteristics of the software development process, and of similar processes, that require additional support. These are that

- Although development is planned to proceed in a set of sequential steps, this does not often happen. Often, the detailed nature of the work means

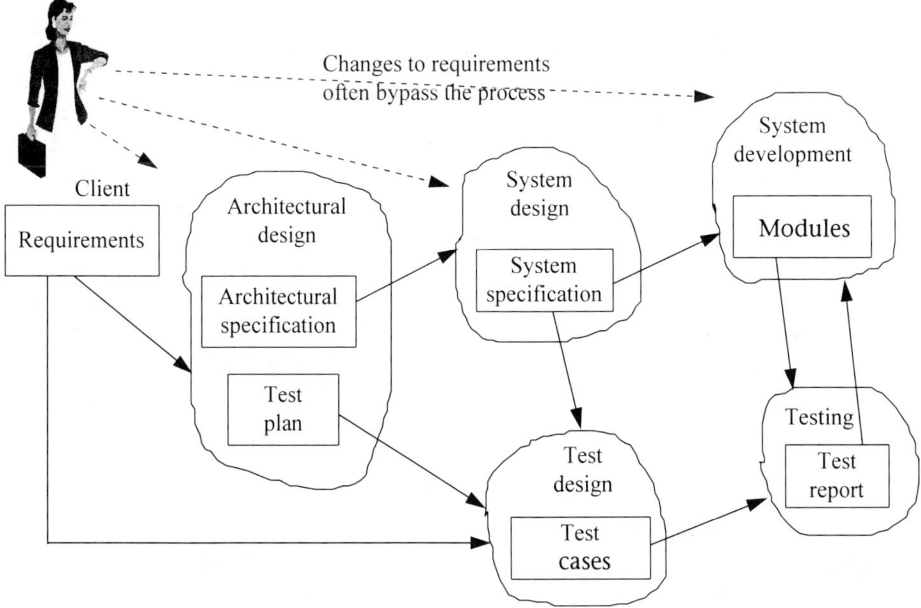

Figure 2.3 Software development.

that misunderstandings can arise between steps and these must be resolved quickly, calling for support for synchronous discussion between the people.

- Requirements will often change, even during development, and everyone has to be made aware of these changes and their impact on their work. Often, change information appears in documents that are not distributed in a timely manner, or simply by word of mouth first and recorded later. As shown by the dotted lines in Figure 2.3, changes are often introduced in late design stages, thus resulting in inconsistencies between the documents. Situations can thus arise where tests are developed using a set of documents that do not include late design changes, leading to uncertainty of whether a test failure means an incorrect program or an out of date test. Support is thus needed to quickly propagate changes to documents. This will require a document structure that allows changes to be quickly traced through the document structure.
- To expedite development, tasks often proceed *concurrently* in an environment of continuous change. This presents enormous pressures on maintaining communications between developers to ensure that all parts of the system are consistent with requirements placing more emphasis on coordination. Typically, this coordination problem is solved by frequent meetings, and people often spend considerable part traveling to attend the meetings.

Workflow support must eventually address all these issues. It should make it possible to reduce the ever-growing amount of time spent in meetings by better coordination support. It is also clear that such support needs more than one service. Of course, a workflow service can move information from one person to another and keep track on project progress. This workflow system must, however, be integrated with document management service to keep track of the documents and allow changes to be easily traced. At the same time, additional support is needed both to keep track of change and keep team members aware of each other's activities. One way to do this is to provide support for people to exchange information informally to iron out many of the problems that arise when building strongly related system components. This can be done through e-mail or discussion databases. Videoconferencing is one approach here, but discussion databases can also be used to keep track of ongoing problems. Still another approach is to use bulletin boards to post new events and notify members of new postings.

The example just described illustrates the problems found in managing many large projects working to deadlines. There are a number of groups of people, each with specialist knowledge. The tasks are closely related and there is the need to maintain coordination between the tasks, ensuring that the latest documents are made available to the people who carry out the tasks. It is here that workflow services can provide considerable support, but these often need to be supplemented both by ways of exchanging information and personal exchanges in resolving complex task interfacing issues. The book will describe how such support can be provided.

2.3 CHANGES TO WORK PRACTICES

The combination of computers and communications can result in major changes in the way that people work. Instead of repetitive simple operations, a person can now have a *portfolio of tasks* and select tasks that need most urgent attention. This book uses the term *task* to define what is to be achieved. Thus our task may be to produce educational material on, say, using computers for cooperation. Another related task may be to write the syllabus for such a course. Another may be develop an exercise showing how to use electronic mail. Tasks can be further broken up into subtasks.

The term *mission* now also frequently appears in the literature. Mission becomes more relevant in rapidly changing environments where goals may change, often quite quickly. Thus it defines a more dynamic target than a well-defined goal. The term mission has a longer time frame and a more abstract meaning than goal. It usually means the image and position that an organization wishes to take in its environment and is very relevant where the environment is changing. Thus, a mission may be to deliver high-quality education. Tasks may be to prepare course syllabi. In software engineering, the mission is to build systems that satisfy user requirements that may change in the course of the project; tasks that must be carried out to carry out the mission are allocated to team members. Team members will be more aware of their context, or situation of their work, as well as the situation outside the organization, and thus contribute better to the overall mission. Thus people in the software engineering team should become quickly aware of changes to user requirements, and adjust their tasks to satisfy these changes.

The ability to communicate through computers makes a number of new work situations practical. However, it usually means that new work practices will have to be adopted to lead to effective work in these situations-both at the *micro* and *macro* levels. At the macro level, there are now a large number of work situations made possible by the combination of computers and communications, and these are described in the next section.

2.3.1 New Work Situations

Collaboration technologies lead to the possibility of getting away from face to face work by providing the ability for people to work together across distances. Such remote work can be between persons in their regular offices or working from remote or mobile offices. This in turn leads to increasing distribution and globalization of work and the ability to work remotely. New metaphors for work are now being identified. The terms being used to focus discussion are telework, telecommuting, and mobile and nomadic workers. Work in these situations is no longer measured by time spent on the job, but defined in terms of mission and tasks to be accomplished by the remote workers.

2.3.1.1 Telework and Telecommuting

Telework refers to a situation where people employed by an organization spend a significant proportion of their time working from locations other than a permanent location in the organization, and carry out their tasks remotely, possibly at home using a computer. The term telecommuting is also common here. There are many kinds of work where the equivalent of telework already takes place. Sales representatives are one example. There is probably no need to have a desk for each representative-and some alternative should be looked at. Thus we may have simply a cupboard with a slot for each person to keep their office papers, with the majority of their work stored on the computer nétwork. At the same time, there may be a number of desks equipped with computers. These may be shared between the representatives, using some schedule. When they come in, they go to the cupboard to take their papers, take position at one of the desks, and are ready for work.

There are an increasing number of people who can work in this way because of the computer communication connection. It implies people traveling to work electronically, and is thus attractive to many city planners and traffic managers as it gives the opportunity to reduce traffic congestion in large cities. Perhaps now it will become possible to choose one's location of work, where people can become more creative and productive working in more familiar and pleasant environments rather than an office. For example, choosing to work in a secluded beachfront home far from the noise of a polluted environment, with a view of the ocean, can provide the environment that opens the mind to new ideas. At the same time, a communication environment that supports communication with coworkers can improve the idea by facilitating discussion.

However, telework will only be organizationally accepted if it improves organizational effectiveness. Early indications are that take-up of the telework concept has been slow. Inhibiting factors are seen to be the perception of loss of management control and maintaining team cohesion. However, significant inroads have been made. In the United States, according to the *EDGE Workgroup Computing Report* - Oct. 30, 1995, 2/3 of Fortune 1,000 executives view telecommuting as productive and about 9.1 million people telecommuted. The majority of executives expect penetration of telecommuting to grow. The Gordon report of May 1995 notes that telework has penetrated 5% of European organizations and that 9.1 million people telecommuted in some way. Most also report that telecommuting has been more prevalent in large organizations, those employing more than 1,000 people.

It should also be noted that telecommuting suits only some kind of jobs. Again, a survey of manager's top choices reported in *EDGE: Workgroup Computing Report*, in Oct. 30, 1995, provides the information listed in Table 2.1. In Table 2.1, the second column lists the percentage of managers that thought that a particular kind of job was suitable for telework.

Table 2.1
Jobs Suitable for Telecommuting

Job	*Suitability for Telecommuting*
Customer service representative	50%
Human resources professional	33%
Information specialist	69%
Manager	24%
Market research analyst	59%
Programmer	61%
Sales representative	68%
Systems analyst	52%

2.3.1.2 Mobile Workers

Mobile workers are a special case of teleworkers. Not only is their location distant from a permanent location, but it is continually changing. The growth of *mobile workers* is also expected to increase. Project managers may spend more time at project sites rather than the office, and sales managers more time in the field. They will need support to keep in touch with developments in their areas of responsibility. Mobile work requires special technical networking support. Imielinski and Badrinath [3] discuss some of these technical issues, especially the idea of wireless support for mobile workers who may not be continually connected to the networks. One criterion here is to reduce the network traffic needed to maintain contact with mobile workers. They suggest the such mobile workers be assigned a home base where their messages are collected for later use, and that such home bases may change with long-term movements of mobile workers.

An often provided tool now for most mobile workers is the laptop computer. There is an increasing demand for services where a laptop can plug into a network and download information from central repositories. The mobile worker can then work on the information in a variety of environments, even while traveling.

2.3.1.3 Nomadic Workers

The difference between nomadic and mobile workers is in the fact that mobile workers usually refer to people who are part of one organization. Nomadic workers are those who may work for more than one organization. Perhaps the best example of such nomadic workers is the growing number of self-employed consultants now found in almost any industry. These people are often employed by an organization on a temporary basis and need similar facilities to mobile workers. However, as these are often external employees, they will have restricted access to enterprise information.

2.3.2 Supporting the New Work Situations

Intranets or other systems developed to support the new structures must support the new work situations at the micro level. To do this requires a better understanding of the work process. Such understanding includes new work practices for sequencing of tasks, ways for remote workers to be kept aware of what is happening in the enterprise, and support for maintaining the interpersonal relationships necessary for people to work together.

2.3.2.1 Maintaining Interpersonal Relationships

Personal relationships arise in any group situation, whether the group is collocated or not. Effective computer support must allow people to maintain such relationships. Furthermore, it should support relationships that have been found socially necessary for people to effectively work together. Benford and others [4] identify a number of key abstractions that can describe such interrelationships.

Perhaps the best way to summarize social requirements is through the notion of *awareness*. Basically, a person's awareness refers to whether that person knows the latest changes and developments in the context. A perfect state is where all group members have the same awareness of the context. Other important factors deal with how person can establish and maintain a unique presence or aura within the enterprise that establishes them as a recognized identity. Ways of drawing attention to issues and focusing on them are also important, and ways of doing this electronically must be found.

There is also another property that is considered important, namely, *group memory*. Group memory has a historic connotation. It refers to reasons to changes to the context and why were the changes made. One reason for maintaining group memory is to make people aware about reasons for earlier changes. In this way, people can learn from earlier practices and choose those that have been found effective. In addition, good group memory means that the same mistake will not be repeated.

2.3.2.2 Social Effects

The trend towards resolving many of these social issues is the formation of electronic social groups, where some of the social factors can be realized through electronic means. Such groups, for example, can then maintain awareness through bulletin boards, news groups, or simply by exchanging messages using e-mail. They can try to focus attention through distributing e-mail messages or maintain nimbus by attaching comments to other peoples work. The formation of such electronic groups or social fields has a number of effects on society, especially on the acceptance of the groups. They can be seen as working outside the normal organization and not within its processes. They can be quite influential and yet their activities are not formally recorded and hence are looked at with suspicion by management. Electronic mail thus provides a *private meeting space* for such members.

The Importance of Trust

Establishment of trust between distributed team members is also seen as an important social issue. Team members must place reliance on information received from people whom they may not have even met. There must be some way to give them an assurance about the quality and authenticity of information received from distant members in networked enterprises.

2.3.2.3 Context Requirements

A further requirement of electronic groups is to maintain a common work space or context. This context, should be easily accessible to all group members and should include the policies and rules of the organization, as well as external factors that may influence decisionmaking. The context must also allow people to associate themselves with certain developments, or perhaps even actively bring their ideas to people's attention. The computer will have to contain more unstructured information such as policies and rules to provide this context, while still continuing to maintain structured data. To date, this has often not been feasible because of the large quantity of information involved. Much of the context has been provided in manuals, which must be constantly updated by their users. The more informal part of the context is developed by regular face to face meeting or visits of personnel to a common location for informal information interchange. However, technical developments in the area of multimedia, together with electronic communication channels, will make it possible to provide such contexts in the near future.

Holding the context in computer storage will ensure better consistency. For example, updates to manuals will be propagated to all personnel, ensuring that the context is up to date. However, such maintenance requires the recording of every action to the computer memory, again often a time-consuming task.

Access to Information

Access to information becomes an important part of gaining access to a context. Thus, as shown in Figure 2.4, a user can look up various services provided by applications and use information obtained from them in their own private workspace. The way that the service is provided depends on the application. Some may open up a new window with a menu, others may deposit standard information into the workspace.

2.3.3 Management Implications

The relationship of remote staff to management is a critical issue, as well as the role of management itself. There are many managerial implications in working from distant locations. Perin [5] notes that management is often unwilling to agree to distant work, as it is more difficult to supervise distant workers. Consequently, an understanding must be reached between management and distant workers, or their representatives, on their tasks and targets. This is again where a clear definition of the task becomes very important.

In any arrangement for distant work, the wider social issue must be considered as well as the work process itself. Socially, the overriding question is whether we can maintain our *presence* in the same way as we do when we are physically at work in an office. Another question is whether we can maintain the same relationships between our fellow workers or even improve them.

Arrangement of office space also becomes important. Where people work from distant locations, in particular from home, there is less need to provide an office or desk for each employee since much of their work is external. Hence, there will have to be more emphasis on shared office furniture and equipment, thus resulting in savings.

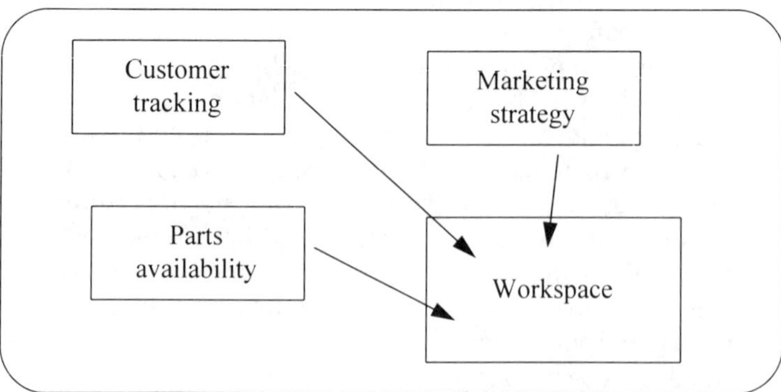

Figure 2.4 Easy access to applications.

2.4 SUPPORTING NEW WORK PRACTICES

One important issue in enterprise networking is the computer interface, which must allow distributed group members to set up and maintain relationships. It will define how people see their context and how their actions are recorded. The interface must support both informal relationships, such as simple interchange of information, or more formal relationships such as joint work on some artifact. The interface should allow group members to get access to information and use it to maintain effective working relationships. Often, the term *platform* is used to describe the interface provided to people. The platform should provide the range of collaborative and desktop services to enable people to collaborate using computers. This book will describe such platforms in detail later in Chapters 6, 7, and 10. In this chapter, we only outline some of the broad platform requirements.

2.4.1 New Interface Metaphors

It is important that the computer interface allow users to see or *visualize the enterprise* in a familiar way. People, however, do not like changes to their perception of work and interfaces should try to preserve views, albeit virtual, familiar to their users. A number of metaphors have been identified for this purpose. Perhaps the most widely known one is the virtual office.

2.4.1.1 The Virtual Office

The virtual office is perhaps the most often quoted metaphor in distributed work. The goal of a virtual office is illustrated in Figure 2.5. It provides an interface that displays a "logical" office. This interface could have icons to represent rooms, people, and all the other objects commonly found in offices. A remote worker can use the system to see who is in each "office" and what they do. Furthermore, the remote worker can initiate an interaction by "moving" electronically to an office. Such a move can open up a video channel and allow the remote worker to "talk" to people in these offices. They can also look at any information in the office. It then becomes possible to do things like have a meeting or work on an artifact. Supporting tools are needed to support such meeting over distance or time.

Virtual offices should be able to provide abilities such as the creation of a "new office" with some defined function. We can assign a number of people to the office and provide it with funds and support services by "dragging" these components from other parts of the system. In fact, we should be able to "drop in for a chat" by dragging ourselves to someone's electronic office and having their face appear on a video display for a discussion. Alternatively, it should be possible to present the equivalent of an open office where communication is more informal. It should be possible to simulate activities such as " seeing things out of the corner of one's eye" or even eavesdropping.

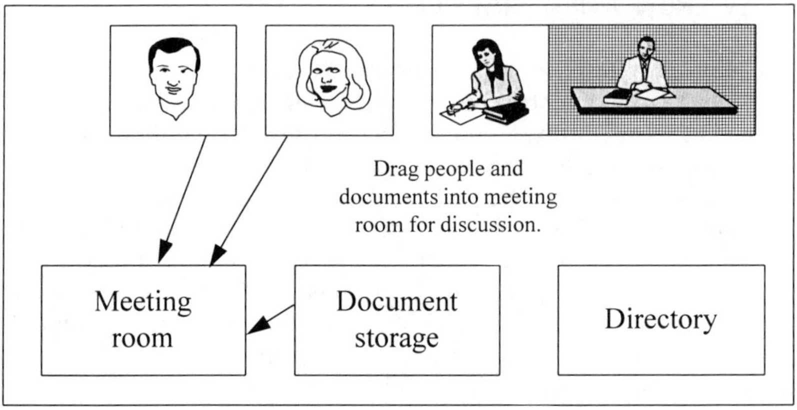

Drag people and documents into meeting room for discussion.

Meeting room

Document storage

Directory

Figure 2.5 The virtual office.

Of course, there is no requirement that people in adjacent offices should also be physically adjacent. Some of them could be in an organization's office, while others could well be teleworkers. The virtual office does provide the ability to "feel" part of the enterprise, even by seeing one's picture on the screen, and it is possible that such offices can be provided with facilities to maintain comments, views, and records of private discussion. Figure 2.5, of course, represents a goal, but research is being directed to create such environments.

2.4.1.2 The Business Process

The business process defines how to coordinate activities into a well-executed process. This process may be to sell a product, install equipment on a customer's premises, or provide any kind of social service. The goal is to execute this process in a way that both maximizes customer satisfaction, while at the same time it is being carried out smoothly within the organization. The term quality is now often used to describe processes that satisfy these goals.

Managers of such processes must be able to quickly get information about the status of process activities. A visualization, like that in Figure 2.6, could be used to do this. It shows the activities and relationships between them. Activity details can be found by clicking on the activity box. Completed activities are those whose output is shaded, and completion records can be found by clicking on this button.

It is now also common to use *workflow languages* to define business processes. The workflow language defines and traces flows of information through the system. One important issue is who becomes responsible for the process. Some organizations are now appointing business process managers to take process responsibility. They must then be directly responsible for the processes. This calls

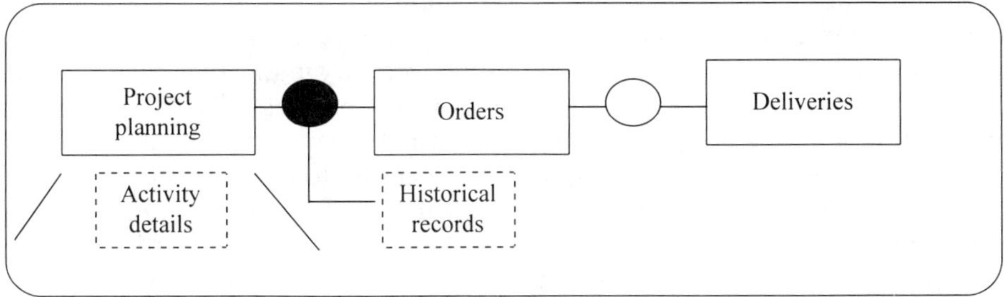

Figure 2.6 The business process.

for special support for businesses processes to be defined, and changed, at the business level rather than by information systems people.

2.5 BENEFITS OF COOPERATIVE WORK

Organizations will only provide support for cooperative work if they see direct benefits resulting from this new way of working. Such benefits must be measured in the terms of the organization's mission, which can often be generalized to get an improved product in shorter times, or improved productivity and better quality. The question is how to illustrate such benefits to management to get their support for group support systems. However, many such benefits can only be specified indirectly. Examples of such measures are the following.

- Better use of people's time by reducing the amount of time spent on routine tasks such as making copies, sending faxes, distributing paper, and in general keeping track of information distribution;
- Better use of information by providing better access to information and ensuring that everyone is aware of the latest information and who is doing what;
- Networking to enable better communication support between people in the organization and outside the organization, using tools such as teleconferencing or videoconferencing;
- Providing support tools such as joint document preparation, joint decision-making, and joint design including assistance in the discussion leading to higher quality decisions;
- Improving customer service by improved facilities to keep track of customer needs;
- Reduced cost of travel through the ability of people to work at a distance.

For managerial purposes, however, one approach is to show the positive impacts of these measures on the organizationn combined with easier communication between people, can lead to higher quality outcomes. Thus, for example, consider an organization preparing proposals to get business. Knowing precisely what a customer's latest requirements and priorities are will mean that a proposal will address the important issues. Similarly, the ability to bring the best expertise quickly to bear on a problem can raise the quality of the proposal. Joint working on documents can further improve quality because of the ability of people to check each other's work. Here as in other cases there is better awareness of what is going on, thus improving the ability to create higher quality products. The formation of informal groups with the ability to quickly exchange complex information further adds to the quality of work.

The *integrative nature* of cooperative work will be seen as attractive to organizations that are seeking to integrate their functions. Cooperative systems are integrative by their very nature because they bring people (and their systems) closer together. Marshak [6], for example, reports productivity gains exceeding 100% obtained from using LOTUS Notes, computed as people time saved divided by cost of software.

2.5.1 Integrating Services

Many benefits can only be obtained with well-designed networks that support enterprise work processes. However, the benefits only eventuate if the networks provide people with the services needed in their work. More often than not such networks will be made up of network services found on the Internet, and designers must work out the best ways to use them. This book makes the following distinction between the level of network services that are provided:

- *Level 1, basic connectivity*—users are provided with network services for message and information exchange such as, for example, electronic mail or file transfer.
- *Level 2, personal connectivity*—users are provided with network services for interpersonal communication and have to adapt them to their work contexts.
- *Level 3, work-specific connectivity*—networks provide users with services customized to the work processes in their context.

Basic connectivity requires the users to work out their own ways of coordinating their work using network services. This requires them to determine their coordination protocols and to maintain them, an effort that can sometimes be difficult for large teams. Level 2, or personal connectivity, provides some personal coordination support such as, for example, videoconferencing. Nevertheless, even here the users must integrate many services themselves, such as, for example, combining videoconferencing with the display of documents. Perhaps what is ultimately needed is Level 3, or work process connectivity. Such support is conceptually shown in Figure 2.7, where a platform of integrated services is provided to each enterprise worker. Here, each such worker is presented with an interface that describes their work situation, including the context and access to coworkers. This is where the intranets differ from the Internet. Effective intranets will provide the enterprise worker with integrated services that support their work processes. However, such integrated interfaces can only be constructed following careful analysis of the needs of the enterprise. Ways of carrying out this analysis and using it to design systems will be the focus of the final chapters of this book. Furthermore, a distinction can be made between two kinds of Level 3 systems-those that support fixed work practices and those that allow users to *adapt the integrated facilities to changing work practices*.

2.6 SUMMARY

This chapter outlined some of the new ways of working across distance made possible by computer communication systems. It described new organizational structures such as telecommuting or mobile work, but noted that their successful implementation requires a better understanding of how people collaborate so that they can be provided with tools to maintain good interpersonal relationships in the new structures.

Thus it is important to keep in mind that benefits cannot be realized by simply making technology available in large amounts and then expecting people to

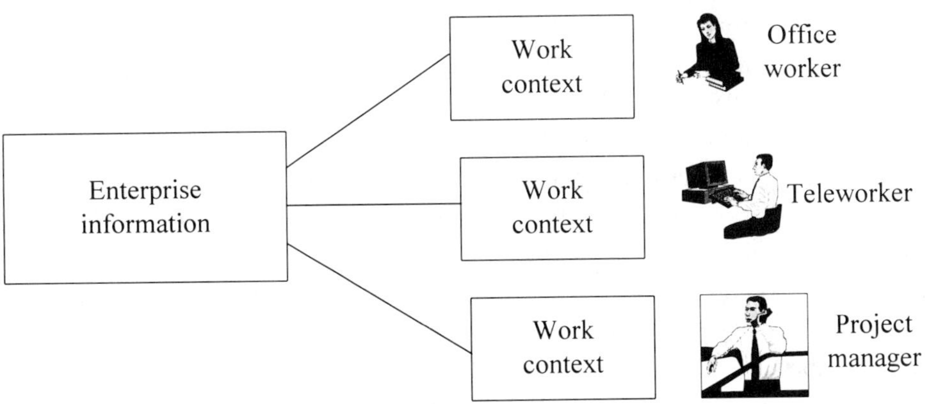

Figure 2.7 Supporting work practices.

use this technology. What is needed is to find out how people work together and supply the technology to support this work. Thus a *"user pull" approach rather than a "technology push" approach* is needed to ensure that selected systems assist rather than impede users.

2.7 DISCUSSION QUESTIONS

1. What are the advantages of collaborating across distances?
2. Describe some group support systems with which you may be familiar. Where would you place them in the matrix in Figure 2.2?
3. Why is asynchronous work more common with large groups?
4. Do you think all work can be carried out asynchronously?
5. Discuss why it is important to ensure that all group members have a consistent view of their context, using the software engineering process being developed by workflow enterprises, as an example.
6. What are the problems of maintaining consistency in an environment of change?
7. What is the difference between a mission and a task?
8. What is the difference between telecommuting, and mobile and nomadic workers?
9. Can you name some kinds of jobs that can be done by teleworkers? And some that cannot?
10. Do you foresee a growth of nomadic workers?
11. What are the problems of maintaining interpersonal relationships in these new work situations?
12. Do you think that all informal interaction can or should be supported in work across distance?
13. Why is it important to provide interfaces with which users are familiar?
14. What do you understand by the term group memory?
15. What would you understand by group memory in report preparation?
16. What do you understand by the term virtual office?
17. Describe how some of the services introduced in Chapter 1 can be used to realize the social factors described in this chapter.
18. Describe some advantages of distributed work.
19. Why are these advantages difficult to justify?
20. Why is user push important in group support systems?

21. What criteria would have to be met to maintain good contact with remote workers?
22. What requirements should a good intranet satisfy?
23. What do you understand by the three levels of intranet support?
24. Name some differences between the Internet and intranets.

2.8 EXERCISES

1. Define the kind of informal interactions that occur in a learning environment and how they can be supported across distance in a virtual university. What are the requirements of the context for the learning environment?
2. What documents would you expect to find in the context for a pilot software engineering project undertaken by Workflow Enterprises, from the viewpoint of a tester?

References

[1] Johansen, R., "Groupware: Future Directions and Wild Cards," *Journal of Organizational Computing*, Vol. 2, No. 1, 1991, pp. 219–227.

[2] Hetzel, B., *The Complete Guide to Software Testing*, Wellesley, Massachusetts: QED Information Sciences Inc., (2nd ed.), 1988.

[3] Imielinski, T., and B. R. Badrinath, "Mobile Wireless Computing: Challenges in Data Management," *Communications of the ACM*, Vol. 37, No. 10, Oct. 1994, pp. 18–28.

[4] Benford, S., et al., "Supporting Cooperative Work in Virtual Environments," *British Computer Journal*, Vol. 37, No. 8, 1994, pp. 653–668.

[5] Perin, C., "Electronic Social Fields in Bureaucracies," *Communications of the ACM*, Vol. 34, No. 12, Dec. 1991.

[6] Marshak, D. S., "Understanding and Leveraging Lotus Notes" Patricia Seybold Group, IN-DEPTH REPORT, Boston, 1993.

Selected Bibliography

Gorton, I., and I. T. Hawryszkiewycz, "Enabling Software Shift Work with GroupWare: A Case Study," *Proc. of the 29th Hawaii International Conference on Systems Sciences Vol. 3,: Information Systems - Collaboration Systems and Technology*, IEEE Society Press, ISBN 0-8186-7330-3, Hawaii, Jan. 1996, pp. 72–81.

Hosmer, L. T., "Trust: The Connecting Link between Organizational Theory and Philosophical Ethics" *Academy of Management Review*, Vol. 20, No. 2, 1995, pp. 379–403.

Kalakoa, R., J. Stallaert, and A. B. Whinston, "Mobile Agents and Mobile Workers" *Proc. of the 29th. Annual Hawaii Conference on System Sciences*, Hawaii, 1996, pp. 354–365.

Luukinen, A., J. Pekkola, and R. Suomi, "Teleworking Arrangements in Finland, " *Proc. of the 29th. Annual Hawaii Conference on System Sciences,* Hawaii, 1996, pp. 366–375.

Mayer, R. C., and J. H. Davis, "An Integrative Model of Organizational Trust," *Academy of Management Review,* Vol. 20, No. 3, 1995, pp. 709–734.

Tamrat, E., T. Vilkinas, and J. Warren, " Analysis of a Telecommuting Experience: A Case Study," *Proc. of the 29th. Annual Hawaii Conference on System Sciences,* Hawaii, 1996, pp. 376–385.

Chapter 3

Groups

LEARNING OBJECTIVES

- ❑ *Group structures and context*
- ❑ *Groups and culture*
- ❑ *Roles*
- ❑ *Types of interactions*
- ❑ *Interactions and activities*
- ❑ *Normative processes*
- ❑ *Conversations*
- ❑ *Social implications of introducing technology*

3.1 INTRODUCTION

The previous chapters described new ways of working made possible by the new computer-communication technologies and the importance of supporting interpersonal relationships within these new working environments.

This chapter looks more closely at how people work together in such environments. It begins by describing the relationships in terms of interactions between people. Such interactions can range from a simple command to more elaborate interactions such as an explanation or decision. The interactions of working together depend on group size. Individual interactions usually concern small amounts of information. It may be two or three people working on a simple design or on one part of a document. Usually, a synchronous joint editing tool can support such interactions across distance.

It is often necessary to go beyond simple interactions when considering systems with large amounts of information. Here, smaller groups of people may work on individual parts of an artifact, but the work of all such groups must be coordinated. Group structure and arranging people into groups becomes important where there are large volumes of information, as in global systems or organization-wide structures. Here, it becomes necessary to subdivide all people into smaller group structures to ensure the smooth flow of information while at the same time ensuring that no one gets overwhelmed by the large amount of information in the whole system.

Another important thing to realize is that groups are not static but, as is often the case, they are very dynamic. They change their composition and structure while carrying out some task. In fact, they often make these changes to meet the requirements of the task. Groups may also change because the task itself changes.

The chapter begins by describing individual interactions and then continues to describe the formation of interconnected groups. It also looks at ways to categorize groups and how they work. This is important for designers because they can provide initial group support using standard techniques for supporting each kind of group. However, the reader should note that such standardization at the moment is still a goal and not reality. The general kinds of groups are understood, but there are so many variations on the group that some adaptation of any group support tool is needed for each group.

3.2 CHARACTERIZING INTERACTIONS

Group or teamwork can be seen to be composed of many interactions between individuals within the group or team. Such interactions can be simple, such as asking a direct question, or they can be more complex, such as choosing one from a number of alternatives. One question that arises concerns ways of supporting such interactions with information technology. The choice of technology usually depends on the type of interaction. A number of ways have been proposed for classifying such interactions. As shown in Figure 3.1, they begin within speech acts, which usually relate to a small group, and then continue on to more complex interactions.

3.2.1 Speech Acts

Speech acts were initially proposed by Searle [1] as a way of describing interactions and are the left-hand part of the spectrum shown in Figure 3.1. Speech acts are seen as the basic actions in communication, and more complex communication patterns can be made up from a sequence of speech acts. The usual model is to

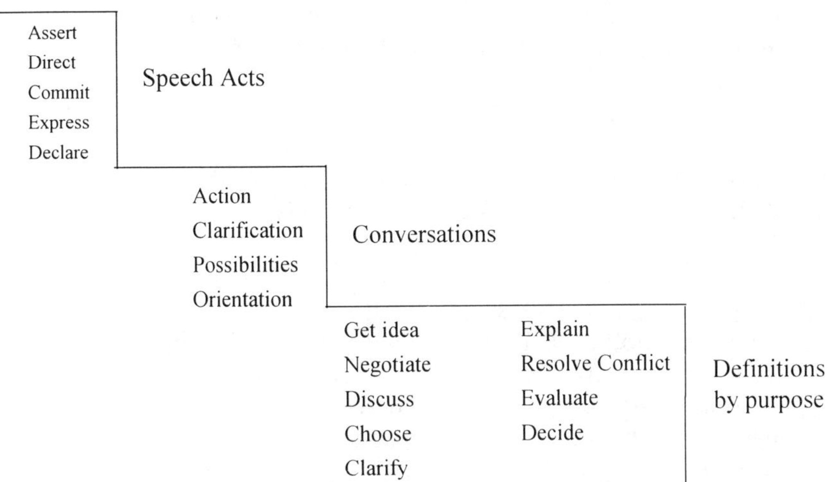

Figure 3.1 The spectrum of interactions.

identify elocutionary acts, which are the basic meaningful units of human communication. The idea here is that every time we say something, that is, perform an elocutionary act, we have some intention-at the same time the listener may respond in a way that is consistent with our intention. The effect on the listener is known as the perlocutionary act. The idea then is that we should be able to describe communications between people as sequences of speech acts. Searle has proposed that five speech acts are sufficient to describe any communicative sequence. These are the following:

- *Asserting:* where the speaker makes a statement of fact and is identified with that statement;
- *Directing:* where the speaker orders or attempts to get the listener to carry out some action;
- *Committing:* where the speaker makes a commitment to carry out some action;
- *Expressing:* where the speaker expresses their attitude or feeling;
- *Declaring:* where the speaker defines some new condition that may impact the listeners.

Describing any activity as sequences of speech acts can become a tedious process. One can imagine that each speech act could be realized as an e-mail message. However, this would mean that both the parties would have to keep track of all these acts and their time relationships to each other. For this reason, it is often more convenient to use standard combinations of interactions, known as conversations, that are what one might call standard combinations of speech acts.

3.2.2 Conversations

Conversations combine a number of speech acts to describe an interaction. Winograd [2] has identified a number of different kinds of conversations. These are as follows:

- *Conversation for action*: This describes the interaction where one party makes a request to another party. It describes the various replies and follow up actions to the initial request.
- *Conversation for clarification*: This describes interactions where one party needs an explanation from another.
- *Conversation for possibilities*: This describes how we decide on a course of action. For example, we may be considering the possibilities for selling a product.
- *Conversation for orientation*: This describes conversations that give a clearer picture of the environment in which particular actions take place. It provides the setting for other activities.

As an example, the process followed in conversation for action is illustrated in Figure 3.2. The process supports an activity where a client requests a provider for some service. The process is defined as a conversation made up of a number of speech acts, hence the term language/action. Figure 3.2 describes the process in terms of state transitions, in a similar way to that used by Winograd and Flores [3]. The states define conversations, and the transitions are speech acts. The top five states (1 through 5) in Figure 3.2 describe the process going through initiation to completion without any interruption. The client makes a request, the provider promises to fulfill the request, the provider reports completion of the request, and the client accepts the completion of the service. The other states and transitions model variations on the process. First of all, a client may not accept the completion requesting further work, as indicated by the transition from state 4 to state 3. There is also the possibility of a counterproposal, leading to the conversation represented by state 6, with a counterproposal accepted by both parties. There can be cancellations on both sides if the provider cannot provide the service, leading to state 8. Cancellations can also occur later in the process, by the provider withdrawing from the arrangement, state 7, or by the client refusing to accept the service.

There is one system, known as the coordinator [3], that has been built based on conversation for action and its use has been reported in the literature. The coordinator supports a process for action that follows a predefined set of steps for one

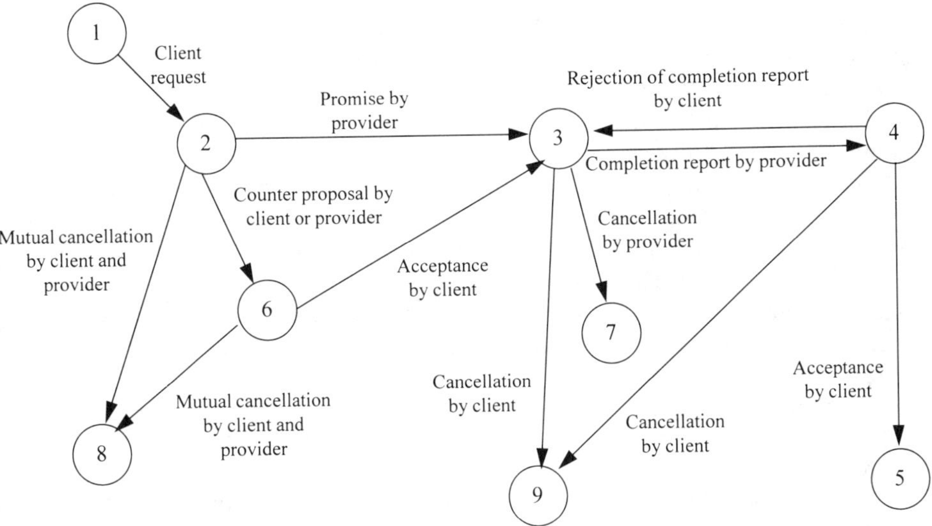

Figure 3.2 The coordinator process.

user to request an action by another. Coordinator requires users to faithfully go through all the steps in the process and it has been criticized for this reason—not allowing users sufficient flexibility in their process.

3.2.3 Defining by Purpose

Speech acts usually refer to two people communicating. One person initiates the speech act, and the other responds to it. Perhaps a better way would be to define broader interactions or social processes that can include a number of people. Some idea of these interactions is given in the following:

- *Get idea:* Generate proposals for a course of action or the solution of problem;
- *Explain:* Describe the characteristics of some object or course of action;
- *Negotiate:* Decide methods of allocation of resources or sharing of resources and responsibilities;
- *Resolve conflict:* Identify reasons for strong disagreement on some proposed course of action and develop a common approach to the course of action;
- *Discuss:* Exchange information about a particular subject;
- *Evaluate:* Evaluate a proposed course of action using agreed-upon criteria and measures;
- *Choose:* Select one of a number of possible courses of action.

It is important to ensure that any group support for these interactions eliminate rather than amplify their undesirable characteristics. Many of these interactions take the form of meetings and suffer from some undesirable meeting properties.

For example, many meeting situations are criticized because of their encouragement of groupthink rather than providing innovative ideas, as most people do not like to disturb a meeting by suggesting some new approach. There is also the assumption that silence means agreement whereas the opposite is often the case, most people preferring not to be identified with opposing what seems to be an agreed course of action. The anchoring phenomenon is also seen as an undesirable factor. This phenomenon means that meetings often start with some assumption of base point that serves as a reference for discussion. Thus starting a meeting with the statement that "I think there will be no demand for this product" will result in totally different discussion when compared to the anchor point "I see everyone owning this product within two years."

Negotiation processes also suffer from many undesirable properties (for example, overconfidence by the negotiating sides, escalation of requirements, failure to consider the judgment of the opposition, and the general feeling that negotiation is a zero sum game; that is, a gain by one side must result in a loss by the other).

Computer support for such interactions or social processes will only be effective if it eliminates or at least reduces some of the undesirable factors. To do this, methods must be devised where people at a meeting are presented with facts with direct bearing on any meeting goal and that arguments are rationalized during discussion. Ways of doing this are discussed in Chapter 10.

3.2.3.1 Case Study: Interactions With the Specialist Designer

Faced with no available design approach, one way to approach designing a support system for the designer is to identify all the interactions and then look at ways of supporting them. Typical kinds of interactions found in design include virtually every interaction in Figure 3.3. This figure uses diagrammatic techniques known as rich pictures that show the roles in the system and the interactions between them. The interaction is enclosed in a cloud with roles involved in the interaction linked to the cloud. The artifacts used by the interaction are also shown by the rectangular boxes. The arrows shows whether the artifact is only examined (arrow to the interaction) and whether it can be changed (arrow from the interaction).

A number of interactions are shown in Figure 3.3. They may be a mix of action and purpose interactions. Arrangement of a contract between the client and designer calls for an "explanation" and a "negotiation" to determine the contract price. Interactions between the designer and assistant include a "generation of ideas" and a "discussion" to improve the design.

What perhaps becomes more important is how to sequence these interactions to smoothly flow from one to the next. Examination of the design processes

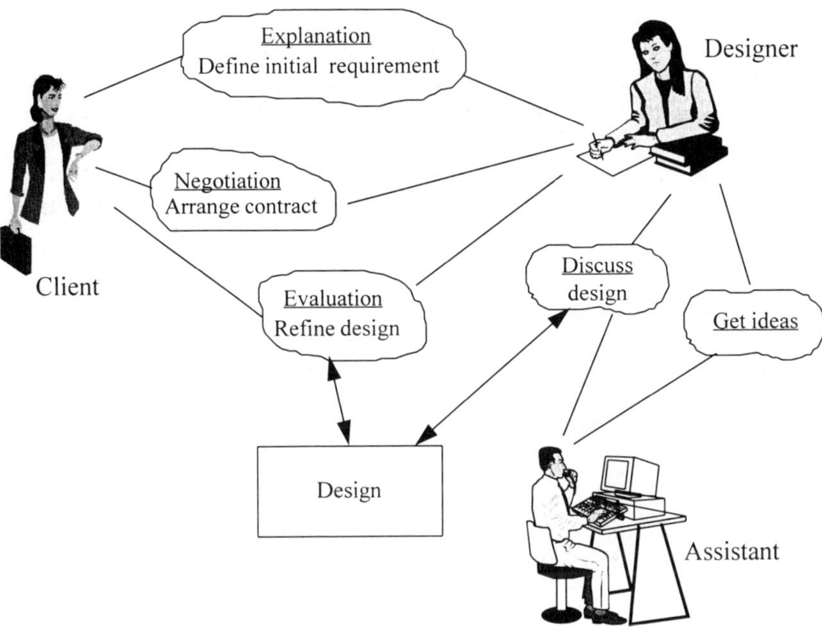

Figure 3.3 Interactions in design.

shows that it is highly iterative and cannot be easily prespecified. Perhaps, at best, what can be done is to describe valid transitions between interactions, and perhaps group them into higher level activities.

Identifying and analyzing interaction is an important part of analysis. Different kinds of interaction often require different kinds of support. Thus interactions that involve information exchange may be supported by different network services than those that require extensive personal interaction. An explanation of a design could, for example, be implemented as an asynchronous exchange of messages, whereas negotiation requires support for synchronous communication, or what is often a face to face meeting.

3.3 DESCRIBING GROUP ACTIVITIES

Interactions are often grouped into higher level activities with some broader task. This book assigns a specific meaning to activities and interactions. Activities are recognized organizationally as having the responsibility of producing some organizationally recognized output, for example an approved design or a budget. Interactions are the steps that make up activities-they are the detailed interactions between the roles in the activity needed to produce the output. Another example is software development, where there is an activity that produces user requirements. This activity can be made up of group meetings about requirements, interviews with users, or comments on system proposals (all of which concern the activity goal, which may be to prepare a requirements specification).

3.3.1 Normative Processes

An activity definition must include all its interactions and the relationships between them. Such a definition may require some time. An alternative is to see if we can define generic activity types, such that each type has the same set of interactions. For example, the activity of providing a service to a client becomes a generic type. This is made up of interactions such as request service, make an offer, agree on service and so on. Standard sets of interactions are sometimes called *normative processes*. The next section describes some well-known normative processes.

3.3.1.1 McGrath's Process

McGrath [4] proposes that processes are made up of the activities shown in Figure 3.4. There are four major activities: idea generation of alternatives, choice of an alternative, resolution of any conflicts through negotiation, followed by the execution of the task. The activities themselves do not necessarily proceed in the sequence shown in Figure 3.3 and teams are likely to switch between them as a task proceeds. Each of these activities are divided into different parts. A distinction is made in idea generation between creative tasks that generate ideas and planning tasks that generate plans. In the choice step, there is the intellectual assessment of an alternative in terms of its positive and negative features and the actual process of selecting it based on an argumentation process. There can be some discussion about the possibilities and perhaps some disagreements that must be resolved. Conflict resolution that often arises when making a choice distinguishes between conflicts of viewpoint and conflicts of interest. Execution itself also calls for resolving conflicts of power and responsibility within a group, often requiring some negotiation, as well as the performance-oriented issues of how to best execute some action.

McGrath also makes a distinction between the cooperative and conflict-oriented activities. Thus idea generation , planning, and selecting criteria for choice are mainly cooperative whereas the process of choice, negotiation, and assigning responsibility often involve considerable conflict.

Perhaps one lesson to learn from this framework is that any tools provided for teams must support all the cooperative and conflicting activities within a team.

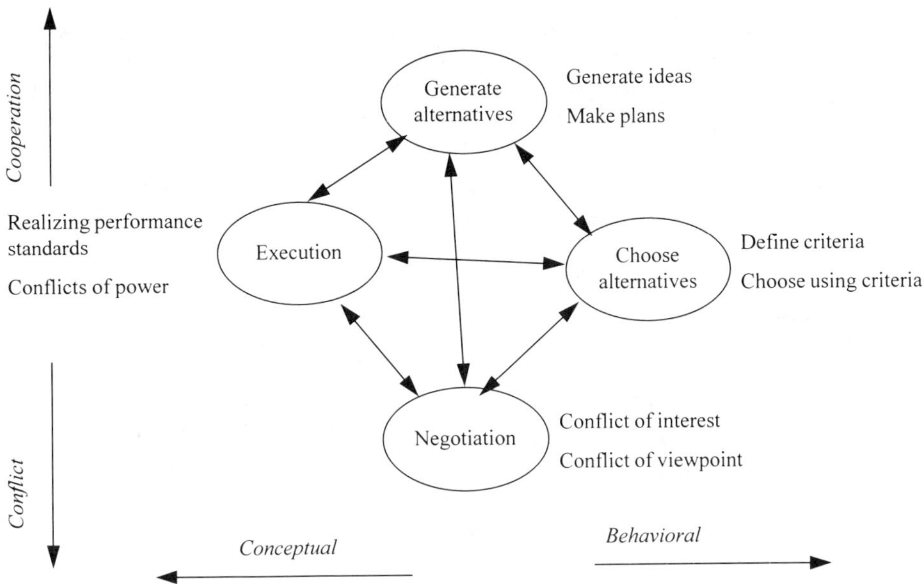

Figure 3.4 McGrath's process.

However, they should not impose a process structure or sets of steps to be followed. On the other hand, team members must be able to select the appropriate tool for the activity in progress. This contrasts with conversational systems that are more strict in the sequence for performing steps, thus detracting from their value.

3.4 DIVIDING WORK INTO GROUPS

A group or team is usually given a task within a higher level mission. However, there are many occasions where tasks can become so large that they cannot be managed by the one group. In that case, tasks are subdivided among groups, with groups themselves coordinated. Information flows between and within groups are an important component of group creation and coordination. Information can be exchanged synchronously or asynchronously, and it can be exchanged through face to face meetings or by exchange of documents. There are two ways of looking at information flows: the physical and the logical. The logical view primarily looks at *what* information is exchanged whereas the physical view looks at *how* information is exchanged.

A group of people is usually made up of people with some common interest or goal. Different groups may organize themselves differently to carry out the tasks needed to achieve their goal. Task characteristics often indicate the support needed by groups. For example, groups with shared decision-making will need different support from those where decisions are delegated to individuals. In other groups, such decisions will be made by group discussion. In still others, they may be made by group leaders. Sometimes the word behavior is used in the context of group work. Behavior itself is difficult to describe, but often means how people react to different situations. Thus technology may be chosen to reinforce behavior that is most beneficial to the group's objectives. Conversely, new technology that does not match current group structure and behavior should be avoided as it will probably not be adopted by the group. Group structure thus becomes important because it determines the kind of support that a group needs. Group characteristics can often be used as an important guideline in choosing the technology for the group. One should avoid simply transferring a technical support system used by one group to solve a problem to another group with a similar problem. If the second group works in a way that differs from the first group, then the transfer will probably be unsuccessful.

Choosing the right technology to match a group requires some understanding of group structures and how groups work. This chapter will propose some frameworks by classifying groups in different ways. Guidelines can then be provided for choosing the right supporting technology for a given group.

3.4.1 Roles

A discussion about groups would not be complete without mentioning roles. Roles are an important way of viewing groups. There are two major views of roles. The traditional view sees individuals in organizations as occupying positions where each position has a set of responsibilities, which may be things like approving some purchase, buying a part, or managing a project. These responsibilities become the roles, which are assumed by the person occupying the position. A position can be responsible for more than one role, and a role can be moved from position to position in an organization.

One question, however, is now becoming important. Should roles be assigned to positions or to people. Assigning roles to positions requires the occupant of the position to possess all the skills to carry out the roles. This has some impact where the role of a position changes but the position occupant does not have the skills to carry out the new roles. This often places a constraint on organizational change, requiring change to be such that it still fits with the skills of existing occupants. Otherwise, it may be necessary to cope with often unpredictable behavior when there is a mismatch between a person's skill and their roles. The idea of task-oriented structures proposed by Drucker, and described later in Chapter 5, solves this problem by getting away from the idea of assigning roles to positions and assigning them to people. This becomes the second view of roles-assign them to people, not positions. The organization can become more flexible if we can reassign roles to people rather than going through the process to change an organization to fit with people's skills.

People's behavior is often influenced by their role and also by what is expected of the role within the organizational culture. Furthermore, a group where roles are assigned to people can often assign roles to its members and change the roles during the life of the group. Clearer definition of roles gives a good indication of the kinds of flows that must later be supported within the system.

3.4.1.1 Case Study: Defining Roles for the Worldlink Consultants

Worldlink Consultants have found that managing the complex relationships that occur when preparing proposals across the globe requires a clearer definition of people's responsibilities. It then becomes possible to define the preparation process in terms of role actions. For example, a role of proposal coordinator is being proposed to coordinate the project and keep track of its status. A set of expert roles can then be designated and assigned responsibility for parts of the proposal. Then, consideration is being given to formally defining reviewers for the proposal to improve its quality. A finer distinction can then be made for the roles. We can have the "technical" expert and the "budget" reviewer. It becomes possible to make statements like "section A is now with the budget reviewer." Once the roles and responsibilities are defined, it becomes possible to define possible interactions between the roles and information flows needed to support these interactions. Once the roles are clearly defined and agreed, then it is proposed to implement a workflow support system to support the process.

3.4.2 Identifying Information Flows

While defining roles, it is often necessary to look at the artifacts in the system to ensure that a role is assigned responsibility for each artifact. Questions like, "Who will look after the budget?" must be asked. What information is needed to accomplish a task? Groups interested in particular artifacts will then emerge and ways of jointly working on the artifact will be identified, giving designers an idea of how information is exchanged in the system. Distinction of joint work through information exchange, personal interchange, or workflow will later be useful in choosing the best network services to support the interactions. Where joint work is possible through exchange information, then e-mail may be sufficient. Where more detailed personal ideas must be exchanged, then discussion databases may be considered.

In such analysis it is often beneficial to get away from current physical activities and look at what it is that is really being done. Once the reason for some action is known, it becomes possible to ask questions like, "Do we really need to do it this way?" or, "Is a better method possible once network support is provided?" For example, if we find that a meeting is being held simply to interchange information, we can suggest that a better way is to distribute information for comment, and then only hold a meeting if conflicts are identified. We can then ask whether meetings have to be held over a fixed time period, or at regular times, or made asynchronous and extended over a longer period using computer communication support. Answers to these questions can lead to networked systems that improve the collaborative process rather than simply automating it.

Thus instead of proposing expensive tools such as videoconferencing that simply replicate face to face meeting, an alternative way may be to have extended meetings. Members of the group perceive themselves to be in a constant state of meeting where they contribute to the topic at any time that they have some useful idea. This is an idea that is strongly put forward by Turoff [5]. The book will return to this topic in later chapters that describe technical tools. Costs are also very important in making the choice.

3.4.2.1 Case Study: Information Flows for Worldlink Consultants

Figure 1.8 in Chapter 1 outlined some of the major activities in preparing proposals. A more detailed examination was then carried out to determine the information flows that occur during proposal preparation to see how they can be expedited with network support services. Examination of the activities has produced the more detailed picture in Figure 3.5. New roles or reviewers have been added and a collator identified to work together with the coordinator to assemble draft sections into a final report.

The development of this picture has shown that most information flows between the coordinator and the experts, who may be at different locations. The collator is most often at the same place as the coordinator, although at times the coordinator may be away, in which case the collator assumes some of the coordinator's responsibilities. A coordinator may be managing up to ten proposals at any one time. A proposal is made up of an average of seven sections, which often require consideration by two or three experts, each of whom may add comments that must be eventually be collated and any differences of opinion resolved. In addition, ad-hoc questions may be made by experts or reviewers.

It is clear that this environment must support both the exchange of information as well as personal communication to resolve differences of opinion. One alternative for information exchange is to maintain a central repository, perhaps based on a document management system, allowing the roles to make copies of their parts and recording their comments back into the repository. This will have to be supplemented by a workflow tracking system to see where all the parts are. Another alternative is to distribute parts using a workflow system itself. An important question to be resolved is how to support the flows for personal relationships

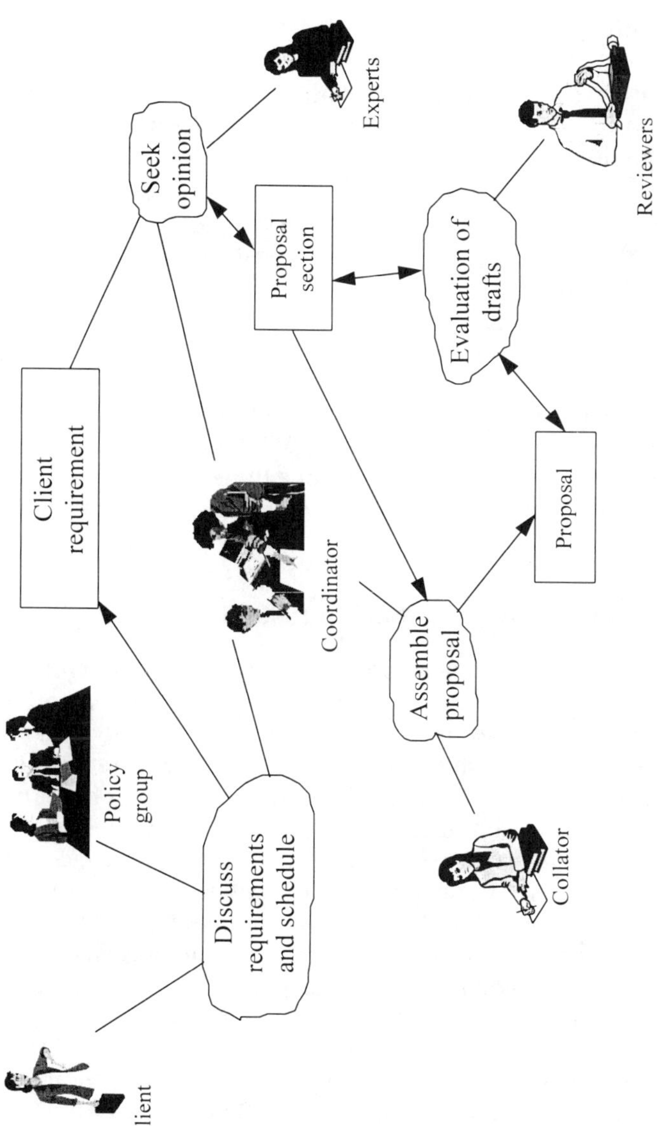

Figure 3.5 Information flows in proposal preparation.

and, in particular, whether some form of discussion database support is to be provided because of the wide distribution of the experts and reviewers involved.

3.5 GROUP CHARACTERISTICS

There is more than one way to describe groups, and it is recognized that group dynamics depend on the combination of many factors. The most common factors used to describe groups are

- The culture of the group environment or its setting;
- The group structure;
- The task of the group;
- The membership of the group;
- The characteristics of people of the group.

This section begins by describing the group culture and then uses a classification of group structures as given by Sprague and McNurlin [6] to describe the other factors. The classification is two-dimensional-one dimension defines group membership and the other group structure, both of which can be used to describe group cultures.

3.5.1 The Setting or Culture of the Group Environment

Group behavior depends on its role within an organization. Important parameters that define the setting are how the group is supported by the organization and what it is authorized to do. Does the group operate autonomously or do individuals have to report to their respective superiors before decisions are made? Does the group have access to all the information that it needs, and does it have the authority to implement any of its recommendations? Whether it becomes responsible for its decisions is another important part of the setting. All of these factors make up the setting of the group, and the group must operate within the constraints imposed by its setting. These constraints and responsibilities together make up what is sometimes known as the *culture* within which the group operates. They also include whether innovation is encouraged, how decisions are made, and how people in the organization are rewarded.

Organizational culture becomes even more important in the international context where it can be influenced by national traditions. The importance of culture in the international context is described by Watson and others [7] and by O'Hare-Deveraux and Johansen [8]. Even more important are the differences in culture as one passes between national boundaries and the methods to support groups that include more than one country. Thus there may be cultures where people work as individuals with relatively loose structures and those that are authoritative and work in strictly hierarchical ways. Group support systems must somehow bridge these cultures.

O'Hara-Deveraux and Johansen [8] identify five major cultural parameters—language, context, time, power relationships, and information flow—as the most useful variables in identifying cultural differences. Language is perhaps the best known of these parameters. Language must be seen beyond its simple syntax, more as the expression of national culture in terms of language phrases or phrases. Thus the word "soon" may have different meanings in different cultures, as do phrases like "its up in the clouds." Simply translating such phrases syntactically into another language will completely lose the intended meaning.

Context is also often related to national culture. It defines what must be taken into account when acting on a particular message or event. Is it possible to act on an event independently of the rest of the context? High-context cultures consider many factors in considering their actions. They often go beyond the present and consider action within a longer historical perspective, considering the historical standing of a corporation or even, perhaps, its family owners. In others, it includes reasons for earlier decisions and the relationship of current personnel to these decisions. Thus, for example, one may not always wish to change some process or activity that was initially set up and is favorably considered by the chief executive. Low-context cultures are those that tend to use information directly related to some actions, and often compartmentalize their activities by relying on information directly related to their task, often separating business and social relationships.

Time also is also an important cultural element. It has two dimensions. One is whether activities are looked at one at a time (monochronic) or whether a number of activities are taken together (polychronic). The other viewpoint of time is its past/present/future orientation. Again, some cultures, and even professions, emphasize one of the orientations more than others. Thus accounting is primarily present-oriented, whereas research and development is future-oriented. It is often a challenge to mix these two disciplines when making a decision.

The distribution of power is perhaps the most important element as it determines the way that relationships are built up within and between organizations. Organizations can vary from highly authoritarian to loosely coupled, and matching support systems must be chosen to support these relationships. Thus one would not broadcast a new idea to all people in a hierarchical structure, whereas one may do so in a flat structure. In some alliances, it is necessary to match the different structures of alliance partners. For example, a contact person may be created in the flat structure to correctly pass information to members of the hierarchical structure. The contact person must first learn the organizational culture to follow the correct channels in information distribution.

Information flow is another factor that considers both the speed and path of communication. Power relationships have an important bearing on information flows, as does the context. Thus information will often flow up and down a hierarchy in authoritarian relationships, with extremes where only downward flows exist. Furthermore, in low-context cultures, information will become compartmentalized and communication tends to be often to the point. High-context cultures rely on a wider distribution of information. It is also often the case that information flows more quickly and freely in high-context cultures, whereas compartmentalization and barriers in low-context cultures tend to inhibit such flows. Once a flow pattern for a given structure is determined, then methods are developed to minimize time taken for information to flow along this pattern.

All these factors have an interesting implication on group formation. Is it, for example, easier for groups with the same time orientation to work together than one where members have different time orientations? Alternatively, what support processes are needed for groups whose members have different time orientations?

3.5.2 Group Tasks

Group behavior also depends on its tasks. In highly structured tasks, a group may define roles for each of its members, define the work to be done by each role, and the relationships between the roles. Thus a group processing customer orders may have a role that looks after enquiries, another to expedite orders, and another for customer liaison. The person with the inquiry role may pass a customer to the person who takes the liaison role, who in turn arranges the customer request to be met through the expediter. The expediter is responsible for meeting the request. The person who takes the liaison role liases between the customer and the expediter to clarify any issues during order processing and, later, to arrange delivery.

The division into roles together with stable role relationships may not be appropriate for unstructured tasks. For example, developing an alliance may lead to unforeseen tasks to be accomplished. For example, a new idea for a venture may call for specialist analysis and agreement of new organizational units. New roles must be created to carry out the analysis and the role assigned to a person suited to execute it. This role will then disappear once analysis is complete, but new and not easily predicted roles may be needed further down the process.

3.5.3 Group Structure

The classification here is more closely related to the structure of groups and the roles of people in each group. A large variety of group structures are found in practice. An indication of such groups is given in the following:

- *Authority groups:* These usually include a supervisory role that assumes responsibility for the tasks carried out by other people in the group. Roles in authority groups are often fixed, as are the people who undertake them. In fact, such roles are often tied to positions.

- *Management relationship groups:* This represents relationships where people in managerial roles provide guidance to people in their groups. It is less formal than an authority group because managerial roles act more in an advisory rather than supervisory role.
- *Clerical processing groups:* Here, the group breaks up a task into simple components and each member becomes responsible for one component. In these groups, roles are often tied to positions and changes require redefining position responsibilities.
- *Peer groups:* Where people with the same interests can organize their work. They may do so by assigning roles to each other with the goal of accomplishing some task from time to time. For example, some people may agree to keep records of minutes, or collect statistical information, or perhaps even lobby political groups for support.
- *Project teams and task groups:* This extends the idea of authority group by adding coordination roles to the group. These roles monitor the activities of other roles to ensure that they fit together in a way that completes a task in a prespecified period.
- *Committees:* Groups here primarily discuss issues and often have a task of deciding on future actions. Here roles, such as keeping minutes, may also be fluid.
- *Information exchange networks:* Where groups try to develop a common view of the world through the interchange of information.
- *Business relationship groups:* Members of these groups usually exchange informal information that may be of interest to them.
- *Client-server relationships:* Where some people offer services to others. Here, some people can assume a service role whereas others assume a client role.
- *Social networks:* These are groups that usually exchange informal information related to their context. Role-taking is infrequent here unless the group tries to organize itself to carry out some specific task, such as organizing a party. Some group members may then undertake to make arrangements for the provision of food, or arranging a location, or some other task.

Group structures range from authority groups with clear reporting relationships to social networks where reporting patterns are uncontrolled and continually changing. There are also structures where roles can be continually reassigned such as those that occur in many peer groups or committees. The kinds of roles vary in these groups. In authority groups, each role would be assigned some specific responsibility and reports their progress to an authority role. With peer groups, people may be assigned roles to deal with specific problems as they arise, and report on progress to the whole group. Group structure again becomes important in design, as different group structures have different communication patterns and require different kinds of supporting networks. Clerical groups, for example, are likely to need work process support based on systems such as LOTUS Notes, whereas information exchange networks can make good use of the World Wide Web. More thought must be given to structures such as committees, task groups, or peer groups whose work emphasizes interpersonal relationships, suggesting greater use of discussion databases or interactive bulletin boards.

3.5.4 Membership Characteristics

Membership factors as those that define the broad dynamics of a group, in particular how easy it is to join a group and how closely together people work in the group. The categories proposed by Sprague and McNurlin [6] are shown in the following:

- *Open groups:* Where new members can be freely added to the group;
- *Closed groups:* Where the authority to add members to a group rests with the owner;
- *Loosely coupled groups:* Where members work relatively independently;
- *Tightly coupled groups:* Where there is strong dependence between the group members;
- *Hierarchical groups:* Where information flows through predefined paths.

They are similar to those described by Constantine [9] for the software development environment.

3.5.5 People Characteristics

The characteristics of people in a group are also important and can affect the operation of a group. Apart from personal characteristics, it is necessary to pay attention to the skills needed for group work. Thus, for example, skills needed for effective collaboration must be developed (for example, ability to resolve conflicts, make presentations to groups with a variety of skills, evaluate alternatives in a rational manner, and deal effectively with people). Some specific skills can also become important such as, for example, leadership skills for leadership roles.

3.5.6 The Variety of Groups

The range of cultural and structural factors simply illustrate the variety of groups. Groups can exist in any combination of cultural and structural characteristics (for example, an open committee or a closed management relationship group). There are further specializations that define processes followed by a group. For example, groups may make decisions jointly or may delegate it to a single individual or small group. In that case, the communication support will be different in each case. In one case, there is a need to maintain continuous contact between members, and in another case this is not needed. Consequently, it is difficult to define a design process that is sufficiently generic to cover all groups. Perhaps we should leave it up to the group to choose one service from a range of available services.

There are also many cases where there are a number of groups interacting with each other. One way to view the globalization example described as *global planning* in Chapter 1 is as a number of open peer research groups that are currently loosely coupled. The goal is to increase the coupling and project orientation of the groups by the definition of specific tasks. At the time, there may be structured project teams that may have to be made more open to receiving of new ideas.

The important factor to remember is that behavior of individuals in a group can change with the introduction of technology, as they may assume new roles using the technology. Thus introducing e-mail can loosen the hierarchical nature of a group, while providing access to new developments such as the World Wide Web can raise the level of innovation within a group.

3.6 SOCIAL FACTORS—ACCEPTANCE OF NEW SYSTEMS

One important, if not the most important, consideration in system design is to ensure that a system will be accepted by its users. There are many cases where installed systems are not used fully by people, who then carry out most of their work independently of the system. This can particularly happen with CSCW systems, which often need to support a much larger variety of work, and in particular work that is often personal, such as jointly editing documents or contributing to a meeting. In addition, CSCW systems often are not seen as critical, and may even be perceived as an overhead that somehow has to be justified. Thus social consideration has a much greater role in CSCW design than in the design of transaction-based systems. The three major social issues in CSCW design are as follows

- How to get acceptance for introducing CSCW systems;
- How to provide facilities that enable growth;
- What support to provide for groups to facilitate teamwork.

This section provides some useful design guidelines or principles derived from research on social issues in CSCW systems.

3.6.1 Principle 1—Minimal Disturbance

One principle comes from the work on Applegate and others [10] on acceptability of new systems. In general, users are not happy about changing their work practices quickly because of the introduction of a computer system unless they perceive some direct and immediate benefit to their work. The resulting guideline is that new systems should *initially NOT CHANGE or at least only have MINIMAL EFFECT on existing work practices.*

Thus, for example, adding e-mail to a person's computer does not affect their existing work practices, but can in fact make some of their work easier. People need not use e-mail, so their work need not be impacted. However, to be accepted, e-mail must support the information flows consistent with the organization's culture.

3.6.2 Principle 2—Adding Value

Following on from the first principle, however, users should be able to perceive some benefit from newly installed systems. Thus there should be some ways that CSCW systems add to the value of a user's work. Adding value can mean many things. It can simply mean making it easier to send a message. It can be giving more options in an application or, in the case of groupwork, giving better access to group members. This should be done by simply providing the communications facilities without requiring changes to user programs. Thus, for example, adding e-mail, enables a user to broadcast some information.

The design goal thus becomes for a new system to *enrich the context.*

However, the process should not be enriched by simply adding new technologies to the workspace and expecting the user to choose methods for using the technologies. Instead, the system should be seen as providing new services to users and not require the user to worry about how to use the operating system to adapt technologies to user needs. Thus, instead of new menus or windows that make new software available to a user, a workspace should be enriched with menus or windows that provide new services directly applicable to a user problem.

3.6.3 Principle 3—Growth

Once users have access to new technologies, they should be able to experiment with them to find good uses for them. This leads to the principle that a new system should *provide the opportunity for growth.*

The principles outline the general strategy that is followed by most designers. The assumption is that people will learn how to use CSCW technologies and thus begin to use more and more of its tools. Design should thus provide a platform of tools and allow the users to select and adapt the tools to their process. Word processors are a good example here. They can simply be used as a typewriter, but also provide opportunities for new ways of doing things (for example, moving paragraphs, rearranging tables and so on, to produce higher quality documents).

3.6.4 Principle 4—No Additional Workload

Finally, CSCW systems are usually justified by giving users the ability to simplify their routine work processes. They should, however, not simply replace one set of routine processes with another. This leads to the next principle, namely, that *a new system should not add to the routine workload.*

This principle becomes particularly challenging when we wish to enrich processes with group memory. Group memory requires reasons for various actions to be stored in the system and made available to the group through the group context. One easy solution here is to simply require users to enter reasons for any actions at the time of the action. This is often perceived as additional workload and resisted by users. Thus design for capture of group memory should ensure such capture as part of normal processes rather than additional work. This is an extremely challenging problem in CSCW design. In fact, new systems should reduce the workload (for example, sending faxes directly from computers, removing the need to walk to fax machines).

3.6.5 Additional Principles

The four principles just outlined are seen as the most important in this book. There are, however, many other principles, guidelines, and suggestions in the literature. Grudin [11], for example, suggests many additional features. This includes paying attention to exceptional conditions, ensuring that there are sufficient users to justify a networked system, ensuring that group support is consistent with the group culture, and integration of group services. Grudin also draws attention to the difficulty of measuring the effectiveness of group support systems to use previous experiences to improve group support system design. The second important requirement is that it must provide a new platform for growth. This has often become known as structural adaptation theory, which has through empirical research found that users always try to adapt computer systems to their most preferred ways of working.

3.7 SUMMARY

This chapter described how groups of people work together. It characterized groups in terms of their culture, task, structure, membership, and characteristics of the team individuals. It then discussed alternate ways of describing how members of a team communicate, outlining the theory of speech acts as well as normative processes such as conversations and decision processes.

3.8 DISCUSSION QUESTIONS

1. What is the advantage of assigning roles to people and not positions?
2. Do you think classification schemes are useful for describing complex systems?
3. Why is it important to match group structure and culture when providing support?

4. Why is culture important when forming international alliances?
5. Try to characterize a group to which you belong.
6. Characterize your group members by their context and time orientations. How do they compensate for differences, if there are any?
7. Do you think the type of group suggests the kind of network services that it needs?
8. Do you think a distinction between tasks and goals is needed?
9. What do you understand by the term normative process?
10. Do you think structured normative processes can be used in all group activities?
11. What is the major difference between McGrath's steps and conversational systems?
12. What kind of problems would be amenable to conversational support?
13. Do you think there is a relationship between group structure and the processes it uses to carry out its tasks? For example, take a group structure and see what kind of activity, in terms of McGrath's model, would be more prevalent in each structure.
14. What conditions should be met to improve chance of acceptance in a group?

3.9 EXERCISES

1. Consider the impasse reached by the expert designer and propose some solutions to resolve it.
2. Define the nature of groups in the software engineering project.
3. Suggest the kind of network services that may be needed by an information exchange network.
4. Consider your experiences at a university. Define the roles and interactions between them in this environment. Can you classify the interactions using the scheme in Figure 3.1, listing the roles in each interaction?
5. Consider global planning of research and try to identify some interactions in this context.
6. The following scenario describes the production of short films. See if you can define the roles and interactions in the scenario.

   Film production goes through a sequence of activities. First there is a story to be written. Then a set of frames must be produced to illustrate how the film that describes the story will evolve-this is often called a storyboard. The storyboard is checked against the original script and then "filled in" by frames between the initial storyboard frames. Much of this work is done by people with different skills. Usually, the people are self-employed workers. The mission, again, is to develop quality films in minimum time at minimum cost.

Again, it is possible that many of the professionals involved in film production can be widely distributed and their work can perhaps be coordinated and facilitated through the use of computers. Coordination requires not only verbal discussion, but also display of pictures and scenes. Again, the same questions arise as in the case of the small trader. How to achieve such coordination with the least travel time. The question now is more complex as communication here is between a large group of people rather than an individual. A process that can include many steps must be established for the group. To define such a process, it is necessary to define how these professionals communicate, how much of this communication needs to be face to face, and what facilities are needed to support either case of contact.

References

[1] Searle, J. R., *Speech Acts - An Essay in the Philosophy of Language*, Cambridge, MA: Cambridge University Press, 1969.

[2] Winograd, T., and F. Flores, *Understanding Computers and Cognition: A New Foundation for Design*, Norwood, NJ: Ablex Publishing Corporation, 1986.

[3] Winograd, T., "A Language/Action Perspective on the Design of Cooperative Work," *Human-Computer Interaction*, Vol. 3, 1987–88, pp. 3–30.

[4] McGrath, J. E., *Groups: Interaction and Performance*, Englewood Cliffs, NJ: Prentice-Hall, 1984.

[5] Turoff, M., "Computer Mediated Communication Requirements for Group Support" *Journal of Organizational Computing*, 1. 1991, pp. 85–113.

[6] Sprague, R. H., and B. C. McNurlin, *Information Systems Management and Practice* (3rd ed.), Englewood Cliffs, NJ: Prentice-Hall International, 1993.

[7] Watson, R. T., T. H. Ho, and K. S. Raman, "Culture: A Fourth Dimension of Group Support Systems," *Communications of the ACM*, Vol. 37, No. 10, Oct. 1994, pp. 44–55.

[8] O'Hara-Devereaux, M., and R. Johansen, *GlobalWork: Bridging Distance, Culture and Time*, San Francisco, CA: Jossey-Bass Publishers, 1994.

[9] Constantine, L. L., "Work Organization Paradigms for Project Management and Organization," *Communications of the ACM*, Vol. 36, No. 10, Oct. 1993, pp. 34–43.

[10] Applegate, L. M., "Technology Support for Cooperative Work: A Framework for Studying Introduction and Assimilation in Organizations," *Journal of Organizational Computing*, Vol. 1, 1991, pp. 11–39.

[11] Grudin, J., "Computer Supported Cooperative Work: History and Focus," *IEEE Computer*, Vol. 37, No. 5., 1994, pp. 19–26.

Selected Bibliography

Chang, M. K., and C. Woo, "A Speech-Act-Based Negotiation Protocol: Design, Implementation, and Test Use," *ACM Transactions on Information Systems*, Vol. 12, No. 4, 1984, pp. 360–382.

Chapter 4

Processes and Workflows

LEARNING OBJECTIVES

- ❑ *What is a process*
- ❑ *Formal and informal processes*
- ❑ *Roles in processes*
- ❑ *Facilitating a process*
- ❑ *Notification schemes*
- ❑ *Workflows*
- ❑ *Process characteristics*
- ❑ *Process and task support*
- ❑ *Process improvement*

4.1 INTRODUCTION

Earlier chapters referred to processes in a relatively loose way. The term *process* is important when discussing systems because it describes *the way we do things* or, more specifically, the way tasks are carried out. It describes the sequence of tasks, the tools used, and who carries out the tasks. If there is a better way to carry out the tasks, then there is a better chance of producing better outcomes, such as a higher quality product with a lower cost. Businesses talk about improving their processes. But in addition, people as individuals can now also do things differently using computer-communication support. This is because they have better access to expertise, they have more contacts, and better access to information. Furthermore, businesses are talking not simply about incremental improvements such as adding another person to speed things up or installing a computer for some task. They are talking about wholesale restructuring of the processes, or the way things are done, including the reorganizing of their business.

There is continuing work to define processes in a general way so that experiences gained in one process can be passed on to another. The goal is to find some general principles and properties about processes and apply them in different situations. One question here is whether to define processes broadly or to define them as detailed steps. A broad description can be "find a customer for our product," while a detailed way specifies every step that must be followed to find a customer. Correspondingly, any definition method must be able to specify both the top-level business process or go into detailed interactions between individuals involved in this process. It is often difficult to describe processes where people collaborate in detailed ways. The steps are very small and often we do not think about how we carry them out. For example, how do you describe the way a meeting goes at a detailed level, or how do you write a report.

This chapter outlines some approaches used to describe specific processes. The chapter also describes some ways of improving or redesigning processes. In particular, it defines a set of measures for processes and bases improvement on these measures. We can then take a specific process, evaluate it in terms of process characteristics, and propose a new structure for the process.

4.2 ROLES AND PROCESSES

It is important to realize that the way arrangements to carry out tasks are made is usually a combination of process, groups, and tasks, as shown in Figure 4.1. In design, it is usually necessary to determine the relationship between these factors and choose the best combination for a given goal. Usually, there are constraints on what can be done, and the current work situation will determine what is easy to change and what is not. Thus a process may have to be designed to fit a given group or a task. Or it may be necessary to choose a group structure for a given process and task.

Chapter 3 described the importance of roles as the responsibilities assigned to people. A design can start by identifying the goal of the process, and then identifying the tasks needed to accomplish the goal. These tasks are then assigned to roles. The next step is to define the sequence for carrying out the tasks and provide the support needed by the roles for their tasks and for the process. *Task support* should make it easier for roles to carry out their work, whereas *process support* should make it easier for the roles to communicate with each other.

The way that roles are assigned and the support that they get can affect the outcome of a process. Thus choosing roles with the correct specialization can mean that each process step becomes more effective. Process support can improve the flow of work between the roles.

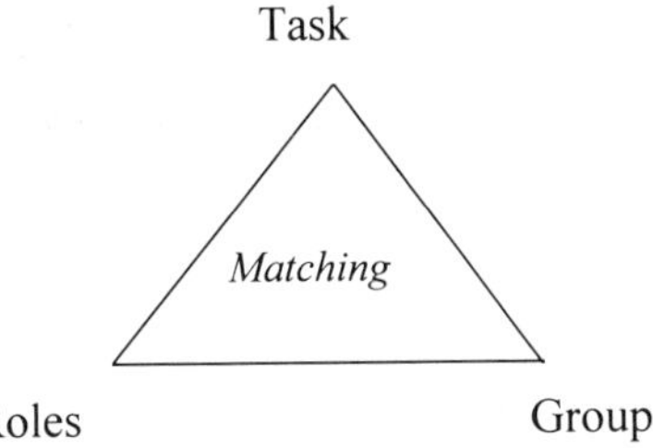

Figure 4.1 Interdependency of roles, tasks, and processes.

The choice of a process is thus complex. One way to manage this complexity is to classify processes by task or group structure. We can say that given a group or task of this kind, we should use a process of that kind. Or for a given process or task, we should choose a group like the following. The previous chapter described a classification of groups, and this chapter will outline a classification of processes and try to find matches between them. These classifications can be used to support designers. Processes themselves must be structured to meet certain goals.

4.2.1 Goals of Process Design

Processes have two goals to satisfy, one is to assist people to carry out required tasks, the other is to carry these tasks in the best possible way. Sometimes the term *quality process* is used to describe processes that work effectively. Where processes support groups, they must do the following:

- *Support collaboration* by integrating people and artifacts into the process and fitting in with agreed upon work practices;
- *Achieve quality* by supporting process improvement, maintaining awareness and memory, and in general raising satisfaction with the system;
- *Improve productivity* by providing better access to information and better coordination support, thus improving *awareness* and *group memory*.

Many businesses are now redesigning their processes to improve their quality. Chapter 5 will describe such redesign, now generally known as *business process re-engineering*.

4.3 CLASSIFYING PROCESSES

Classification is an important factor in design, and one design goal is to classify processes and match them to groups. A process classification can then also be used to choose the best collaborative technology for process support. There is a wide variety of processes used in group work, and often many different ways to do the same thing. Processes can be structured or informal. They may require synchronous contact or may be done asynchronously. This chapter classifies processes by their general philosophy and structure.

4.3.1 Process Philosophy

Constantine [1] has described one way to classify processes in software engineering. The classification is by its philosophy, and although it was defined in the context of software engineering, it is generally applicable. In this classification, processes are classified by the way they are carried out. This can be

- *Orderly*: Where things are done in a predetermined way;
- *Openly*: Where the sequences can be changed to improve the process;
- *Zealously*: Where users have a mission and all action is directed to that mission;
- *Creatively:* Where users discover and work on problems as the process proceeds.

In a way, orderly implies a formal structure where tasks in the process are clearly defined, as are the roles who must carry out these steps. Each role knows what they must do and the next task to be carried out. At the other end of the spectrum, the process becomes more informal. The next step depends on the outcome of the previous step and may be totally unpredictable.

Another reason for classifying processes is to determine the kinds of support collaborative technologies must provide for a process. Thus, orderly processes require support for predefined workflows; open processes require support for varying slightly from a predefined set of steps, with more emphasis on interpersonal relationships; zealous processes require constant awareness of the mission; and creative processes must allow the process itself to be defined as work proceeds.

4.3.2 Process Structure

Another way to classify processes is by their structure. The simplest structure is where the process is a sequence of sequential steps as, for example, process A in Figure 4.2. Here, step 2 follows step 1. Step 2 is then followed by steps 3 and 4. This process is quite predefined as its outcome can often be predicted. For example, a bank transaction goes through such a predefined process. We can of course have

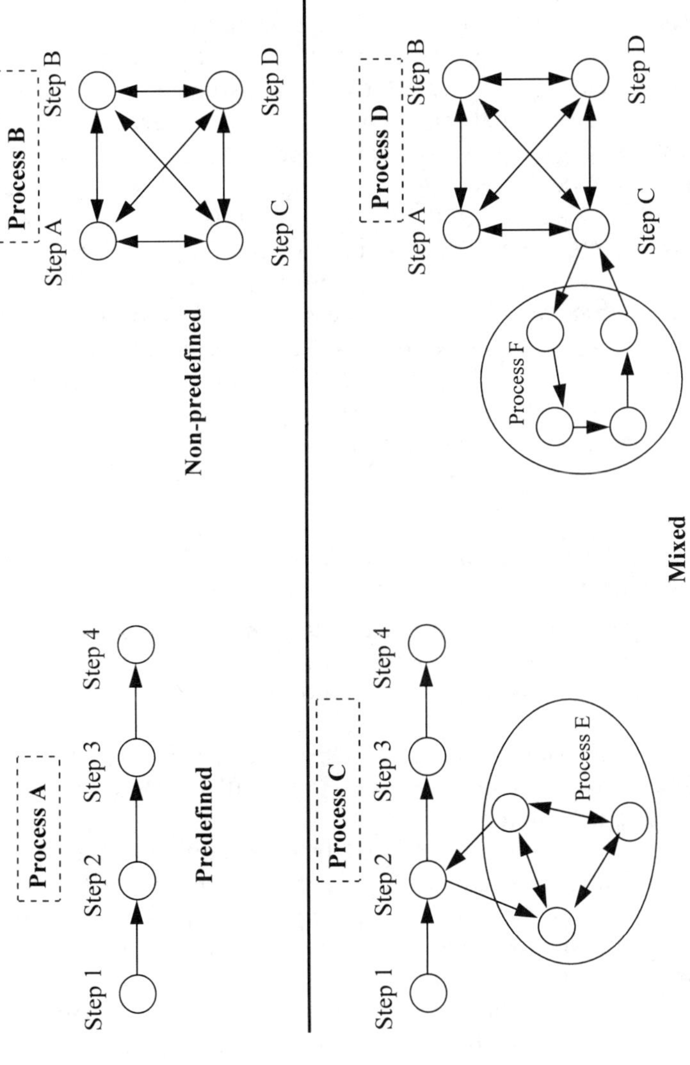

Figure 4.2 Process structure.

variants on a predefined process, such as allowing it to be interrupted, replacing one of its steps, or adding an alternate path.

A non-predefined process, on the other hand, is quite different. Thus in process B, step A can be followed by any of the other steps. The next step is often decided following the completion of the previous step and the selection can require considerable specialist knowledge. It is not possible to predict the exact sequence of steps, but the interactions in each step can be defined. An example may be a meeting. It can be made up of interactions, like a discussion, a vote, an explanation, and so on. It is known that these kinds of actions are taken during the meeting, but, in most meetings anyway, you cannot predict their actual sequence.

Then there are mixed processes. For example, each of the four steps in the predefined process C, can itself initiate a non-predefined process, as for example step 2 initiating process E. Thus, for example, a step in a predefined process may initiate a meeting. Similarly, steps in non-predefined processes can start predefined processes, as for example, step 3 of process D starting process F in Figure 4.2. For example, a meeting can start a process to carry out some estimate using an agreed upon set of steps. *It should be noted that most processes, in practice, are mixed processes.* Hence, more attention must be paid to the detailed steps and how they can be supported to improve the whole process.

4.3.2.1 Case Study: Midvale Community Hospital—Process Structure for Health Services

No area has generated more discussion on its provision than health. This discussion predominantly centers on how to reduce costs while maintaining a quality of service. Central to this discussion are hospitals and the way they provide their services.

The provision of health services in hospitals often requires the support of teams that may not always be physically close to a patient. These can provide expertise that may be crucial in many cases.

Hospital operations are characterized by a mix of processes. Usually, as shown in Figure 4.3, a process starts with patient admission. Following admission, the patient is examined. Doctors, on examining patients, request tests to be carried out, often by independently run laboratories. Results of these tests are then used by doctors to perhaps order further tests until a diagnosis is complete.

The actual running of the tests usually follows a fairly structured process, whereas the process of defining the sequence of tests is usually quite indeterminate and variable. Thus a test process may schedule a patient, arrange an appointment, get details required of the doctor and any samples from a patient, carry out the test, write a report and then return the report, perhaps with comments. At the same time, special situations may require changing of one of the static processes, as for example, an urgent case that must bypass scheduling.

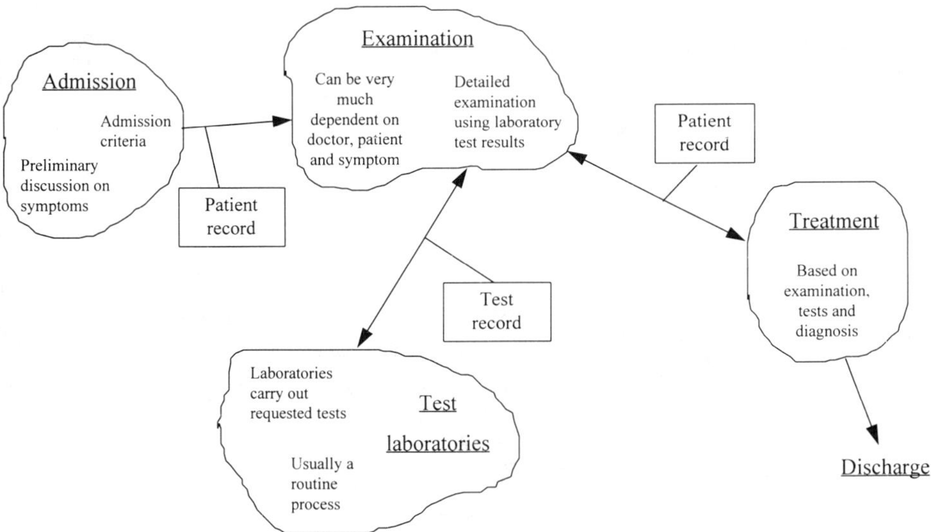

Figure 4.3 The hospital activities.

The tests that are carried out, however, are determined by the doctor. Thus following the result of one test, the doctor may decide the next, although there are some requirements that must be met given certain symptoms. The context here is very varied, some of it being the doctor's expertise. The information provided by the patient also forms part of the context. It can only be accessed through discussion with the doctor.

Test results are then used to determine the treatment, following which the patient is discharged. The treatment itself may uncover more symptoms that require further examination and tests. The usual process here starts deterministically by admitting a patient. The flow at the top level can be quite structured, as shown in Figure 4.4. A patient is admitted, examined, there is a course of treatment, and then the patient is discharged. However, each of these top-level steps can initiate additional processes. The examination itself is non-predefined in that it requires tests to be carried out on the patient. The sequence of tests are usually dependent on the judgment of doctors and outcomes of previous steps. The process of selecting tests is thus non-predefined and is often the subject of discussion and negotiation. Furthermore, each of these tests can be carried out by a different hospital unit. Each test, however, often is carried out as a predefined process.

The treatment can be a structured process that follows a set of prespecified steps. However, it is always possible that given some unexpected symptom in the treatment, some additional tests may be carried out. The discharge step is usually

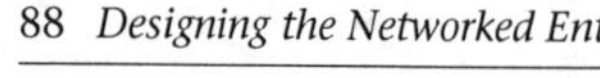

Admission
Examination
Treatment
Discharge
Test A
Test B
Test C
Step A
Step B
Step C
Step X
Step Y
Step W
Step Z

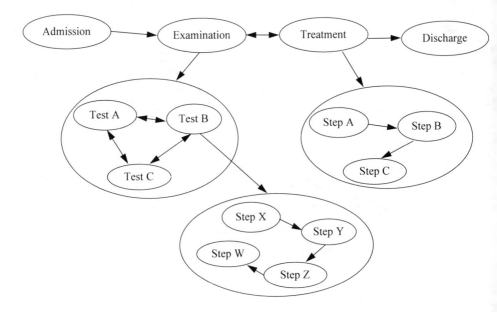

Figure 4.4 A process for hospital patients.

quite structured, resulting in a record of discharge, and perhaps the generation of an invoice.

4.3.3 Process Commonality

Sometimes it is thought that every process is different. However, it is also noted that there is much commonality between processes. For example, take Figure 1.9 in Chapter 1. This describes a loop in global planning. A group of researchers are kept aware of research needs, and the results of their work are fed into practice. But this kind of loop can apply in many other situations. Apart from planning for health-care delivery as was the case in Chapter 1, the same loop can be used in planning innovation of products in the market. The project group side of Figure 1.9 would define the needs, whereas the research side attempts to construct innovative products that meet these needs, and a strategy is developed to market the new products.

There is also much commonality in processes concerned with building systems, be they buildings in construction, products in manufacturing, or software systems. All of these have a life or product cycle that starts with requirements, which is followed by design, then construction, then testing in use. All these processes usually have a further common characteristic: they are accompanied by change during development. Software requirements can change as software

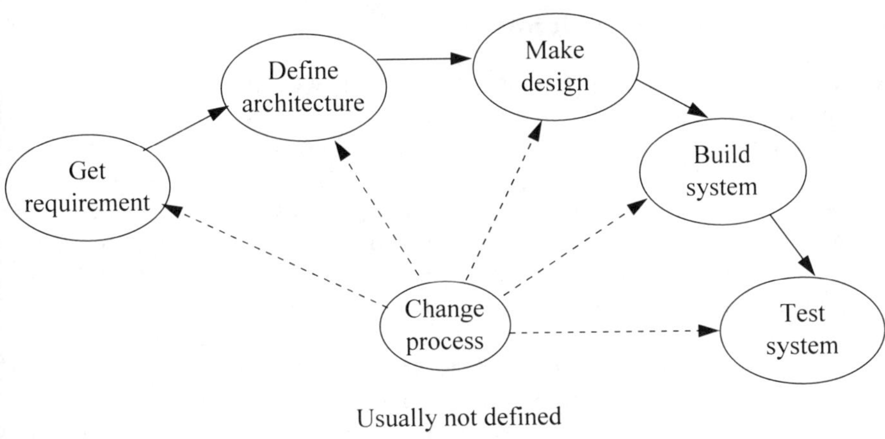

Figure 4.5 A linear software engineering process

development progresses, as can building requirements (especially changes to floor layouts).

4.3.3.1 Case Study: Workflow Enterprises—Product Cycle for Software Engineering

Much software engineering is premised on the fact that software can be designed by sequentially following a set of stages. The workflow for these stages is illustrated in Figure 4.5. Thus in the first step, users define their requirements and record them on a user requirement document. These requirements then produce an architectural design, which in turn is used to produce module designs. The module designs are then used to construct the modules, and once this is done the system is tested. The whole process is, however, not as simple as it sounds because some additional work such as testing is carried out throughout the whole process.

There are other problems, also. One is that the design process is continually affected by changes to requirements, that affect the process. Thus the process is continually interrupted and earlier document changes invalidated. Detailed workflows often become cumbersome here because of the unpredictable ways that people can sometimes respond to changes. Change in Figure 4.5 is illustrated by a change process linked to each workflow step. The organization is studying whether to incorporate the change process into its workflows or to entirely redesign the process. One idea is to make greater use of notification schemes in the current process to integrate change management.

4.3.4 Can Anything Be Predefined?

There are many advantages to systems made up totally of structured processes. In such systems, everything can be predicted, costs can be estimated, and the process easily monitored and controlled. Many organizations seek to automate the processes-and in fact make them predefined. The goal also extends to governments, which are trying to contain their budgets in costly areas such as health or education. The pressure always is to control costs by attempting to institute predefined processes whose steps can be precisely measured. Can health costs be controlled by requiring practitioners to carry out a totally predefined process given an identified symptom, or should they vary treatment depending on their judgment? But a predefined process usually means lack of creativity as it discourages new avenues of thought, lack of responsibility, or identification to the process as the process determines what is to be done. So, who is responsible if a patient does not respond but a predefined procedure was followed in the treatment? Here is the dilemma—to control costs, the goal is to make processes predefined, but this in turn can discourage creativity can eventually lead to processes that produce low-quality outputs unresponsive to change.

Is it thus better to invest effort to see how non-predefined processes can be organized and controlled rather than making everything predefined?

4.4 IMPORTANT QUALITY PROCESS PROPERTIES

One important design goal is to create a quality process. To do this, it is necessary to have some definition of what a quality process is. There are various ways to describe process characteristics. Many of the descriptions are general, but there are now a number of special generic properties that are recognized as being important to the process. These properties can then be used to measure processes and thus determine whether we have any improvement to the process. Properties that are generally recognized as necessary for good processes are described in the following.

4.4.1 Maintaining Awareness

Awareness is another term important to the process. It means that participants in a group are always aware of the current process state. Thus all doctors involved with a patient should be aware of all the tests that have been ordered to date. Awareness provides the *context* for people's activities, ensuring that their contribution is relevant to the group and that they do not take any unnecessary, or even counterproductive, steps. A more detailed elaboration of the context is given by Dourish and Bellotti [2]. Awareness is important because wider knowledge can be used to design roles with reduced repetitive work, but enables them to exercise more initiative given wider knowledge. Process awareness includes the following:

- Awareness of the current process state and who is doing what now;
- Awareness of what is needed and expected of the process;
- Awareness of the plans of other participants.

Awareness can also be associated with group memory, as for example:

- Awareness of what has happened in the past;
- Awareness of who has done what;
- Awareness of why certain things were done and what should not be repeated.

The ways of providing awareness often differ in synchronous and asynchronous collaboration. For example:

- In synchronous work, awareness is maintained during interactions between the roles. It may be maintained through observing other people's attitudes through their gestures, gaze, or expressions. However, this may not often be sufficient as reasons for taking actions may not be recorded, resulting in loss of group memory that may be needed in later interactions.
- Annotation schemes are used widely in asynchronous work to describe a role's action to other roles.

One important criterion in awareness is to integrate it into individual work as much as possible so that it is not seen as an overhead by group members. This is one criterion that must be important in designing notification schemes.

4.4.2 Providing Group Memory

Group memory, or process memory, is now being recognized as an important process property. It refers to the ability to retain memory about decisions that were made during the course of a process so that we

- Do not have to go over some actions again;
- Remember actions taken in similar circumstances and can either avoid them if they led to problems, or reuse them if they were effective.

Or more commonly, we "do not go over things again and again" and "do not reinvent the wheel."

Conklin [3] describes some ways of retaining memory and distinguishes between memory about documents and memory of the process. He suggests that most memory in organizations centers around artifacts such as documents. Document memories often do not retain information about how the information was obtained and whether the outcome was satisfactory. There is little memory kept about processes. As a result, we cannot reproduce why certain document states were reached or difficulties encountered in process steps. Such information is needed if we are to improve the process and not repeat earlier mistakes. A quality process should keep records about the process steps used to produce a document, especially which steps produced outcomes that needed rework and that produced acceptable outcomes.

4.4.3 Some Other Desirable Properties

A number of additional desirable properties for processes have also been identified. These are the following:

- *Adaptability to change.* A process must easily respond to its environment. It must include features that leave it up to the process roles to make decisions on what they should do as well as allowing the process steps themselves to be changed.
- *Be useable and friendly.* The actions expected from people in the process must be clearly defined. This means that all the task support needed by people is always there and the support presents a friendly interface to the user.
- *Provide assistance to users.* The necessary information needed by users to take an action should be readily available. Steps in processes are always explained and users are assisted or advised of what to do where there are unexpected outcomes.
- *Encourage creativity.* Users should be able to deal with unusual circumstances and vary their actions in light of these circumstances. This often requires an understanding of the mission of the organization and the latitude for individual actions by individuals within this mission.

4.5 NOTIFICATION SCHEMES

One way to maintain awareness in both synchronous and asynchronous work is to use notification schemes. Notification schemes keep track of changes and inform interested roles about these changes. Notification schemes depend on the type of process and often revolve around documentation. Each step usually results in a document change, which in turn requires further changes to this or some other document. These changes are usually carried out through another role, which must be notified that it must do these changes.

Of course, another way to manage a process is for each role occupant to scan their context regularly to see if any changes have occurred. This approach, however, may result in considerable unnecessary work by each role occupant, especially if changes occur irregularly. Alternatively, there may be urgent changes that

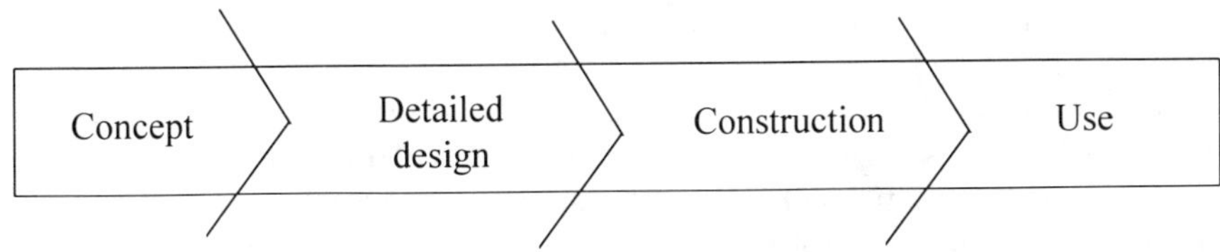

Figure 4.6 Construction phases.

are not checked in sufficient time by the role occupant. Thus, for example, in a hospital there must be a notification when a patient test is completed.

It has been suggested that notification schemes are needed in industries where close coordination is required between independent people. One example that is often quoted is the case of contractors in the construction industry.

4.5.1 Case Study: Independent Contractor Support

If any industry is characterized by a combination of small traders closely linked together, it is the building construction industry. The major phases, shown in Figure 4.6, usually start with a concept, often carried out by close liaison with architectural firms or individual architects. This is followed by building design, carried out by a variety of professions that produce detailed design plans, carried out by draftspeople. This is then followed by construction that includes a variety of jobs such as painting, wiring, and concreting, which are often subcontracted. There is a continuous need to cooperate to get construction completed on time. All this work requires considerable coordination between these contractors. Deadlines usually mean that many of these jobs are carried out concurrently, adding further pressure on coordination.

Coordination between contractors can be particularly difficult because of the amount of detail involved in the specification and the requirement that all persons conform to that detail. Thus, for example, builders must prepare the right ducts and cavities for electrical wiring and electricians must be consulted before building work commences. Doors must be placed in the right position for furniture requirements, which also must be related to electric outlets, and so on and so on. However, it is often the case that it is impossible to define all the detail in design, and specifications will evolve as building progresses and as requirements change, as is often the case. Some changes are even made when use commences, to fine-tune office layouts, for example.

There have been suggestions that computers can support such concurrency through keeping the traders informed of status changes. An important question here is how to use computers both to manage the detailed drawings and changes made to them and to ensure that everyone involved is aware both of drawings relevant to them and, in particular, any changes to them resulting from requirements changes. It is thought that computers and communications can assist this coordination.

Notification schemes have been suggested as a possible solution to the construction industry. It is suggested that construction be defined as a set of tasks, using a project management package, with a set of roles associated with each task. Task progress can then be monitored and roles notified of any changes made to planning that can affect their work.

The support system in this case will contain a project management part that defines the tasks and the relationship between them. The project management system must then be integrated with a notification scheme that notifies roles by following paths of activities that are of interest to them. Often, this calls for the use of a workflow management system to monitor project progress and detect changes to task completion times. However, there is one important criterion that such a system must satisfy. Many of the people in this process are highly mobile, and any system must allow for such mobility.

4.6 REPRESENTING PROCESSES

Considerable attention is now being paid to representing processes. The term *workflow* is often used to describe process representations because these representations describe the way that work flows in a group. Figure 4.2 illustrated one way of representing processes. Representations like Figure 4.2 are also representations of workflows, as for example, the representation of process A showing that work flows from step 1 to step 2, then steps 3 and 4. Figure 4.2 is a relatively simple representation of a workflow. It only shows the steps that can follow other steps, Such representations are also sometimes known as *transition diagrams*. There are other and more detailed representations. One is where we have non-predefined processes. Another is where it is necessary to study the detailed *interactions* between people, *not* between people and the computer screen.

4.6.1 Workflow Models

Workflows are a common way to represent processes. A way of representing workflows was illustrated in Figure 4.2. A workflow defines the process steps with arrows showing the sequence of steps. A complete workflow model also contains more information about each step, including the roles in the step, the artifacts they use, and a description of their task. Most workflows are a mix of predefined and non-predefined parts. Figure 4.4, for example, illustrates a process with a top-level set of predefined steps, where each step itself is implemented by a sequence of detailed non-predefined steps.

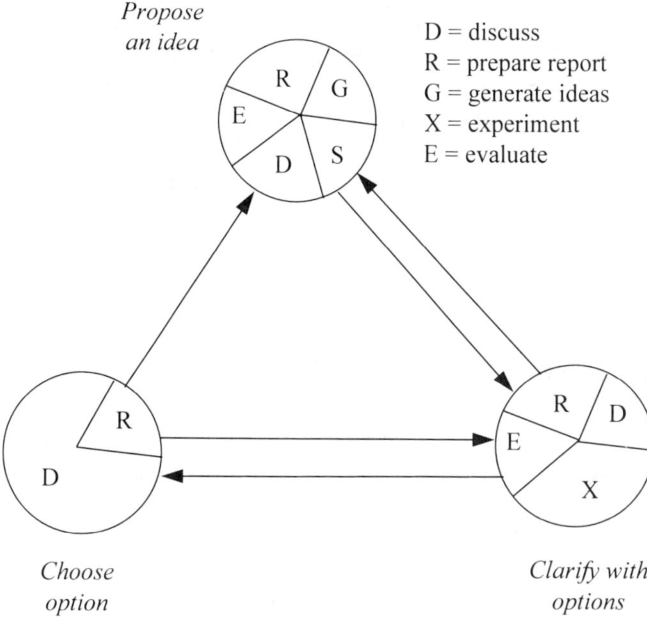

Figure 4.7 A graphical representation.

4.6.2 Describing Non-Predefined Processes

It is not possible to define steps to describe non-predefined processes although we usually know the kinds of interactions in these processes. The amount of time spent on each kind of interaction may be needed and useful to provide collaborative technologies to support the interactions. One way of doing so is shown by the transition diagram in Figure 4.7. Here, each circle is an activity. Here also, each such activity is subdivided into interactions in the activity. The proportion of time in each interaction is indicated by its proportion on the activity circle. The arrows show possible transitions from one activity to another. It is also understood in this diagram that one can freely move the activities.

Figure 4.7 defines a design process where an idea is proposed, is evaluated, and a decision made on whether to adopt it. The process is described in two levels. There are the transitions between the activities as well as transitions within the activities. Thus there can be an idea, it is evaluated, and a request is made for some change. It is then reevaluated, goes on for decision, and so on. The idea generation itself goes through a number of interactions such as discussion, looking up information, generating a design, and so on. It should be noted that a representation like Figure 4.7 gives an idea of what people do but not the sequence of doing this.

An important field of study of groups is transitions between interactions and activities and, what is more important, what are good transition paths. Figure 4.7 describes these transitions in terms similar to McGrath's [4] normative model, described earlier in Chapter 3. Figure 4.7 includes three of McGrath's activities: generating ideas, clarifying alternatives, and choosing an option. It also shows paths that correspond to McGrath's process between the activities. The arrows illustrate transitions between activities. No transitions are shown between interactions within those activites as these are highly volatile.

4.7 IMPROVING THE PROCESS

A process can be improved in four major ways:

- Improve the process structure by more clearly defining the process steps and responsibilities. Often, this results in the creation of roles with specific process responsibilities.
- Improve process support by making it easier to follow the process. A description of the process itself and support for scheduling process tasks is an example of process support.
- Improve task structure by making it easier for users to organize information in a given task and thus organize their work better. Thus, spreadsheets are an example of support task structure in the accounting field as they allow users to structure financial information.
- Improve task support by providing tools that can easily maintain and operate on information in a given task. Thus, a word processor is an example of task support in document preparation as it makes it easier to edit documents.

The design can then look at various tools and make a judgment on how they contribute to the process.

4.7.1 Obtaining Process Gains

The term process gain has been used to describe an improvement to the process. Usually, it is found that introducing a new technology while introducing a gain can also add a loss. One loss, for example, is the time needed to learn how to use the technology. The idea behind the design is to improve the process by having the process gains overshadow the process losses. Improvement can be incremental or it can be structural. In the first case, we may reduce process times by increasing resources. In the second, we may look at a totally different way of doing something, a step sometimes called radical redesign. A chart like that shown in Figure 4.8 can be used to identify process gains and losses.

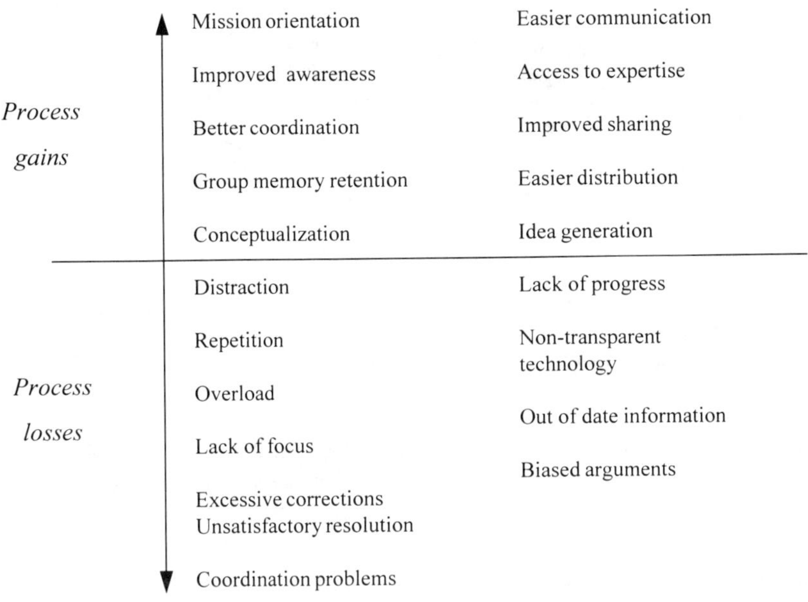

Figure 4.8 Process gains and losses.

Typical improvements include the following:

- Making it easier to get extra relevant information;
- Making it easier to retain memory about the system;
- Making it easier to involve others in our work;
- Providing a better knowledge of the working environment;
- Better ways of creating a new solution;
- Making it easier to find out what others are doing;
- Ability to communicate within the work context.

Later, this book will describe tools that can be used to obtain such gains.

4.7.2 Process Facilitation

Often, some facilitation is needed to ensure that a process proceeds smoothly, and a facilitation role has been proposed for this purpose by Viller [5]. A facilitator is a special role with responsibilities to improve process quality. This may be someone who assists a group to use the network to exchange information or to conduct an electronic meeting. The need to distribute facilitation has also been proposed by

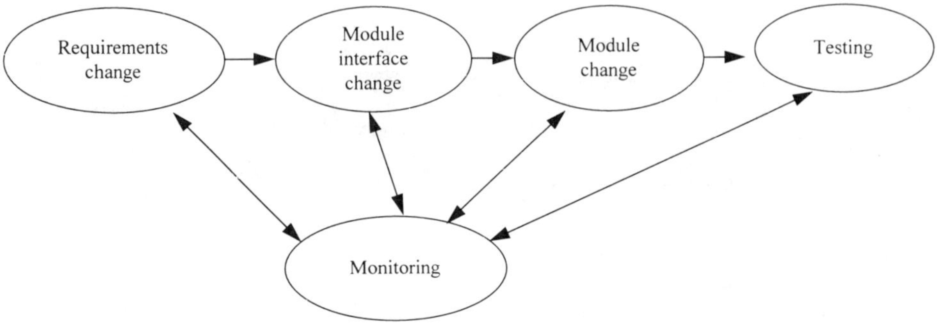

Figure 4.9 Making change the driver.

Dubs and Hayne [6]. One responsibility of a facilitator can be to provide advice on the technology needed to support groupwork and how to use the technology.

4.7.3 Radical Redesign

Processes can be improved incrementally or by radical change. There are now many writers and practitioners that claim that radical change is often needed because many processes in organizations do not address the critical business of the organization.

4.7.3.1 Case Study: Workflow Enterprises—Improving the Process

Workflow enterprises have noted that the software engineering process is characterized by continuous changes to user requirements. However, it is managed as a strictly sequential construction process. The most pressing problem in terms of the table in Figure 4.8 is making people aware of changes in the systems that may affect them. Such awareness will ensure that everyone is always working towards the same goal. This can be done by simply broadcasting each change to everybody. However, such broadcasts can themselves result in voluminous information having to be sent out and absorbed. Some of the information flows could be reduced by using selective notification schemes. The possibility of using a facilitator to help other roles to coordinate their activities given a change has also been considered.

However, somebody has come up with a radical approach to redesigning the process to emphasize change in the way shown in Figure 4.9.

Now each change results in the execution of a process that centers on implementing that change. The life cycle now has to be redesigned with more emphasis on testing for change, with more emphasis on regression testing. Testing often repeats earlier tests to ensure that any changes do not invalidate earlier functionality. Some people have noted that this approach is similar to prototyping and perhaps should be seriously considered.

4.8 SUMMARY

This chapter described the importance of the process. It described a process classification and ways of representing processes. It defined properties that should be possessed by most good processes and described how process improvement can center around these properties. The chapter also made a major distinction between predefined and non-predefined processes, suggesting that the goal is to make all processes predefined, although its achievement poses the dangers of discouraging creativity and innovation.

4.9 DISCUSSION QUESTIONS

1. What do you understand by the term *process*?
2. Why is it difficult to define some processes that frequently occur in collaborative work?
3. Make some suggestions of the kinds of processes where the planning cycle of Figure 1.3 in Chapter 1 would be useful.
4. Do you agree that the production cycle can be applied in many situations?
5. Do you think all processes can be made predefined?
6. Can teaching and learning become predefined?
7. What is the importance of notification schemes to processes?
8. Would you expect notification schemes to be more useful in predefined or non-predefined processes?
9. What do you understand by the term *process gain*?
10. Do you think that the process driven by change shown in Figure 4.8 can be generalized into the construction and manufacturing industries?

4.10 EXERCISES

1. Define processes or workflows that would be commonly found in learning environments. Can you characterize them into predefined and non-predefined processes?

2. Can you define a process for joint document production as described for Worldlink Consultants?

References

[1] Constantine, L. L., "Work Organization Paradigms for Project Management and Organization," *Communications of the ACM*, Vol. 36, No. 10, Oct. 1993, pp. 34–43.

[2] Dourish, P., and V. Bellotti, "Awareness and Coordination in Shared Workspaces," *Proc. of CSCW 92*, Toronto, 1992.

[3] Conklin, E. J., *Groupware '92*, Morgan Kaufmann Publishers, 1992, pp. 133–137.

[4] McGrath, J. E., "Time, Interaction and Performance (TIP): A theory of Groups," *Small Group Research*, Vol. 22, No. 2, 1991.

[5] Viller, S., "The Group Facilitator: A CSCW Perspective," in Bannon, L., M. Robinson, and K. Schmidt, (Eds.) *Proc. of the Second European Conference on Computer-Supported Collaborative Work*, Amsterdam, 1991.

[6] Dubs, S., and S. C. Hayne, "Distributed Facilitation: A Concept whose time has come," in Turner, J. and R. Kraut, (Eds.) *Proc. of the Conference on Computer-Supported Cooperative Work*, Toronto, Canada, 1992.

Selected Bibliography

Davenport, T., "*Process Innovation - Reengineering Work through Information Technology*," Boston, MA: Harvard Business Press, 1993.

Davenport, T. H., and N. J. Aquilano, "The New Industrial Engineering: Information Technology and Business Process Redesign," *Sloan Management Review*, Summer 1990, pp. 11–27.

Hammer, M., "Reengineering Work: Don't automate, obliterate," *Harvard Business Review*, July–Aug. 1990, pp. 104–112.

Chapter 5

The Changing Organization

LEARNING OBJECTIVES

- ❑ *Evolution of computing in organizations*
- ❑ *Changing way of using computers*
- ❑ *Flat and hierarchical management structures*
- ❑ *Group empowerment*
- ❑ *Extended organization*
- ❑ *Electronic commerce*
- ❑ *Networks and alliances*
- ❑ *Enterprise directories*

5.1 INTRODUCTION

The trend toward networked enterprises is not an unforeseen event but a part of the process of enterprise evolution. Change is not new to enterprises; many traditional organizations have been in a constant state of change almost since their conception. Much of the recent change has arisen because of a strong relationship between computers and organizations—they affect each other. New computer structures provide opportunities for new ways of working, which in turn place demands for new computing systems and so on. What is different now is that change is becoming more rapid and that this change is made possible by the combination of computers and communication. Many people work with or will work in enterprises that extensively use computers and will be affected by this change. These people will be affected by all of the things described in earlier chapters including culture changes, processes, and trends to group work. Support for groups rather than just individuals will add new opportunities—the ability to work across distances and the ability to more easily form alliances. This chapter describes how organizations have changed over time and the relationship of this change to computing. It also looks at possible future changes emphasizing the trend to networking.

5.2 A HISTORIC PERSPECTIVE

Changes to organizations have had a tremendous impact on management structures. A historic perspective is sometimes useful in trying to evaluate future developments. It also gives some insight on existing legacy systems. One of the earliest models was that popularized by Anthony [1].

5.2.1 Anthony's Model

The traditional organization was portrayed by Anthony as made up of three levels. The three levels were as follows:

- The strategic level that defined what the organization was to do;
- The management level that is responsible for obtaining and organizing the resources to meet the organization's goals;
- The day to day tasks that had to be carried out to achieve these goals.

Anthony's model was strongly related to the idea that all decisions are made at the management and strategic levels. Actions resulting from the decisions are then carried out at the operational level. A natural extension is to *integrate* the levels through the database. It is assumed here that the database can store all of the organization's operational data, which at times can be unrealistic. In that case, we should be able to summarize this data for higher level decision making. Thus if we have a record of each sale, we can develop a chart of sales trends for management. The integration approach, however, has not always been successful because operational data is usually *historic* whereas decisions often use *futuristic* assumptions. Change in the environment is putting pressure to change the nature of strategic planning so that it produces plans that can manage unanticipated change rather than simply extrapolating the past-we will return to this later in the chapter.

5.2.2 Nolan's Stages Of Growth

In 1974, Nolan proposed a model that describes how computing is introduced into organizations. This study, although reported in 1974, still holds for new organizations embarking on computer use, as it defines the ways computers are introduced into organizations and how computer use matures. The first four stages are the following:

- *Stage 1, Gee-Whiz:* Where an organization becomes familiar with ways of using computers to assist them in their task. For example, in a small shop, computers can be used to balance accounts or keep track of invoices, thus improving efficiency. Once the power of the computer is understood and it is realized what computers can do, people or organizations begin to look at ways to use the computers in almost every possible situation, leading to stage 2.
- *Stage 2, rapid proliferation of computers:* Where everybody in the organization wants to use them, but there are not enough resources to satisfy all needs. Development often proceeds in an uncontrolled manner using ad-hoc techniques and absorbing considerable funds, leading to stage 3.
- *Stage 3, management control:* Where managerial and technical methods are introduced to control the development of individual applications. The controls include organized development methodologies and ways to monitor development costs. Such methodologies follow a formal process that requires priorities to be defined, projects to be identified, cost estimates to be made, and user requirements to be agreed upon before programs are written. Such overall enterprise control leads to stage 4.
- *Stage 4, integration:* Where an emphasis is placed on integrating some of the resources provided for individual applications for the benefit of the business as a whole rather than improving the efficiency of individual operating units. Orientation is changed to one of sharing, especially sharing the organization

In 1979, Nolan revised his model with two additional stages. Stage 3 in the new model still emphasizes a more professional approach to system development. Stage 4 recognized a shift from management of the computer system to management of the data resource and this in turn results in the next phase.

- *Phase 5, data administration:* Where enterprises realize the benefits of administering data centrally, thus even increasing the power of the information systems department. The goal then becomes to reach phase 6.
- *Phase 6, maturity:* Where an enterprise develops a portfolio of applications that mirrors the enterprise structure. Users now begin to take more responsibility for system development. This trend to a more cooperative environment also addresses the issue of the balance of responsibility and power in organizations. In earlier phases, the information systems department had the power of application of computing whereas the users had the responsibility for organizational performance. Hence, statements from users such as

"the problem is in the computer system" in case of problems were often heard.

5.2.3 Drucker's Predictions and Flat Structures

The most recent paradigm is that described by Drucker [2] in a famous paper, which identified the fact that organizations are becoming information rich, making it difficult for all decisions to remain at management levels. He then proposed that new forms of organizational structure will be needed and that such new forms will have the following characteristics:

- Flatter structures with fewer management levels.
- Knowledge is at the bottom of the organizational structure, requiring operating level personnel to make many decisions previously made by management.
- There is a trend toward creation of task forces made up of specialists with wide ranging objectives and responsibilities in carrying out their task towards a common mission.
- Departments become standard setters rather than decisionmakers.
- Everyone takes responsibility for information.

Drucker draws an analogy of the new organization to an orchestra. In an orchestra, there can be many specialists each versed in one instrument. Each is highly skilled in playing their instrument. However, the goal is to take these highly skilled players and combine them to work together to produce a symphony. The organization should thus work like a symphony orchestra-specialists on playing individual musical instruments working together to a common mission, the musical score. To do this, everyone must be aware of what everyone else is doing-something that must be achieved by networking in business.

5.2.3.1 The Change to Information—Rich Structures

Changes to flatter organizations have a number of important social effects. One is that it is no longer possible for all decisionmaking to be concentrated at only the higher levels, because of the growth of information and its complexity. Expert teams are needed to process such information. The trend illustrated in Figure 5.1 is thus to information-rich structures that reduce the role of management while placing more emphasis on task-oriented groups to develop information for decisionmaking. The implication is that most work will be carried out by closely coordinated groups. These groups will be made up of people with all the needed specialist skills for the task. The task-oriented approach also calls for a changed role for management. Instead of a controlling and approving role, management's role now becomes one of providing support for the task groups. Management must

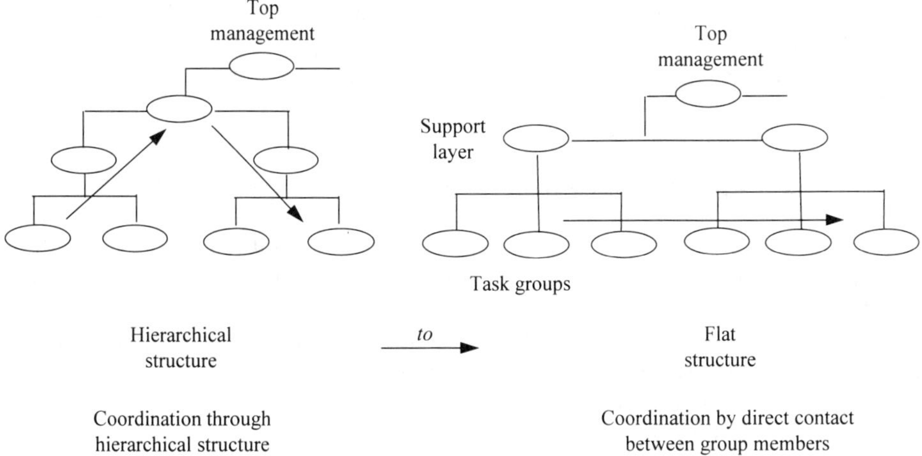

Figure 5.1 Moving to flatter organizations.

thus ensure that groups (or task forces) have the right tools, space in which to work, and methods for communication. This is a very significant change from direct responsibility.

It further leads to what is known as the *empowerment* of groups. This means that whereas in hierarchical organizations all decisions were made by people at higher organizational levels, now a significant amount of this responsibility is delegated to individuals or groups. Secondly, there is considerable enrichment of work that people do. Before, everyone might have had a task allocated to them, without even knowing the purpose of the task. Now, people have much wider knowledge of what is happening and can often use this knowledge to adjust their work to better contribute to the wider goal. In summary, people now have a much wider context in which they work and more scope to change their activities, often leading to increased commitment to their work.

The move to a group-oriented paradigm introduces a number of changes to the way that people think and behave-a change of emphasis. The group paradigm is thus characterized in the following ways:

- There is more attention paid to process, or the way we do things.
- Shared database becomes a jointly owned one, as now people have to agree on changes made to the database.
- We must get away from prespecified processes to dynamic processes as our next step often depends on what just happened.
- Interactions now become more focused on social rather than task-oriented issues and revolve on what is to be done to accomplish a mission.
- We now work with evolving rather than structured artifacts.

The evolution of computing technology itself tends to substantiate Drucker's prediction-again illustrating the strong interrelationship between technology and organization. The trend of technology is towards improved communication through networking.

5.2.4 Impact on Information Systems

One characteristic of the early stages was that the information systems (IS) department was in a position of considerable influence. Thus in Nolan's stages 3 and 4 the IS department often prioritized applications to be developed and placed constraints on the functionality provided for operating divisions through arguments based on cost. Often, computer systems determined what could be done in operational units and operations managers could not be held accountable for unit operations. Thus operational managers had the responsibility for their units but not the power to change their operation, since such changes required agreement to changes in priorities in the IS department. Statements like "we cannot help you because our computer systems are not designed for this service" were and still are common in such environments.

However, the growth of networks and cheaper computers has shifted power back to the users. Operating units can now purchase smaller computers for their tasks and are no longer constrained by IS department priorities. Furthermore, they can arrange for their systems to be developed outside the organization to expedite their work. In this culture, people become responsible for their processes and must be able to control all the components needed by the process. Operating units are now required to provide a service to the business, and they must do so at a minimum cost.

In the service culture, computers are no longer seen as a miracle product, but are viewed as something to provide effective business support. Most businesses thus no longer see computers as having some favored status, but require computer systems to work in a business-like fashion to achieve a firm's objective. Thus, the mystery has somehow faded, and there is little interest by users in the details of computing technology but greater emphasis on its contribution to the organization's mission. Perhaps computers are going the way of the telephone or automobile. All people want to know is how to use them easily to go about their business, but not their internal technology. Thus most of us would not know or even care about the complexities of a telephone switching network (and contemporary telephone networks are very complex). We just want to be sure that when we dial a number we are connected to the right person.

The ability of operating managers to make decisions in light of their own needs rather than IS department preference places enormous stresses on the internal IS department. Its culture now must be to become competitive and business-oriented rather than being technology-oriented. This cultural change has come as somewhat of a shock to many IS groups. They can no longer choose what they want to do, but must be competitive in meeting the business demands of operational managers. In fact, if they are not competitive, the operating groups may get their computing services from elsewhere, making the IS department redundant. The power has thus shifted significantly from information systems to operating system departments. IS departments are now viewed as a cost to be reduced or even outsourced to another organization.

The impact on the IS department structure has also been significant. Recent reports from the United Kingdom indicate a 54% reduction in the number of programmers, with a small growth of 20% in the number of analysts, and a huge growth (almost doubling) of user support personnel. There are other implications, such as the gradual loss of control of data, a cornerstone of stage 4 of Nolan's second model. Thus, instead of a tightly controlled central database, the control over data is now distributed over an ever-growing network. Some people feel that this loss of control over data can lead to serious problems in the way that the organization operates. Time will tell if this is the case.

The trend towards greater and greater responsibility for computers by users has recently almost achieved avalanche proportions. This shift of responsibility is consistent with the culture of lean management. In this culture, people become responsible for their processes and must be able to control all the components needed by the process.

5.3 FUTURE TRENDS

The changes described in the earlier sections are going to continue and further impact enterprises. This section discusses some such changes. One is the impact of rapid change on strategic planning. The others are the trends to lean systems and growth of business alliances.

5.3.1 Strategy and Chaos

One issue that is increasingly being challenged has been the need for strategic planning. Strategic planning assumes a steady environment, where a given set of actions can produce predicted outputs of benefit to the organization. Thus, for example, strategic planning often emphasized increase market share of its products. This could be done by extending the customer base and increasing production facilities. Given a steady requirement for a product, this plan would, in many cases, work.

However, it is found that such an approach assuming a steady state is no longer valid. Customer preferences change, new laws are enacted, and production methods can quickly change. A lot of organizations cater for such change by including contingencies in their planning. These are actions to take if the strategic plan cannot be met. However, if the contingency plan tends to dominate strategic plans, then strategic planning is no longer valid.

There are now some people, who are proposing an alternate to strategic planning-known as chaos theory [3]-whose goal is to design systems that can easily respond to a range of possible events. Although chaos is perhaps a strong word to describe environments, this theory implies less predictability and equilibrium in the environment and looks for methods of organizing organizations to deal with this dynamic change-methods for organizing chaos. The general idea here is that do so, it is necessary to emphasize the ability to react quickly to change and define processes to deal with change. Another suggestion is that strategic planning should change its emphasis and cover new planning concepts, such as building up capabilities, improving links to the external world, and facilitating the learning process, thus improving adaptability to change.

5.3.2 Lean Systems

At the same time, faced with increasing competition, organizations and businesses are trying to build better and better products. They find that because of increasing specialization, it is no longer possible to be best, or to compete in everything. Perhaps, it would be better to concentrate on its main line, which may be electricity production for an electric company, or account management by a bank. This becomes the *core business* of the organization. But once you have defined your core business, should you also worry about the other activities that go along with producing your product? For example, should management worry about how to run its computers effectively or about the best media for its advertising? Why, then should senior executives of a corporation spend a significant part of their time deciding how their computer department should be run, thus neglecting their main area of revenue? Perhaps this should be handed over to external specialists, while the organization concentrates on its main business, what it is best at. The other functions should then be provided as services by other organizations that have appropriate specialist skills. The computer systems can be contracted to software houses, and selection of advertising media to advertising agencies. As a result, many organizations are now changing-they are identifying their core businesses and shedding their other components, and they are becoming *lean organizations*.

5.3.3 Extending the Organization

A further trend is to give outsourced units or personnel easier access to the organization. This includes both access to its information and integration into its business processes. For example, form processing is often outsourced to external workers for checking as part of a business process. These external workers in effect become part of the organization, thus "extending" the organization (in fact, becoming its teleworkers). This requires supporting services to integrate the workers into the business process through a secure firewall.

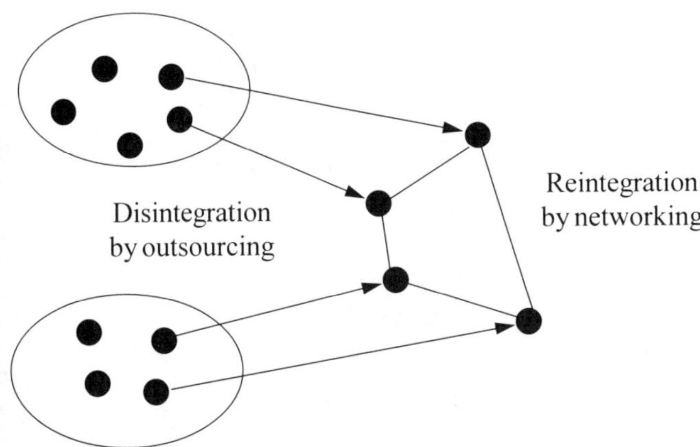

Figure 5.2 Disintegration and integration.

5.3.4 Outsourcing and Networking—The Future Direction

In this environment, the model shown in Figure 5.2 has some relevance. Here, the organization outsources functions least related to its core business. These functions then become smaller organizations, which may then form alliances to get the benefits of scale. One such alliance is often with its parent organization.

This leads us to predict the following changes within the organization:

- *Internal independence:* Functional units become autonomous, calling networking within the organization towards a common mission;
- *Loosening the ties:*The autonomous units either become subsidiaries or independent enterprises;
- *Networking: New strengths* of the newly independent units to form larger, loosely structured enterprises to give them strength.

Such changes will increasingly depend on networks to support them.

5.4 TECHNOLOGY AND ORGANIZATIONAL CHANGE

There has been a close link between computing technology and organizational change. Figure 5.3 illustrates the changes in the way computers are used over the last few years. These started with batch systems; then there were online transaction systems, corresponding to Nolan's early stages. Database technology and decision support systems supported Nolan's fourth stage and emphasized strategic decision-making using the organization's database. The availability of personal computers and networks began a trend towards decentralized decision making with a

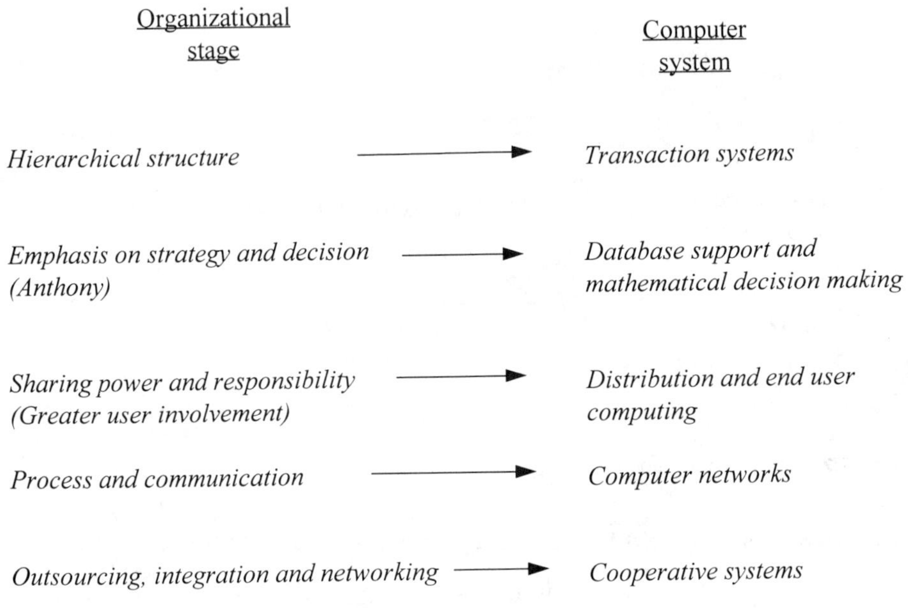

Organizational stage
Computer system
Hierarchical structure
Transaction systems
Emphasis on strategy and decision (Anthony)
Database support and mathematical decision making
Sharing power and responsibility (Greater user involvement)
Distribution and end user computing
Process and communication
Computer networks
Outsourcing, integration and networking
Cooperative systems

Figure 5.3 Changing computer paradigms.

corresponding trend to end-user computing. This was also associated with greater sharing of power of computer use by user departments, with a lesser role for the information systems department. Greater networking led to more emphasis on process and communication, with a consequent emphasis on open systems to add process flexibility. The continuing trend is to greater support for business processes and a greater role for computing to provide collaborative support between people in different business units. A consequence is expected to be the growth of intranets within organizations, commencing with level 1 systems to provide initial support for information exchange, but gradually evolving to level 3 systems to support the newly integrated processes.

The whole evolution is characterized by continuous *re-engineering* or *redesign* of systems. Existing systems are converted from one paradigm to the next. This continues to create a demand for personnel and equipment.

One conjecture here is that increasing support for teamwork by computer technology will result in organizations that are more technological in nature. Computers in this environment will now become tools in the hands of every user rather than the computer specialist. This will have considerable impact on the way interfaces are designed. They will have to be more natural and easily understood by users. Such users will not be expected to know details of computer or operating system commands, but will work with commands strongly related to their everyday

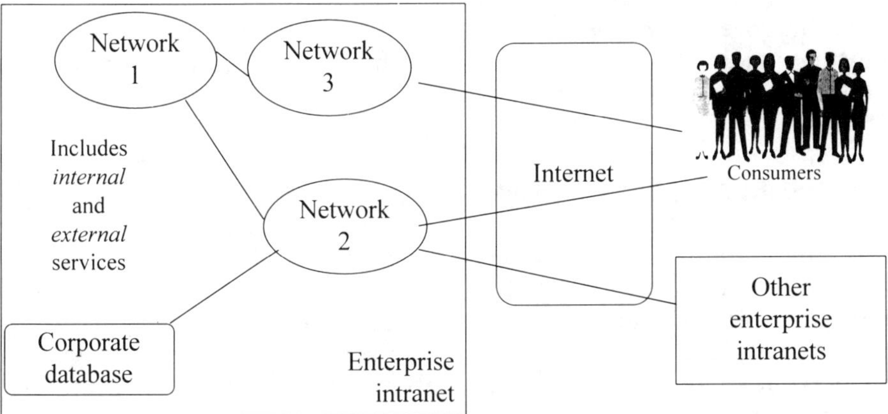

Figure 5.4 Evolving networks.

work. This is already happening with the use of the World Wide Web and will strongly effect interface design, which is discussed in Chapter 9.

5.4.1 Future Technical Structures

Computer communication systems will play a major role in the new networked enterprises. The trend may be to the kind of architectures shown in Figure 5.4. This is towards loosely connected networks where each network serves one enterprise unit, supporting the business processes in this unit. At the same time, the network is interconnected to other units and the external world. As units devolve, so the connections become more controlled with increased emphasis on privacy for each individual network. This, as shown in Figure 5.4, can be seen as a trend to secure intranets connected through security "firewalls" to the external world through the Internet. Where the intranets are in the same enterprise, they may have special connections. Otherwise, they may be connected through the Internet to their clients and other enterprises. Appropriate services must be provided to the external world to support interaction. Furthermore, as independence increases, so will the level of networking support. Level 1 systems maybe sufficient initially to maintain contact, especially initially where coworkers know each other. However, greater integration as implied by level 2 and 3 systems will become essential to maintain the more complex coordination patterns that are needed in wider distributed networks.

However, at the same time there is a force often inhibiting this change. These are the organization's existing systems. Organizations are thus faced with the problem of converting these existing legacy systems to meet the needs of the new culture. At the same time organizations must improve the way they do business-they must improve their *business processes*. Most organizations face this problem by constructing new systems to meet new needs while letting legacy systems run on until their services are no longer needed.

5.5 RE-ENGINEERING THE BUSINESS

Many organizations, faced with the new environment, are now looking at changing the ways that they carry out their business. Thus there is now a situation where business processes are changing. At the same time, computer systems are being changed or re-engineered to meet the needs of the new processes-a very dynamic environment. Re-engineering has recently become a very important issue in computing and is often referred to as *business process re-engineering*.

There are now many writers [4] who advocate total and radical redesign, not simply incremental changes. Others [5] see a major role for information technology in this redesign. Organizations are being urged to:

- Rethink what they are doing, thinking about outcomes not tasks and redefining their mission;
- Radically simplify the way they do things, eliminating unnecessary steps, and concentrate on their main line of business;
- Use the business as a driver, identifying their central business mission and concentrating on supporting this mission;
- Clearly defining the process followed in their critical path and smoothing out the process workflows rather than the operational units that make up the process;
- Define a migration path from their existing processes to newly defined processes.

Re-engineering not only changes the computer programs, but also the way we do business, and changes to existing computer systems in an organization should look more closely at business processes rather than at individual units. Defining a mission more precisely is often of value here. Perhaps once we clearly know what the mission is we can align our activities and tasks to the mission. The mission helps people rethink what it is that they are actually doing and provides the focus for business redesign.

5.5.1 Case Study: Homesafe Securities—The Growing Business

Homesafe Securities began by producing home security systems and their success led them to diversify by purchasing other companies with related products, in particular products related to office security such as video observation systems, card-

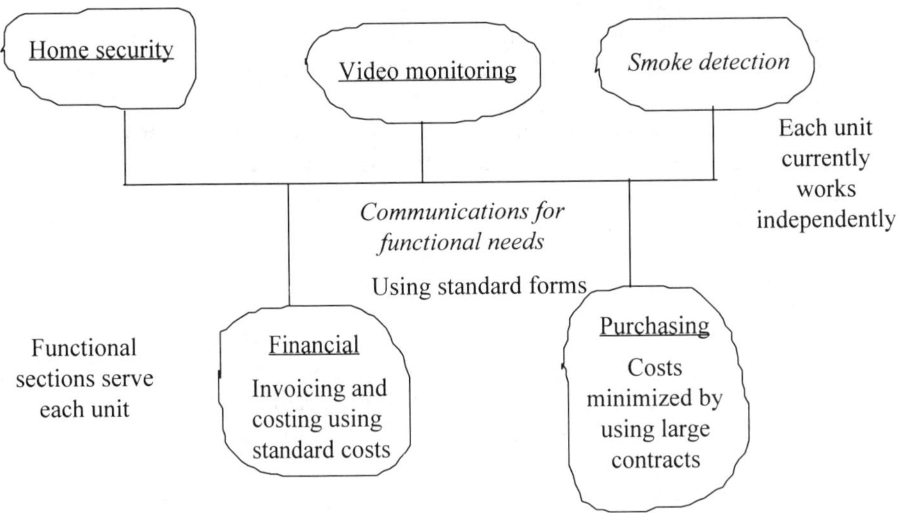

Figure 5.5 Homesafe Securities.

activated entry to buildings, smoke detection systems. The result was that home security systems received less attention. The company concentrated on obtaining savings by integrating common operations; in particular, computer systems and accounting, as well as purchasing and inventory. It now has well-developed computer support for its financial and purchasing sections. The sales and installation of individual products remained separate. This resulted in a standard purchasing section where production sections were required to use the same purchasing procedures, with savings obtained through price reductions by bulk purchases from suppliers and common inventory. The structure of the expanded organization is illustrated in Figure 5.5.

Changes in customer demand have meant greater customization of integrated services, requiring proposals that integrate different systems. The requirement to respond quicker often required purchasing nonstandard parts. Special permission has been given for this, but lately more parts are purchased through special permissions rather than the standard. Furthermore, approval procedures are now holding up purchases, as approval is required from both the production and purchasing part before any action is taken. Online access facilities have been developed to give operating personnel direct access to computer files to facilitate such purchases. However, such online access to individual functions has not improved its ability to monitor the overall installation processes.

The company is thus faced with a problem. All its business units are functioning well. The parts are bought at minimum cost, the computer system is producing information for the organization, but the company is losing its market because of its inability to respond quickly to requests (because it is concentrating on standard units while demand is increasingly for more customized products).

Customers have also begun to express some dissatisfaction with keeping track of orders that may involve more than one product. Because of increased customization, they need to put more inputs into the process while security systems are being produced, but often find it difficult to find the right person to contact. This is because the order passes from unit to unit and most people in the organization are unaware of its present step.

The organization is looking at ways of addressing these problems so that it can remain viable in the increasingly volatile environment.

5.5.1.1 Homesafe Securities—Beginning Re-engineering Through Redefining the Mission

In examining the security organization, we find that what has happened is that the environment has changed. When the systems were set up, the organization produced a high-quality home security system that was widely accepted in the market. Its mission in that case was relatively straightforward. It was to supply the market with this standard product. On purchasing additional companies, it still used the same approach, addressing what it perceived to be a number of distinct markets-smoke detection, observation, and key entry systems. However, over time the market changed with a drift away from the standard to the ability to produce products that integrate a variety of security systems to achieve an adequate security level at minimum cost. To survive in this environment, a company must match the needs of the environment and redefine its mission accordingly. In the case of our organization, perhaps a better mission would be to "provide integrated security systems that satisfy specific customer needs." This is significantly different from the "supply of high-quality standard home security systems" and perhaps will require a change to the way the organization works. It now becomes possible to examine how to restructure the company in the light of this new mission.

Re-engineering will have to look at the steps followed to produce customized systems, which now require coordination through discussions between production units making special systems, and will perhaps require changing workflows by reducing the number of steps needed to coordinate these units.

The security organization has found that the production of video is a highly specialized skill and is attracting more than its justified share of management attention. One possible alternative is to divest oneself from this part of the business, setting up the function as a separate organization, or perhaps selling it to an existing company that produces this kind of product.

Then, there is an even more radical alternative. Sell all three parts and simply concentrate on the process of producing integrated security systems, using parts from the most suitable producer to construct integrated systems for customers. Now the organization primarily becomes an assembly organization using the functional units to obtain the component parts.

5.5.1.2 Homesafe Securities—Redefining the Process

Looking back at its redefined mission, the company sees that it must place considerable importance on providing an integrated service. Thus, a customer may have a special security request, which usually means that the customer will need to be in contact with the organization while the request is being processed. Such personalized service is difficult to provide in the current structure. In the existing process, the customer product passes from one operating unit to another and units can lose contact with the product once it has passed to the next unit. A customer often has considerable difficulty in finding out about the progress of their order to make further requests or clarify some issues.

The company is thus considering an alternative scheme. That is to create a number of product groups and assign a customer order to a single group. All group members will be aware of the state of processing of the order. One of the group members will have the responsibility of maintaining contact with the customer and coordinating customer requests within the group. This requires a complete management change, from its current hierarchical structure to empowered groups. The group will need access to a much wider range of information to carry out its tasks. It will also need coordination support so that everyone knows what everyone else is doing.

The new process shown in Figure 5.6 is proposed. It has five steps:

1. Obtain customers.
2. Once an order is obtained, a group is formed.
3. The group creates a plan
4. The group then supervises production using services provided by the operating units. One member of the group is the one contact point with the customer. To do this, the group has the authority to request the three operating units to design systems to its specifications.
5. The product is then delivered.

The iterations that now take place between the operating units in the existing process are thus removed.

Much of the information used in this process will be provided through computers. For example, the orders and their current progress can be recorded. But in addition, changes to customer needs will have to be captured and members of the group notified of such changes. This will require databases that detect changes of documents and actively inform group participants of such changes. Again, groupware is being considered to provide these services, especially to support the

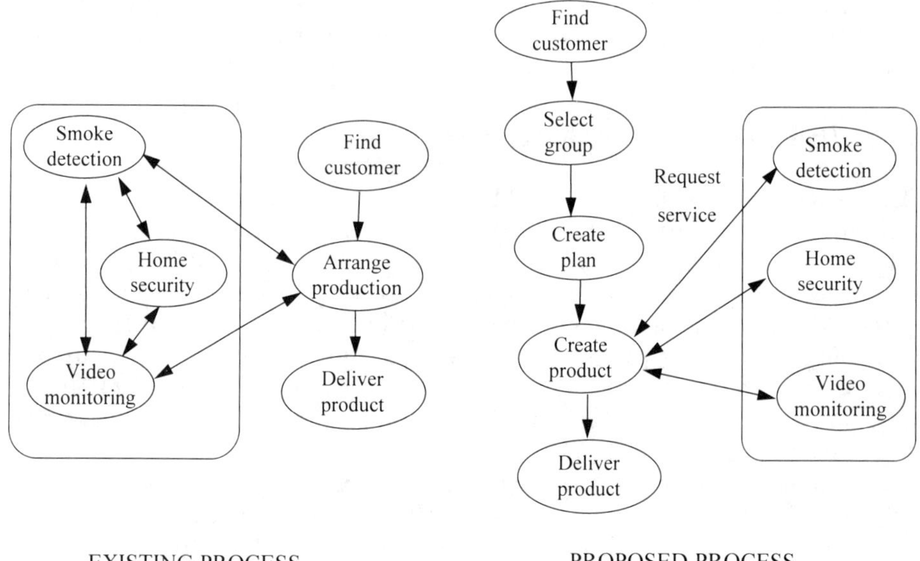

Figure 5.6 Redesigned process.

workflows in the organization. However, the existing systems are placing a constraint on what can be done because it is difficult to integrate new groupware products with existing systems.

5.5.2 Lean Management

Management within the lean organization also changes, placing more emphasis on facilitating work. Thus, rather than providing direction, managers must now concentrate on providing support. An organization may be seen to be made up of a number of highly skilled teams that need continual support to do their work. Furthermore, such teams may rapidly change, again requiring management support in identifying and facilitating such change.

Leaner management usually means fewer levels of management, and consequently less need for managers. Organizations are beginning to only concentrate on their area of expertise and look for services for work outside this area. Thus organizations may need fewer managers that were earlier responsible for work outside the core.

5.5.2.1 Case Study: Midvale Community Hospital—Health Services

One significant illustration of the extended organization is the health system. Anyone who reads the popular press cannot fail to see the constant reference to rising health costs. Much of this cost concerns stays at hospitals, either for testing or for postoperative care.

One suggestion is that hospitals be primarily used as cores for critical processes of intensive care, whereas some of the other processes such as testing can be done outside the hospital or perhaps even outsourced to allied clinics [6]. This kind of outsourcing is made possible by new health technologies that allow many tests to be carried out in less obtrusive ways than exploratory surgery, as was the case in the past.

The process in this extended hospital is still the same as shown in Figure 4.3 in Chapter 4, although the different parts of the process may be separated in time and distance. Thus the testing services may be outsourced, perhaps even electronically. Specialist skills may then be brought onto the problem across distance through collaborative technologies. Although the solution may sound simple, the technical and social problems are enormous. The patient's fear of being outsourced to an outside clinic, the impact on hospital administration, and the need for doctors to communicate across distance all become important issues. The collaborative technologies themselves must be improved to become more ubiquitous, so that communicators feel as those they are using a phone while displaying complex health information. All of this poses considerable challenges and promises for the future.

5.6 ELECTRONIC COMMERCE

The trend of re-engineering to networked enterprises is almost inevitable. Apart from networking individual businesses, possibilities are now arising for such enterprises to include many businesses as well as individuals. Networking can now include many such entities and the term that one hears more and more in this context is that of electronic commerce that can now include

- Business to business networking;
- Consumer to business networking;
- Business alliances.

5.6.1 Business to Business Networking—From Electronic Data Interchange (EDI) to Information Exchange

Computer communication services have been utilized for some time in business to business networking, especially in transaction interchange. The earliest form of electronic commerce is EDI. It has now been practiced for many years, primarily by large corporations for business to businesses transactions. This concerns passing transactions electronically from one organization to another. Thus an invoice may be sent electronically from organization A to organization B. When received by B's computer, the invoice is verified and payment returned by electronically sending a credit to A's account.

Electronic data interchange (EDI) has a number of constraints that limit its use in networking. One important constraint is that it is point to point. Two organizations must agree to establish a physical link before they can commence to trade electronically. Wider networking requires more flexibility in setting up wider and more flexible communication structures, where it should be possible to "dial up" organizations.

New network services can support a greater variety of information exchange than simply business transactions and will have a large impact on the growth of electronic commerce. For example, some organizations have been transferring their office work to other locations with lower wage costs. Sending routine forms for checking and processing is one such example. Activities now commonly found on public networks include the following:

- *Electronic mail,* especially the distribution of reports or delivery of forms to trading partners. In the United States, it is employed by 15% of corporations.
- *Electronic catalogs* to exchange information about products, parts, and so on.
- *Electronic forms* to facilitate business transactions and route information between workgroup members.
- *Electronic funds transfer* to facilitate the exchange of funds resulting from business transactions.

5.6.2 Using Networks to Extend Electronic Commerce to the Consumer

The availability of network services, however, adds a new dimension to trade-extending it to include the consumer. Recently, the growth of networks such as the Internet has also created the opportunity for consumers to participate through such a network in purchasing. The Price Waterhouse report estimates that in 1994, 5% of all retail, wholesale, and mail order sales in the United States were conducted electronically. This figure is expected to triple by the year 2000.

The discussion now is how growth of such electronic trade will be affected by such public networks. The Internet and especially its extension the World Wide Web has been often suggested as a vehicle for electronic commerce. However, there has been considerable concern about the security of the Internet. It is thus more likely that until security is improved, businesses will only use the Internet for public information, while keeping confidential information in secure private

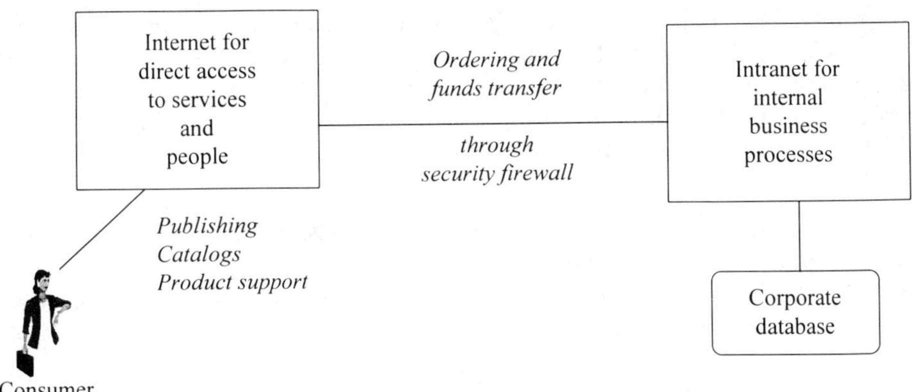

Figure 5.7 Services in electronic commerce.

networks. Thus, possible applications often found on public networks are those that include information that can be generally known rather than being confidential to a business. These include the following:

- Getting business by marketing and advertising on the Internet;
- Distributing catalogs to existing clients or to interested parties in the public;
- Product updates to customers;
- Customer support, including tips on how to get the best out of a product or how to deal with problems;
- Providing maintenance information and advice.

Governments are also increasingly making use of networks to provide information about their programs.

The way of using networks in business to consumer communications is broadly illustrated in Figure 5.7, which is a more detailed elaboration of Figure 5.4. Here, many external services are provided through the Internet, including publishing, catalogs, and product information. In addition, services such as ordering or funds transfer are made either through the Internet accessing the enterprise intranet through a firewall or as a direct connection between networks.

We will again raise this issue in Chapter 9, when we discuss in more detail the way that the Internet can be used to facilitate electronic commerce.

5.7 NETWORKS AND ALLIANCES

There is no doubt that computer support for business alliances will go beyond electronic data interchange. The term alliance itself usually means something stronger than simply trade supported by the exchange of electronic transactions. It implies strong reliance of one organization on another for its operation, and often the sharing of resources and even combining their operations. In that case, network support must go beyond simply supporting information exchange and place greater emphasis on supporting the following:

- Work specific routing of information between businesses to support trade or joint projects between them;
- Interpersonal relationships for interacting in joint project work or discussions leading to such work.

In addition, support is also needed for the more informal but equally important things like tips and news items that are often important to people working in a business or network. These are now increasingly appearing on bulletin boards set up for special interest groups. It serves to keep them aware of any major events, policy changes, or just handy tips about to proceed in their work.

Interpersonal relationships are growing at all levels. For example, Sauter and others [7] report that the use of services such as videoconferencing and e-mail is on the increase with Swiss executives, whereas most organizations now are providing their employees with e-mail.

5.7.1 Kinds of Alliances

Organizational networks are now becoming common, in many cases crossing national boundaries. There can be different kinds of alliances. Typical kinds of alliances include the following:

- *Distribution:* An alliance formed for one organization to distribute the products of another—of course, for some fee.
- *Sales promotion:* Created to advertise products, which may not be competitive or be complementary through each other's marketing network.
- *Market information exchange:* One organization suggests ways of marketing a product of another organization.
- *Product service and maintenance:* One organization makes an arrangement for another organization to service its products.
- *Contract manufacturing:* One organization contracts another to develop some product. Often one of these organizations may develop a design (for example, a fashion design) and contract another to manufacture the product; that is, actually sew the clothing articles from the design patterns.
- *Licensing with title transfer:* Giving an outside organization the power to sell a product.

- *Joint product development:* Two or more organizations may make an arrangement to pool their resources to produce a product that contains components from each organization.
- *Multipurpose joint ventures:* More complex joint ventures that may cross a number of products.
- *Collaborative research:* Here, widely distributed researchers can directly interact by distributing results of their research and entering into discussions electronically on these results. The globalization case indicated the way that this can be done.

5.7.2 Supporting Alliances

Computer support for alliances that goes beyond EDI must consider both social and technical needs. Alliances usually proceed through three generic stages. The first is to identify a potential partner or partners, the second is to develop a business plan for collaboration with the partner, and the third is to set up the collaborative processes. The first stage usually requires considerable searching and information exchange to find potential partners. The second concentrates more on interpersonal relationships. During this second stage, an important social requirement is to build and maintain trust through evolving personal relationships that promote better awareness of each other's activities through information sharing.

The third stage may require different supporting services, which often depend on the kind of alliance. For example, alliances based on sales promotion will emphasize information exchange to facilitate distribution of promotion material. The Internet, or perhaps a private network, may be very useful for this purpose. A joint venture is more likely to require workflow tools to monitor joint project progress. Hence, workflow tools and the intranet may be more useful here. Collaborative research will stress collaboration and possibly emphasize discussion databases. Product distribution can benefit by keeping track of product transfers between organizations and also keeping track of sales, calling for well-defined reporting processes. Furthermore, the nature of support may change as a network itself goes through a number of phases, particularly as they proceed from soft to hard networking.

Some people make a distinction between soft and hard networks. Soft networks are those where the partners trade but do not share resources (for example, long-term contracts to supply large quantities of some product). In hard networks, partners can actually share resources in the sense that these resources are used to produce a product, with all partners sharing the income from the product. Soft and hard networks are also formed in different ways. A hard network thus requires more agreement on pooling resources and on the value of each partner's contribution.

5.7.2.1 Case Study: Business Network Support

Business networks are not just mandated-they usually arise through lengthy processes of negotiation to identify ways of exploiting opportunities. Thus network formation is itself a process that often needs support. Work carried out on network formation for small to medium enterprises has defined a process for network formation (see [8]) that goes through the following three phases:

- *Phase 1:* A business that understands its needs and identifies potential market areas, strategy, and the opportunities in the marketplace seeks and finds partners;
- *Phase 2:* The goal of the business network is clarified and a business plan drawn up between the partners;
- *Phase 3:* The network is established, operated, and eventually expanded or terminated.

Each of these phases concludes with a formal document. Initially, at the end of Phase 1, a letter of intent is drawn up. If this is supported by the program, then more detailed planning commences to produce a cooperation agreement and a business plan. Each phase has different activities that will require different kinds of communication support. Analysis has indicated that during the network formation process information is gradually reduced from seemingly unrelated items of information, into firmer and firmer structures as a network is formed as one proceeds through the network formation stages. Initially, the environment is made up of a collection of seemingly unrelated information about opportunities and businesses. The goal is to search through this information, reducing it to a number of working networks based on structured information.

The changing nature of information has a considerable impact on the kind of service needed in each network formation phase. Phase 1 services can be based on public networks such as Internet and require descriptive databases that are relatively unstructured and include collections of opportunities that often appear on bulletin boards. They can be simply verbal statements or include references to marketing databases. As phase 1 proceeds, it requires increased support for interpersonal relationships in discussions with potential partners, along with the drafting of initial agreements. These become more confidential as issues are raised. They also become more structured as discussion points are made on the issues calling for services to support such structured discussion databases. This becomes even more so in phase 2 as business plans are prepared.

The third level of support, however, is dependent on the particular network as it must support the processes particular to that network. Usually, such processes require workflow systems such as LOTUS Notes, and many telecom are planning to provide value-added services in this area.

One can postulate that support services will gradually evolve to a structure like that shown in Figure 5.7. The Internet provides the descriptive data describing

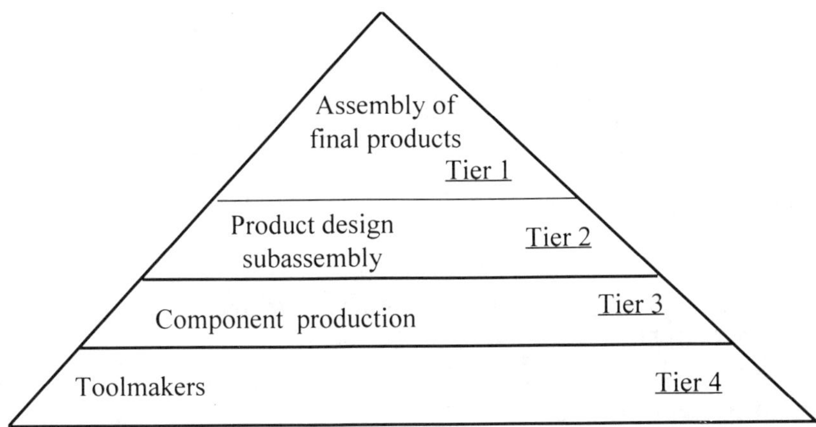

Figure 5.8 The manufacturing pyramid.

the business network program, which provides contacts to its existing members. More confidential services can then be added to support the confidential discussions in phase 2. These will stress services for interpersonal relationships, using either a secure intranet or private pages on a public network. Such private networks can then evolve to support agreed upon business processes at the final phase of the formation process. Some services for business network formation will be described in Chapter 9.

5.7.3 Manufacturing Pyramids

There is not generally enough known about alliances to make general statements about them. Often they are industry dependent. For example, alliances are very common in the manufacturing business and usually industries are made up of businesses that span a number of tiers, as shown in Figure 5.8. In this manufacturing pyramid, the pinnacle (tier 1) is the organization that produces the final product. For example, this may be a motor vehicle. The finished products are made up of many parts. Their main line activity is the assembly of the final product from a large number of parts. These parts are often produced by independent operators that develop the necessary skills to develop the specialized parts.

The top level may utilize designers and other specialists to develop the final product design. These designers (that make up tier 2) are often a few large firms that can carry out this task. Often, quite close integration is required between tiers 1 and 2, and tier 2 is part of the pinnacle organization. As more detailed parts are defined, such as spark plugs or valves, the pyramid spreads out with more and more organizations able to provide the service. These organizations form the next

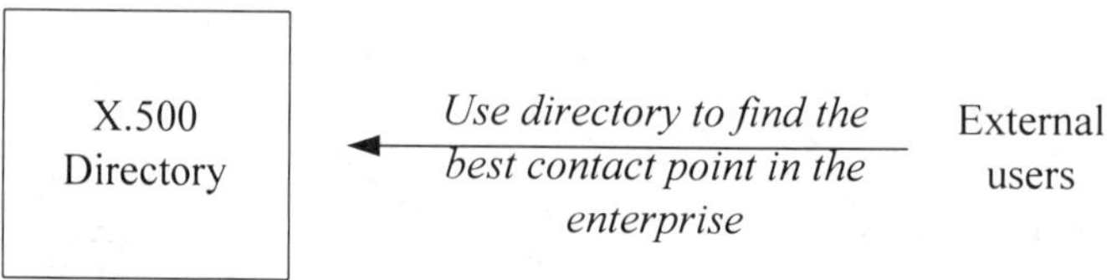

Figure 5.9 Using the directory.

tier, tier 3. The fourth tier usually provides the tools to facilitate production, like machine tools or various kinds of jigs.

5.8 THE ENTERPRISE DIRECTORY

Earlier chapters have stressed the importance of context, or what people see when they are doing their work. This book has stressed the importance of making any task group member aware of all developments relative to their work. This awareness must extend to knowledge about the enterprise itself. We all now obtain this knowledge through things like telephone directories, project plans, part directories, and so on. In networking, this requirement extends to maintaining awareness about partner organizations. Such information is often known as an enterprise directory as it can be used to obtain contacts to the people or groups within an organization. One goal is to provide the organizational directory to users through the computer interface in the way shown in Figure 5.9.

5.8.1 Directory Contents

The question is, what should be stored in the enterprise directory? The directory should not only provide information about people or the enterprise structure, but also include the enterprise policies and services provided by the organization. One early proposal for storing information about organizations is based on the work of Kaye [9] and shown in Figure 5.10. Here, the structural knowledge describes the organizational structure. It includes the main organizational entities such as projects, departments, or people. The procedural knowledge describes the procedures in the organization as well as its mission, plans, and budgets. The knowledge here, again, is mainly descriptive rather that describing the current status of ongoing activities. The user knowledge describes particular people in the organization together with their responsibilities and roles in the organization's procedures.

As you might imagine, with all this information the directories can become quite large. It is thus important to assist users to easily access those parts of a directories needed by them at a given instance of time. Good directories will thus provide their users with assistance through help commands. Because of the directory's potential size, most enterprises only include the structured part of the directory

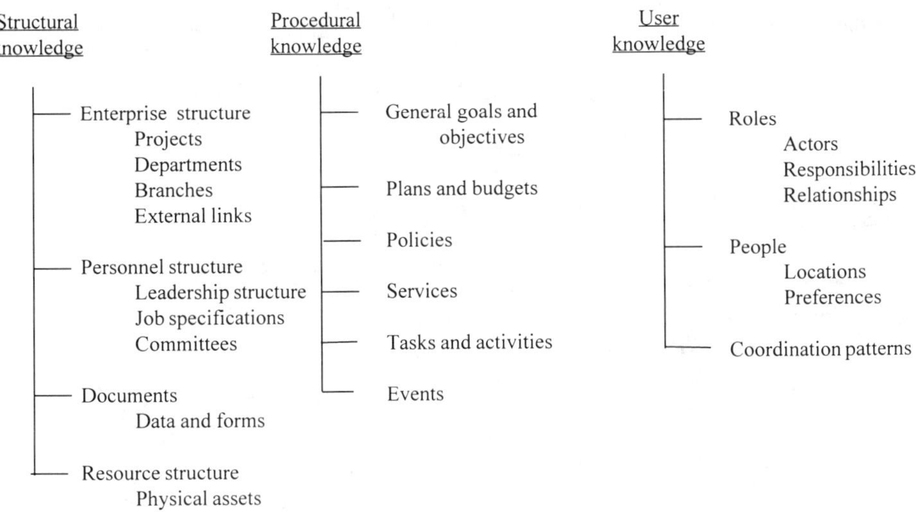

Figure 5.10 Enterprise directory components.

such as accounting information, project budget data, personnel structure and location, or the parts held in warehouses. The database can then be used to identify location of individuals or objects as well indicate responsibilities of individuals and organizational rules and procedures.

However, much of the important information about the enterprise is often not generally available through computers. Missing information often includes policies, descriptions of services, people's responsibilities, or descriptions of tasks. Often, this information is maintained on personal computers and not readily accessible. As time goes by, however, it is expected that more and more information about enterprises will be made available through computers.

5.8.2 Extending the Directory to Collaboration

Figure 5.10 described a context as a static set of objects. However, in actual practice the context is highly dynamic and constantly changing. It is thus necessary to include this more dynamic nature in the context by extending the organizational knowledge shown in Figure 5.10 with knowledge of how particular processes work, and integrating this with the organizational cultural, as shown in Figure 5.11.

Here, the enterprise directory contains information about the organizational structure. The collaborative knowledge understands the flow of information in the processes, and how to access the information from the organizational knowledge and distribute it between the organization's roles. It also uses the cultural knowledge to ensure that information is distributed consistently with existing relationships between the organization's personnel.

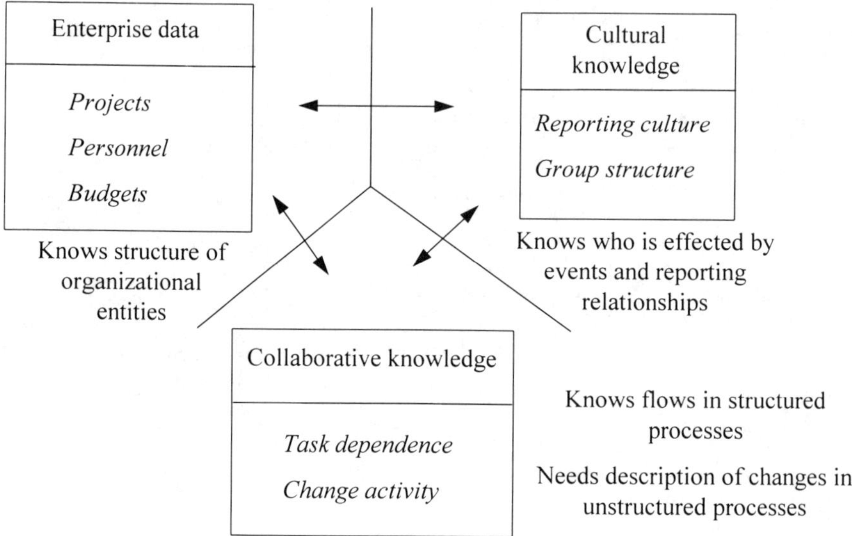

Figure 5.11 Expanding computer context into collaboration.

5.8.2.1 Standards for Directories

Cooperation between organizations requires easy access between people in the organizations. This calls for access to their organizational knowledge, often through computers. Such communication can further be simplified if organizations store their organizational knowledge in some common standard way, thus making it easy for people from other organizations to access it. A number of standards have now been proposed for this purpose, perhaps the best known of these being the X.500 standard, whose goal is to provide worldwide directory services to people within organizations. It is the equivalent of a "world yellow pages," electronically allowing people to be found by referring to an organization's personnel directory, thus simplifying cooperation between trading partners.

5.8.2.2 Case Study: A Directory for Worldlink Consultants

Worldlink Consultants are becoming increasingly aware of maintaining easy access to their experts, under an expert skills heading, as well as access to status documents on critical areas of their activities. This will enable consultants to obtain the latest information about subjects being raised by their clients. In addition, access to standard costings for proposals are also needed. Thus standard and up-to-date costing guidelines would be extremely valuable in expediting budget parts of proposals. Similarly, standard legal phrases related to company policy would be of great value to the group member responsible for legal aspects.

A directory that contains references to this information is being developed. It will be accessible to their employees through an intranet. A number of questions have been raised here. One is how to store information such as legal requirements and clauses, and, what is more important, how to easily copy them into new documents. Entering of all of the requirements into a word processing package has been ruled out because of the amount of work involved, and alternatives are being considered. Another question is whether to support direct links to experts through the directory. Still a third is how to provide the ability to download any information into laptop computers for use by the consultants, many of whom are often traveling.

The context may be different, or at least specialized, for different kinds of organizations or for different kinds of activity. For example, a hospital may need a context specialized to its activities.

5.8.2.3 Case Study: A Proposed Enterprise Directory for the Midvale Community Hospital

The context that supports a health network may contain information such as available procedures, where they are located, and what conditions must be satisfied before they can be requested. The location of testing services and their status can be used to determine the fastest way to get a service. This kind of information could be provided as guidelines for practitioners.

In addition, information about hospital services would be of great value. People to contact about various tests as well as waiting lists and times could assist the planning of patient examinations. All of this context could eventually appear on a computer screen, enabling quick decisions to be made. However, much of the context is a doctor's personal knowledge and experience and cannot be easily captured on computers. So doctors themselves are an important part of the context.

One problem here is that the users are very mobile. A doctor may schedule tests while at the hospital, but will require results at the local surgery. Again the question arises as to how such requirements can be met. Providing a mobile computer usually restricts the screen display and may not be able to reproduce pictorial data. Placing computers at all locations is considered too costly.

5.9 SUMMARY

This chapter described how organizations have evolved over time and the relationship of this evolution to computer developments. It described the impact of these changes on the management structure and the importance of groups in the new structure. The chapter also pointed out that the changing environment, with less predictable components, will call into question many of today other. The chapter concluded by describing enterprise directories that help people find their way around such networked enterprises.

5.10 DISCUSSION QUESTIONS

1. Do you think Nolan's stages of growth are still appropriate now?
2. Do you think the power shift in computing from specialists to the business units will benefit organizations?
3. Explain what is meant by *Drucker's predictions*.
4. Identify some managerial responsibilities in Drucker's new organization.
5. What do you understand by the term *empowerment of groups*?
6. Can you name any manufacturing industries with a pyramid structure?
7. Describe some ways of extending an organization.
8. What is the purpose of strategic planning?
9. What are the implications of chaos theory?
10. What is electronic commerce?
11. Describe some common forms of electronic commerce. Describe the kind of business activities that could be placed on the Internet.
12. What commerce applications are commonly found on the Internet?
13. What are the possible security risks for commerce on the Internet?
14. What do you understand by the term *alliance*?
15. What are the usual steps in forming alliances, and what kinds of services do they need?
16. Describe some common business alliances.
17. What kind of alliances would you find in the construction business?
18. What kind of information should be stored in an organizational database?
19. Do you think it will be possible to store the entire context on computers?

5.11 EXERCISES

1. Is Figure 5.8 in Chapter 1 applicable to planning for innovation in any organization?

2. Define the kind of context you would require in a distance university. Identify the artifacts, policies, and processes that would be needed in this context.

References

[1] Anthony, R. N., *Planning and Control Systems: A Framework for Analysis,* Boston, MA: Harvard Graduate School of Business Administration, 1965.

[2] Drucker, P. F., "The Coming of the New Organizations," *Harvard Business Review,* Jan.–Feb. 1988, pp. 45–53.

[3] Zimmerman, B. J., "Chaos and Nonequilibrium: The Flip side of Strategic Processes," *Organizational Development Journal,* Vol. 11, No. 1, Spring 1993, pp. 31–38.

[4] Hammer, M., "Reengineering Work: Don't Automate, Obliterate," *Harvard Business Review,* pp. 104–112.

[5] Davenport, T. H., and N. J. Aquilano, "The New Industrial Engineering: Information Technology and Business Process Redesign" *Sloan Management Review,* Summer 1990, pp. 11–27.

[6] Braithwaite, J., R. F. Vining, and L. Lazarus, "The Boundaryless Hospital," *Australian and New Zealand Journal of Medicine,* Vol. 24, 1994, pp. 565–571.

[7] Sauter, C., et al., "CSCW for Strategic Management" in Swiss Enterprises: An Empirical Study," Marmolin, H., Y Sundblad, and K. Schmidt, (Eds.), *Proc. of the Fourth European Conference on CSCW,* Sept. 1995, pp. 117–132.

[8] Mjelve, O., "The Norwegian Network Programme - a brief presentation," *Proc. of the International Conference on Cooperation and Competitiveness,* Lisbon, Oct. 1993.

[9] Kaye, A. R., and G. M. Karam, "Cooperating Knowledge Based Assistants for the Office," *ACM Transactions on Office Information Systems,* Vol. 5, No. 1, Oct. 1987, pp. 297–326.

Selected Bibliography

Badrinath, "Helping small and medium size firms to enter export markets," *International Trade FORUM,* Feb. 1994, pp. 4–29.

Bernard, R., *The Corporate Internet,* New York, NY: John Wiley & Sons, 1996.

Economides, N., "The Economics of Networks," *International Journal of Industrial Organization,* Vol. 14, No. 2, March 1996.

Golden, W., "Electronic Commerce at Work: Kenny's Bookshop & Art Galleriers, Galway, Ireland," *Proc. of the Ninth International Conference on EDI-IOS,* ISBN-961-232-000-4, Bled, 1996, pp. 291–303.

Karste, H., "Organizational Memory Profile: Connecting Roles of Organizational Memory to Organizational Form," *Proc. of the 29th Annual Hawaii International Conference on Systems Sciences,* Hawaii, 1996, pp. 188–196.

Loshin, P., *Electronic Commerce,* Rockland, MA: Charles River Media, 1995.

Nolan, R. L., and C. F. Gibson. "Managing the Four Stages of EDP Growth," *Harvard Business Review,* Jan–Feb. 1974.

Nolan, R. L., "Managing the crises in data processing," *Harvard Business Review,* March–April 1979.

Rao, H. R., K. Nam, and Chaudhury, (Eds.), "Information Systems Outsourcing," *Communications of the ACM,* Vol. 39, No. 7, 1996, pp. 27–54.

Swanson, E. B., "The New Organizational Knowledge and its System Foundation," *Proc. of the 29th Annual Hawaii International Conference on Systems Sciences,* Hawaii, 1996, pp. 140–146.

Chapter 6

Interfaces for Collaborative Work

LEARNING OBJECTIVES

- ❑ *Mental and conceptual models*
- ❑ *Interfaces for group work*
- ❑ *Metaphors and icons*
- ❑ *Multimedia displays*
- ❑ *Maintaining awareness*
- ❑ *Visualizing change*

6.1 INTRODUCTION

The first five chapters of this book described groups and their environment. The following chapters begin to describe the technologies that can support groups in this environment. The link between the group and the technology is the computer interface. This chapter will describe the role of interfaces in supporting teams, stressing the additional interface requirements placed by group work compared to interactive work. In interactive work, interfaces support one user performing a task, and interfaces concentrate on a natural presentation of objects used in the task as well as the operations on those objects. Interfaces in collaborative work must go one step further and represent the personal interrelationships between team members.

As a result, the research goal is not only to identify easy ways for users to retrieve and enter information from the computer, but also to make it easier for them to visualize the group situation. *Visualization* is becoming more and more of an issue because of the increasing complexity of users' problems. Good visualization means that users can see their problem in a natural way. Presentations that clearly show the relationships within the context, and what is happening in the context, make it easier for users to both understand these relationships and to manage the complexity of their work.

Many writers make a further distinction between *private* and *shared* workspaces in interfaces that support collaborative work. A private workspace only contains information that can be used and changed by one person, and usually concerns an individual's tasks. Information in a shared workspace can be used by many people. It then becomes necessary to define how information can be moved between these two kinds of workspace and how two or more people can work in the same workspace.

6.2 CRITERIA FOR INTERFACE DESIGN

Most work on designing user interfaces to date has concentrated on the interaction between an individual and the computer, but there is now interest in interfaces for groups [1]. Sometimes the term mental model is used when talking about interfaces. The idea is illustrated in Figure 6.1. Here, the mental model is the way that the user sees the world using terms with specific meanings to the user (for example, accounts or balance sheets). The computer interface presents its information using its terms, often known as the *conceptual* model. Thus, for example, entities and relationships are two conceptual terms often used in databases to define data. The user must then see the data in these terms to effectively use computers. For example, they must see accounts as an entity. The goal is to reduce the cognitive gap between these two models by choosing terms and operations in the conceptual model that closely correspond to the user's mental model and *visualizations* that enable users to see objects in natural ways. To do this, the conceptual model must present the context in ways that closely approximate the user's mental model.

The role of the interface becomes even more important in collaborative work. The interface must have all the properties needed for good individual support, such as being robust, friendly, and easy to understand and the other properties discussed earlier in this chapter. In addition, for group work, the interface must maintain user awareness of developments in their context and their role in the context. This includes the following:

- Providing easy access to information;
- Maintaining awareness of what others are doing;
- Supporting group interaction through the interface;
- Keeping track of documents.

Good computer networking requires all these characteristics.

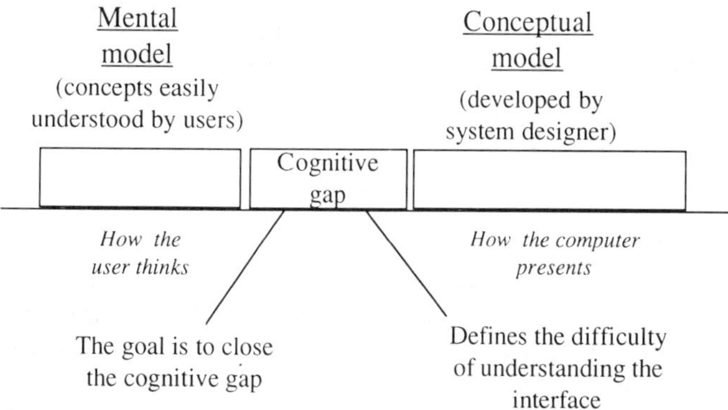

Figure 6.1 Reducing the cognitive gap.

6.2.1 Providing Access to Information

Earlier, this book stressed the importance of context and defined it as all the information available to a user. The context defines the situation in which the user works and must satisfy important criteria, such as always being an up to date and correct representation of events in the user's environment. Some of a user's context is provided through the computer screen, whereas other context parts, including personal contact and reports, are found in different ways.

Designers must consider what part of the context to include on the computer screen and yet make the computer screen not cause cognitive overload. There has been a trend to include more and more of the context in the computer. In fact, there is the possibility that the computer can now virtually represent all of the organization's information, including its people, through video.

However, the question still remains as to whether it is possible, and indeed desirable, to obtain the entire context through the computer screen. The advantages are that everything is at the user's fingertips, information can be easier to integrate, and assistance can be provided in following the process. The disadvantages can be excessive dependence on the computer, possible suppression of creativity, less flexibility, reduced informality, and less personal contact. This is especially so if expert computer technical skills are required to obtain the information. Interfaces should be designed in a way where people do not have to worry about low-level technical components, but only have to concentrate on their problem.

6.2.1.1 Integration With the Corporate Database

One important way to increase the amount of context provided through the interface is to make the enterprise data readily available to the user. There are many ways for providing access to the enterprise data. The simplest is to let the user access the database to retrieve needed information. However, this again may require specialist technical skills or knowledge of retrieval languages. Another way is to simply present icons or menu objects that represent enterprise objects important to the user's context. Users can then select these icons to retrieve information.

6.2.2 Improving Usability

Visualization is only one of the things used to make interfaces easy to use. There have been many others. Much of this work has concentrated on defining the usability of interfaces and most of it has addressed individuals interacting with computers, but not with groups using computers. The Music project at the University of Loughborough, for example, has identified the following important measures:

- *Analytical metrics,* which describe direct objects, such as whether all the needed information appears on the screen;
- *Performance metrics,* which describe how long it takes someone to carry out a task, and issues like system robustness or how easy it is to make the system fail;
- *Cognitive workload,* which is the mental effort needed to understand the information on the screen and use it effectively;
- *User satisfaction,* which include things like how helpful the system is and how easy it is to learn.

These are all important for group work, but group work also has additional requirements. Some important requirements here are

- To establish and maintain one's *presence* through social activities, such as nimbus or focus and the other important factors described in Chapter 3, in virtual teleprocessing environments;
- To ensure that computer use does not establish unnecessary boundaries on the user's workspace;
- To maintain *awareness* of activities and processes in the environment, especially being aware of *changes* and how they effect an individual's work.

6.2.3 Maintaining Awareness

Awareness of other people's work is one of the most crucial factors in collaborative environments. It sets the scene and its priorities (what is considered important and unimportant). The design issue becomes one of keeping group members aware of each other's work through the interface. To do this, a system should provide the following kinds of information to its members:

- Who the workgroup members are and what their current status is;

- Team members' attitudes and feelings about the work being done;
- What the artifacts of interest are and who is currently working on the artifacts;
- The issues associated with each artifact;
- The current process state.

Other important issues in interface support for groups are

- Easy reference to group memory;
- Support for artifact management;
- Independence of the technical platform.

The shared workspace must often maintain the agreed upon changes by all participants. In synchronous collaboration, the workspace is often used to create new changes through spontaneous discussion. Changes to the workspace in asynchronous work may be made by making a change and then passing messages between participants describing the change. Usually in asynchronous collaboration, it becomes important:

- To have a clear definition of roles and their responsibilities;
- To have clear assignment of tasks;
- For group members to maintain versions of their work and discuss them with others;
- To distribute versions and allow annotations to versions to support feedback;
- To add attachments to workspaces offering suggestions and distributing them to group participants.

6.3 METHODS FOR SUPPORTING GROUPS AT THE INTERFACE

The three most common approaches used to date to support cognition at the interface are metaphors, multiwindow display, and multimedia. All of these, often used together, must improve user's cognition of their workspace and their ability to work in that workspace. This not only covers the presentation of artifacts, but also all personal interrelationships within the group.

6.3.1 Metaphors

Cognitive workload can be reduced by presenting information in an easy to understand way using symbols and groupings of information. The idea of *metaphors* becomes important here. Metaphors are to some extent icons that have specific meanings associated with them. Thus there may be a set of metaphors for groups

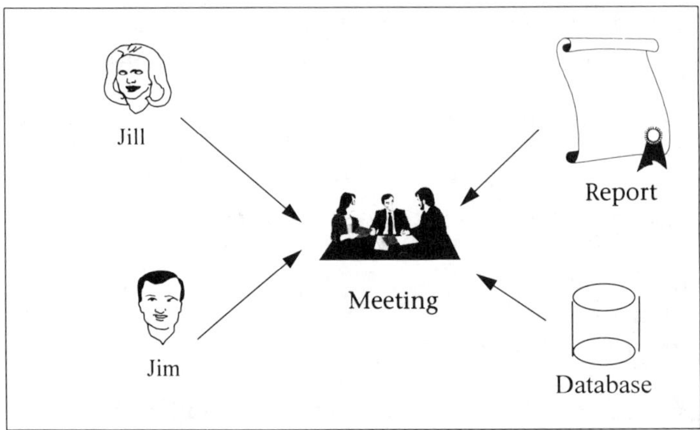

Figure 6.2 Metaphors.

themselves, another set for artifacts within the workspace, and still another for interpersonal interactions. Metaphors can also have another role-by serving as the adaptors to the interface by amplifying important issues and drawing people's attention to system events. Figure 6.2 illustrates the idea behind metaphors.

Here, there are four icons that are representations of user contexts. Two of these represent people, say Jim and Jill. The third represents the idea of a meeting, and the fourth consists of objects like a report or a database. The idea is that if you want a meeting about the report, you simply link the meeting, people, and report icons, and the system arranges the meeting (which may be a video meeting). The important aspect is that the metaphor must be easily recognizable by the casual user of computers and must be almost natural to use.

6.3.2 Multiwindow Displays

By allowing a multiwindow display, users can compartmentalize the information in front of them. This is particularly useful in representing corporate data, where each window represents some object in that database. An example of such a multiwindow presentation is illustrated in Figure 6.3.

The presentation in Figure 6.3 includes four windows. One window is the order, another is a window of project budgets, a third window shows prices quoted by the suppliers, and a fourth shows supplier records. The user is presented with all the information needed to make a decision about placing the order. They can look up the prices for the parts, select the supplier, and make an entry in the order. Windows can be selected, usually by clicking on them with a mouse, one at a time, when the menu for the selected window appears on the screen. A further extension is to include a work window, used for simple computations such as computing the

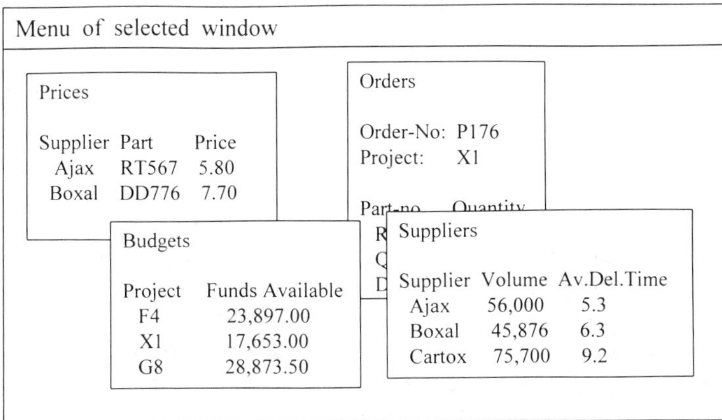

Figure 6.3 A multiwindow presentation.

total value of an order following supplier selection. In that case, the screen would be the total workspace for the user. Finally, it should be possible to move information from one window to another, thus supporting simple integration of information from different functions (for example, to move a supplier price to an order).

Multimedia adds another dimension to multiwindow displays. Now, each window may use a different medium, leading to the possibility of including people within the display.

6.3.3 Multimedia Displays

In a multimedia display, the screen may contain some text, an image, a video, and even speech. Each of these may use their own window. This provides more options to the screen designer. For example, a user may be controlling tools in a machine shop. There may be a video that monitors such machine tools and displays them on a multimedia display. The user can then activate such tools through the screen.

The variety of media gives a much better opportunity for visualization of the user's context. The most common media are voice, image, video, and text. Interactions at the interface can use combinations of media. For example, people may be concurrently talking using voice while discussing sketches transmitted by video.

Multimedia gives designers even more choices in interface design, thus making design more complex. Often, more than one medium can be used in one screen with some advantage (for example, using a shared screen to display a drawing and telephone for verbal discussion). Careful design of CSCW interfaces requires use of media in ways that satisfy the social design principles defined in the previous chapter. Thus, for example, capturing design rationale in voice can mean that we can preserve group memory while not increasing a person's work time. The list of possibilities is almost endless.

6.4 SYNCHRONOUS COLLABORATION

Synchronous collaboration replaces face-to-face meetings or discussions with an equivalent computer-supported system. One of its goals is to maintain the same kind of contact between people who may be far apart as they would have facing each other. Thus social issues, such as facial expressions, movements, or methods of speaking or maintaining attention become very important. Another important consideration in synchronous collaboration is support for shared workspaces, where more than one person can be updating the same document. Lauwers and Lantz [2] describe some of the important issues in supporting synchronous collaboration, in particular:

- Supporting a window layout that is easily understood by all users;
- Ensuring that participants can interact spontaneously and quickly on artifacts displayed in a shared window by supporting interpersonal communication aspects such as focus or nimbus;
- Allowing access for users to the "floor;" that is, to be the ones that are currently controlling the shared workspace;
- Allowing new persons to join a meeting and explaining to them any previous work carried out to make them fully aware of the group status;
- Combining work of the shared window with other media so that a participant can be pointing at a part of the window while explaining it to other participants.

Perhaps one goal is towards interfaces like that shown Figure 6.4. This figure shows the members of the group and those who are currently participating in the session. Group members, with videos, can also be displayed on the screen. There is a shared work area. Individual users can gain control of the shared area and propose possible changes. Users who gain such control are said to *have the floor* at the meeting. There can then be some discussion on whether to adopt a particular change. During the discussion, the floor can be passed from user to user to discuss alternatives.

Interfaces like those in Figure 6.4 cannot be commonly found with commercial products and are experimental. Seamlessness is also important in such interfaces, and it must be possible to compose objects or integrate work using information from more than one application to another. Users should not have to revert back to operating basics, by closing one application and opening another, to do this, but should simply move the data from one space to another on the screen.

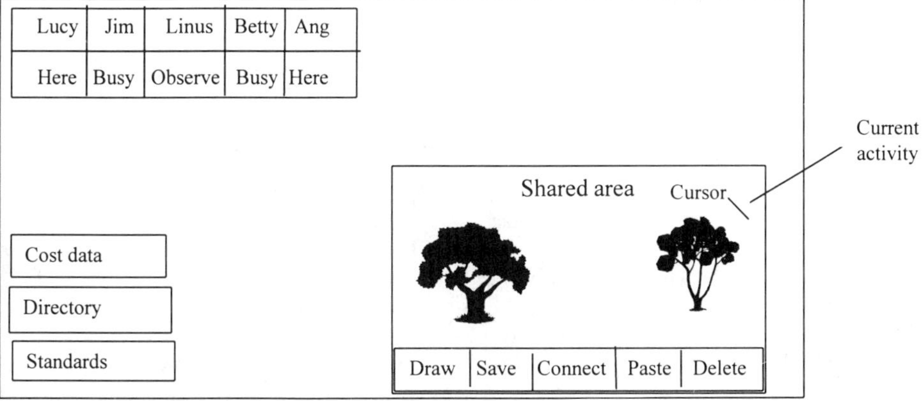

Figure 6.4 Synchronous collaboration.

6.4.1 Supporting Social Factors

Multiwindow display can be programmed to indicate personal factors through color or flashing displays or icons. However, the user still has to interpret the meaning of these events on the screen. This requires considerable familiarity with screen icons, which may be initially difficult for the new or casual users. There is some advantage to supporting these social factors through personal gestures and expressions, which many claim are an essential part of collaboration. Ishi [3] has developed an experimental system known as CLEARBOARD that supports, what is known as *gaze awareness*.

CLEARBOARD, shown in Figure 6.5, maintains eye contact by having a collaborator's face appear as background in a screen. It is important for the face to seem to be "behind" the screen or else people will not draw on it. Ishi [3] makes a distinction between

- *Interpersonal space*, which maintains a sense of Telepresence;
- *Shared space*, which contains common information.

6.4.2 Granularity

The term *granularity* is often important in synchronous collaboration. This identifies a piece of the shared workspace that is available to one person for a defined period of time. For example, in Figure 6.4, a person may only be able to get access to one tree at the same time. This is common practice in shared workspaces. It is not usual to allow two people to change precisely the same object (for example, one word in a sentence). Thus for a short period of time one collaborator gets sole access to that object. The size of objects that can be allocated to one collaborator is

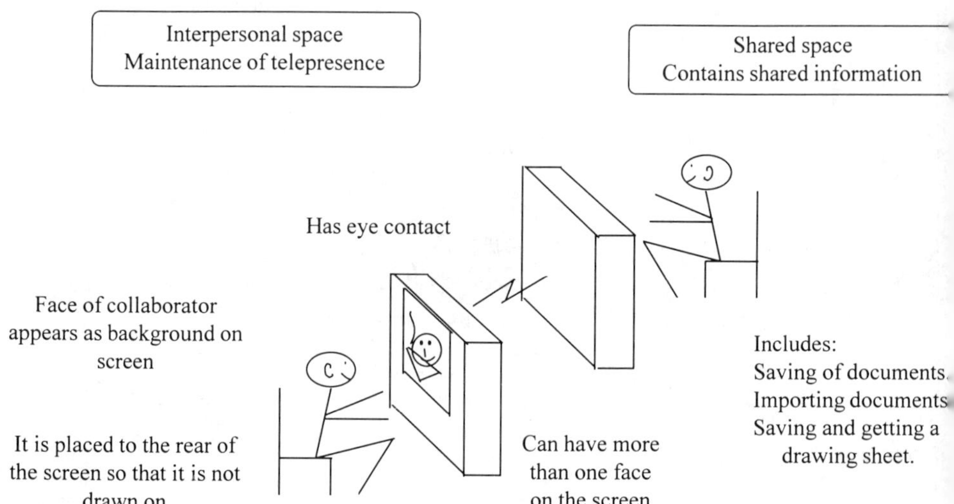

Figure 6.5 The idea of CLEARBOARD.

known as the granularity of the screen. In joint editing, this may be a letter, a word, a sentence, or a paragraph.

The *granularity of cooperation* determines how a shared object is partitioned allowing people to work on different partitions at the same time. In many cases, such as editing documents, it is necessary to know what each user is doing almost at a word (and certainly paragraph) level. Shared workspaces can thus often contain every participant's cursor and what they are doing to what part of a shared object; that is, changing or reading or annotating.

6.5 ASYNCHRONOUS COLLABORATION

Asynchronous collaboration differs significantly from synchronous collaboration. Interpersonal factors must now be represented by screen symbols. Representations of these social factors must be carefully chosen so that even casual users are aware of their meaning. Especially important here is how to make a user aware of what everyone else is doing. In synchronous collaboration, this is done through discussion.

There are a number of ways of doing this in asynchronous collaboration. In asynchronous collaboration, awareness can be maintained through methods like annotations and notification schemes, which are described in detail in Chapter 11. Annotations indicate people's current position on some document, whereas notifications serve to focus people's attention on some document or event. People can leave notes on their work to serve as reminders for future actions. Particularly important here are the methods used to create awareness information and how to distribute it. We do not want all users to continuously broadcast what they have just done, as this may lead to information overload on the group members. Instead, awareness information must be created by the system independently of the user, but it must describe what a user is doing. It should be captured as a natural part of the process without requiring additional work on part of the user. This raises an important question, which is still to be generally resolved, "How to maintain awareness and group memory without imposing high overheads on users? One important approach here is the use of notification schemes.

6.5.1 Maintaining Awareness

Awareness is one of the most important issues in asynchronous collaboration. Imagine the teleworker or mobile worker who needs to make a quick decision in their work. It can be argued that such awareness can be quickly built up by looking at a well-organized database. However, this is not always practical. Firstly, where a decision must be made quickly, lengthy database searches may not be practical. Secondly, awareness goes beyond factual data and must consider social issues such as other people's feelings and attitudes and what are seen as priority issues in one's collaborative space. This cannot be always done by simply searching through a database. Awareness of the environment must be built up gradually and continuously so that people are at ease and confident of what they are doing.

6.5.1.1 Awareness Through Notification Schemes

The book introduced notification schemes earlier in Chapter 4. They keep people in a process informed about any process developments. A notification scheme is almost essential in asynchronous collaboration to maintain awareness. The notification scheme must keep group members informed about things like:

- Who is doing what?
- What has changed since the last time that a user signed on?
- What is happening that a user should know about?
- What is everyone's attitude to a proposed development?

However, such notifications should be provided without requiring users to do additional work. For example, one should not be required to make a change and then also include a note about the change for notification purposes. On the contrary, the notification systems should detect any changes and automatically inform other interested parties about them. For example, a document system should automatically detect changes by one user and inform others without requiring the first user to do anything other than an intended change.

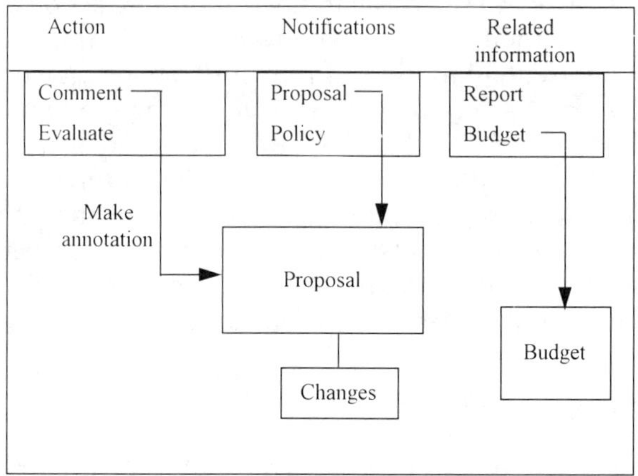

Figure 6.6 An interface goal.

However, it is important to remember that such notifications and presentations should themselves be presented in easily read forms. It is necessary to get away from the usual way of producing serial lists of status of particular events and scan them by eye on a regular basis. Instead, the goal is to try to visualize a user's workspace and the events within it in a graphical or even pictorial way.

6.5.1.2 Goals for Simple Interfaces

It is perhaps worthwhile to state a possible goal for interface design for asynchronous collaboration. The approach here is to *keep things simple for a user.* A user, on signing on, should be able to easily become aware of any changes to their context, get any additional information that they need, and then take an action. Thus we look at an interface like that shown in Figure 6.6.

Using this interface, the user, on signing on, can select notifications to see what has changed in the context and then select changes to a proposal. The user can then add further annotations to the proposal using any of the available options. At the same time, information relevant to the proposal can also be selected through a related information menu selection that has been customized to the notification. Of course, this visualization of changes can be quite complex if there are many related documents. Good design should show both the changes and their effect on other documents, including people responsible for the documents.

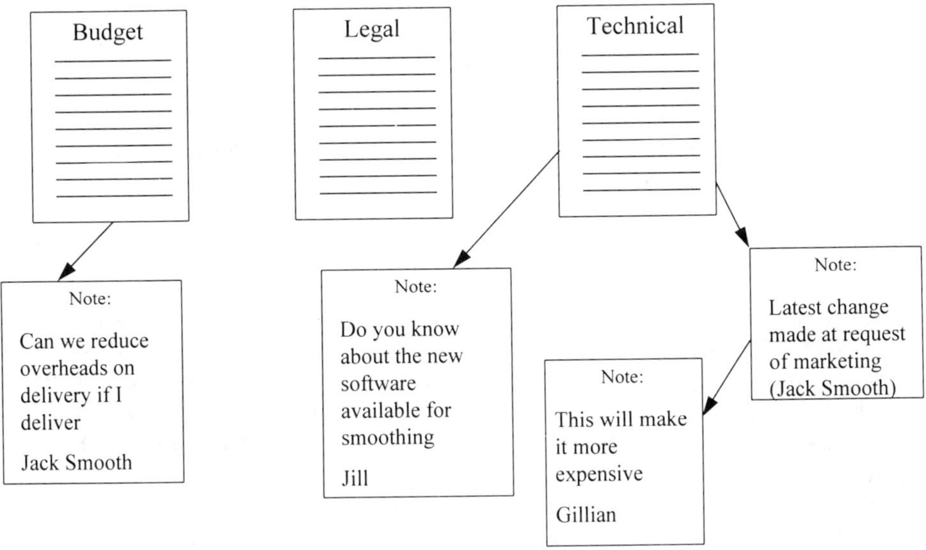

Figure 6.7 Document status.

6.5.1.3 Case Study: Proposal Preparation in Worldlink Consultants

Worldlink Consultants found that there are often different groups working on different parts of a proposal. Each such group is responsible for a different part of the report. It is proposed to develop an interface in this case that shows the current status of each part. Currently, an interface that may look like that shown in Figure 6.7 is being considered.

In Figure 6.7, the presentation is divided into parts that include the suggested change, other people's comments, and overall plan. Here, people can leave notes about various parts of the documentation. For example, a display shows that there has been a change initiated by Jack Smooth of marketing that may increase costs. But there is also a proposal that some of the cost increase can be kept down by reducing some overheads. The report writers thus have information about proposed changes and can act on it. Often, such action requires discussion between the members.

It is of course possible to try to go further to attract particular people's attention to particular parts of the document structure. Thus, following a change, the *technical* part may flash on the display of the person who is responsible for maintaining that part. This part need not flash for other users, who only have a passing interest in this part of the document.

Interfaces for groups introduce further requirements because they must support all the collaboration information described in earlier parts. Thus an important goal here is for an interface to support awareness about current work so that group members can make a relevant contribution to the group. An earlier section in this chapter described many awareness requirements. Such awareness should attempt to visualize the change and its effects.

6.5.2 Visualization of Change

Most group support interfaces usually support the idea of a shared workspace accessible to all participants and private workspaces for each participant. Awareness is maintained through the shared workspace. Participants can then move information between the workspaces, engaging in discussion when making a change. In many cases, the first thing that a person does is to look at what has changed in the shared workspace since the last time. One important consideration is to present information about change in a way that allows the user to quickly recognize developments in their context and act on them. This requires special attention to be paid on how to visualize change.

6.5.2.1 Case Study: Proposed Visualization of Change in the Software Process

Documents frequently change in the software process because of changes to requirements and designs. Users must be aware of changes to documentation that may affect them to ensure that their work is based on the latest requirements. One possible design is shown in Figure 6.8.

Figure 6.8 shows documents used in the software process on the screen and marks the ones that have been changed in some way, either by color or having them blink. Then, the screen would also show those documents that are affected by the change and the people responsible for them. A user seeing this screen can select the documents that affect them and, if necessary, change them. They can make this change by selecting a collaborative action that supports discussion with other members before making a change permanent.

6.6 SUMMARY

This chapter outlined the importance of interfaces. It described additional features needed to support group collaboration, making a distinction between synchronous and asynchronous collaboration. The interface must both maintain awareness of what is happening in the context of the group and allow group members to easily contribute to the group. The chapter outlined the importance

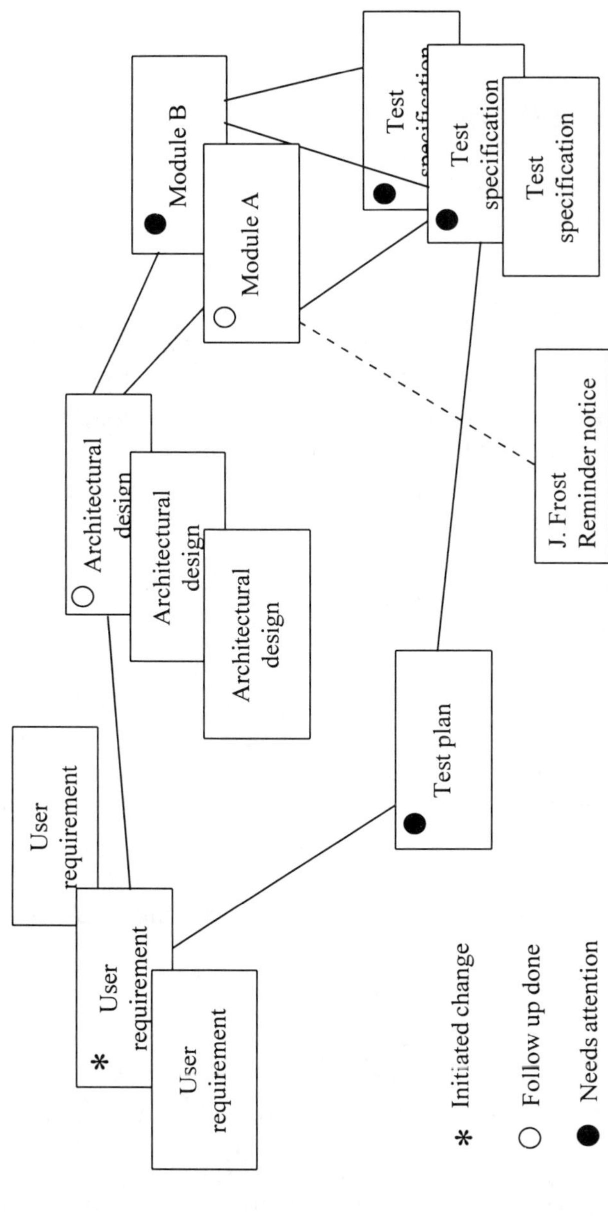

Figure 6.8 Documentation changes.

of maintaining awareness in asynchronous collaboration and described the role of notification schemes in doing so.

6.7 DISCUSSION QUESTIONS

1. Do you think the interface could contain the entire workspace in your job?
2. Do you think the use of many media on one display can be distracting?
3. Discuss some ways of representing nimbus in displays.
4. What are the major interface differences between synchronous and asynchronous collaboration?
5. Why are notification schemes important in asynchronous collaboration?
6. Discuss whether the three interface components of action, notification, and information form a good basis for interface design.
7. Describe some possible options of representing other people's activities when working on a joint document.
8. Why is visualization important?
9. Can you improve on the interface shown in Figure 6.6?

6.8 EXERCISES

1. Do you think Figure 6.8 is a good visualization of change? Can you think of a better one?
2. Propose some interfaces for the learning environment.
3. How would you maintain awareness in global planning of case B?

References

[1] Shneiderman, B., *Designing the User Interface*, Reading, MA: Addison-Wesley Publishing Company, 1992.

[2] Lauwers, J. C., and K. A. Lantz, "Collaboration Awareness in Support for Collaboration Transparency: Requirements for the Next Generation of Shared Window Systems," *Proc. CHI'90*, April 1990.

[3] Ishi, H., "Integration of Interpersonal Space and Shared Workspace: Clearboard and Experiments," *ACM Transactions on Information Systems*, Vol. 11, No. 4, Oct. 1993.

Selected Bibliography

Brooke, J., "User Interfaces for CSCW Systems," in Diaper, D., C. Sanger, (Eds.), CSCW in *Practice: An Introduction and Case Studies*, London, UK: Springer Verlag, 1993, pp. 23–30.

Chapter 7

Platforms

LEARNING OBJECTIVES

- ❑ *Collaborative services*
- ❑ *Service classifications*
- ❑ *Repositories of information*
- ❑ *Combining services into platforms*
- ❑ *Seamless integration*
- ❑ *Middleware*
- ❑ *Software agents*

7.1 INTRODUCTION

The previous chapters described ways in which people collaborate in networked enterprises. Those chapters introduced the services needed to support teamwork and networking within enterprises. This and the following chapters will describe these services in more detail. The ultimate goal will be to produce an environment like that shown in Figure 7.1. Here, each team member is provided with a platform of services. Team members can use these services, through an interface, to collaborate with other team members. These services may be e-mail, videoconferencing, and the other kinds of services introduced in Chapter 1 and now being provided by Internet or intranet networks.

Designers must choose the best services for the enterprise, paying particular attention to the social context determined by the enterprise culture, with the chosen services supporting the information flows to satisfy this culture. To choose the best services, it is necessary to identify the relationships and interactions between

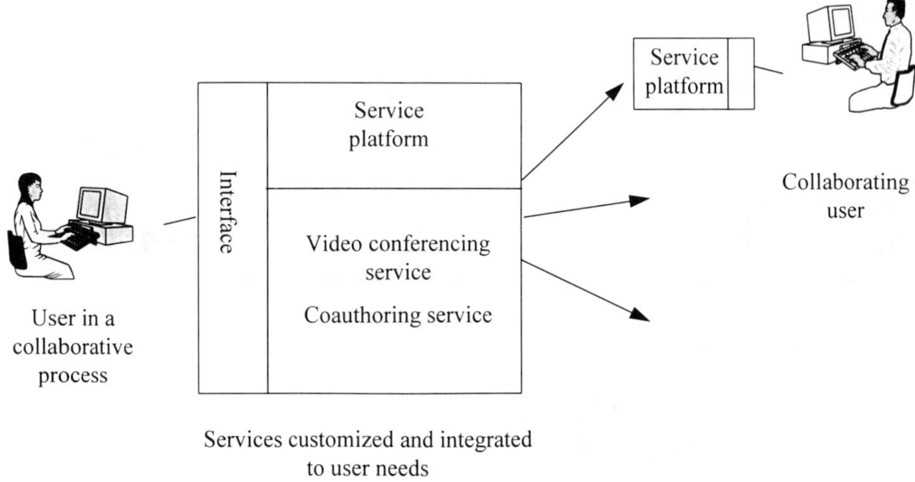

Figure 7.1 The design goal.

people in the enterprise and then specify a platform of services to support these interactions. This chapter will also describe ways of specifying such services.

7.2 PLATFORMS FOR NETWORKING

Generally, platforms are combinations of the two kinds of services shown in Figure 7.2.

Desktop services provide the tools needed for people to collaboratively work on their tasks, whereas operating software, which is described in Chapter 8, provides the connection needed to transfer information between them. Thus, word processing may be a desktop service, whereas the network provides the operating system to move documents produced by the word processor. The Internet or intranets are a combination of the two kinds of services. Both of these can be based on the same kind of network services, but often Intranet services are chosen to meet specific enterprise needs, whereas the Internet serves more general needs.

7.2.1 Desktop Services

Chapter 2 defined three levels of collaborative services. Most desktop services are at the first level—they allow users to construct documents and exchange them. It is common to classify desktop platforms into broad types. Thus platforms can be seen as primarily:

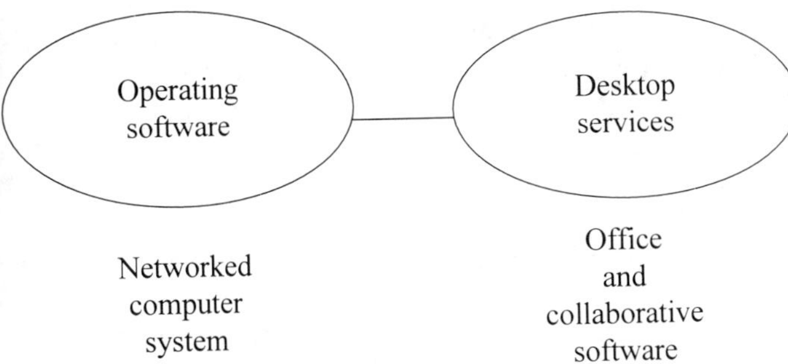

Figure 7.2 Combining networks and desktops.

- A document library that allows shared access to electronic documents usually stored in files;
- Workflow systems that route information through a prespecified procedure from one user to the next;
- Discussion databases needed for decisionmaking or for keeping personal information;
- Office support services such as word processing, spreadsheets, or preparation of multimedia documents.

Of these, the office support services are the most common. There are now a rapidly growing number of commercial systems that provide platforms of integrated desktop office support. Each of these has its own set of services, which are evolving rapidly over time. One of the more popular systems is Microsoft Office, which provides a platform for the office environment primarily developed for local area networks. It includes a number of services for carrying out office tasks as well as methods for moving data between the services. The most important services are

- The word processor for the preparation of documents;
- A spreadsheet for the preparation of financial documents;
- A database, known as ACCESS, for storing structured records;
- A drawing tool, known as Powerpoint.

Details of these and other services can be found in relevant manuals.

7.2.2 Collaborative Services

Collaborative services go beyond supporting one user and allow users to share their work. This set of services must cover the entire spectrum of networking activities within the enterprise, including interpersonal communication and work processes, so that the platform can support any kind of collaboration. The question is, "What is a complete set of facilities?" Chapter 1 introduced a set of services classified into messaging, meeting, document, and workflow services. Rodden and Blair [1] have provided a more detailed framework for classifying meeting services, as shown in Figure 7.3. The framework in Figure 7.3 spans the dimensions of time and location. These include

- *Meeting rooms* that support face to face meetings. Here, people sit as if they were in an ordinary meeting, but their deliberations are now supported by technology, which provides task support to improve meeting outcomes.
- *Messaging systems*, such as, for example, e-mail, that support exchange of information, through messages.
- *Conferencing systems*, where users meet electronically, most often using videoconferencing. Such meetings can range from direct face to face conferences across distances or to *extended conferences* where people can contribute to the conference at any time or place.
- *Coauthoring systems*, which to some extent are a special kind of conference that includes additional support for jointly working on artifacts. Here, people will discuss some artifact, such as a design document, that is displayed on each participant's screen.

Designers of collaborative systems must identify the kind of support needed by a group and choose the appropriate supporting system. Classification schemes such as that shown in Figure 7.3 can be used as a guide to choose the most appropriate software.

7.2.3 Integrating Seamless Application Platforms

Collaborative services must be seamlessly integrated in the platform. Seamless integration means that is easy to move from one service to another. For example, people may hold an electronic conference and at the same use a coauthoring service to work on a document while holding the electronic meeting. Or, a word processing service may be integrated with e-mail to allow documents to be electronically mailed.

Integration can be further extended to include corporate applications. This idea is illustrated in Figure 7.4. Here, there are three applications: customer preferences, item availability and prices, and customer mailing. A salesperson can display a combination of customer preferences and item availability in separate windows to decide what may interest the customer. The information can then be moved to the customer mail application using cut and paste methods and mailed to the customer to see if the customer wishes to purchase any of the items. This again depends on the availability of standard formats so that information can be easily moved between the applications.

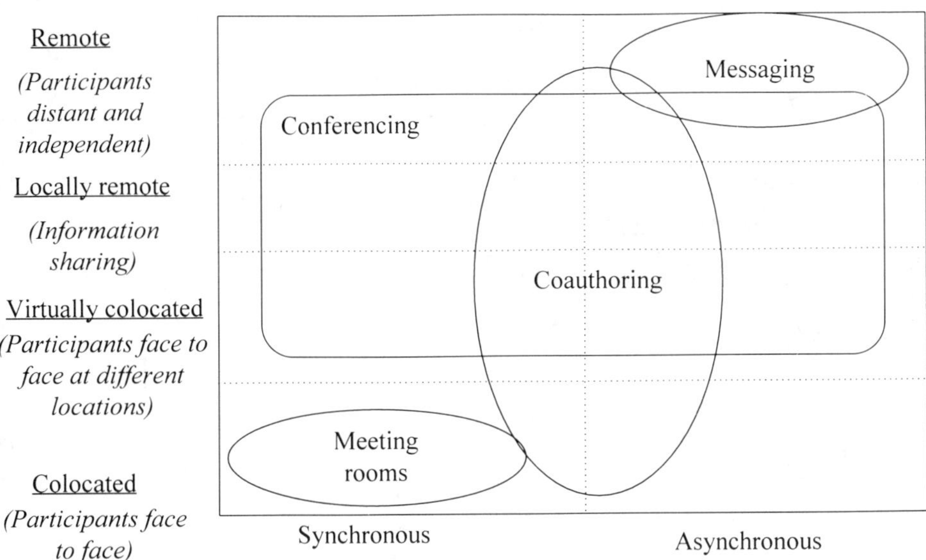

Figure 7.3 Classifying by generic services.

Systems that support such integration require all services to use the same standard. Often, this means buying the entire set of services from the same provider or building interfaces between the different applications. Many suppliers now provide such platforms, as for example, Microsoft Office that includes a word processor, database, drawing tool, and electronic mail service amongst others.

7.3 DESCRIBING PLATFORMS

Design, which is described in a later chapter, requires a method for describing a platform, especially platform requirements. Such requirements will be determined following an analysis of networking needs. There are two ways of describing platforms: the *physica*l description or the *logical* description. The physical description defines the actual software that makes up the platform (for example, the specific workflow package or document management package).

A logical description concentrates more on the requirements side, stating what the system must do. In design, it is often better to start by stating what the system is to do and then choosing the best physical services to achieve this. This approach is particularly important when setting up Intranets to suit particular enterprise needs. Here, the requirements must be specified in terms of what services the platform must provide, and then specific physical hardware and

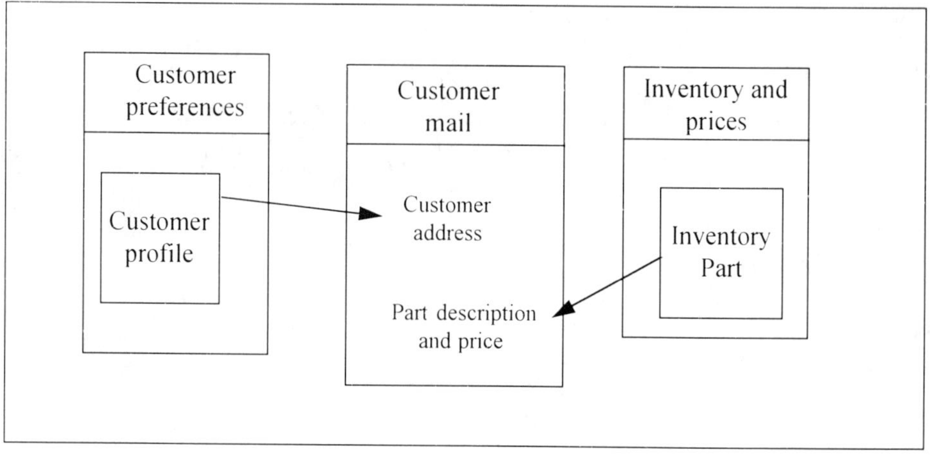

Figure 7.4 Providing collaboration services to existing applications.

software can be chosen to realize these services. This design method will be described in Chapters 16 and 17. The remainder of this chapter outlines ways to describe platform requirements.

7.3.1 A Logical Service Classification

A number of writers have suggested that classifications separate the distribution of information from the processing of information. This book proposes a semantic classification that looks at what it is that people do when setting up collaboration and provides a method of putting a number of components together to directly support the collaboration. It sees a classification in terms of three orthogonal dimensions, shown in Figure 7.5.

In this orthogonal object classification, services forms the semantic dimension, components the detailed dimension, and there is the physical dimension that adapts the services to the location of people. At the same time, each of the orthogonal dimensions can be defined in different levels of detail, thus providing another level of abstraction. We now broadly outline these services.

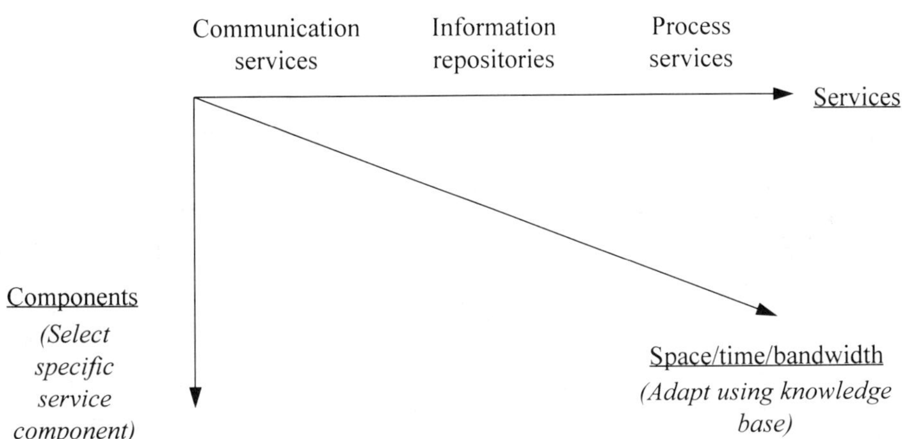

Figure 7.5 Classification dimensions.

7.3.2 Repository Services

Repository services, as shown in Figure 7.6, are objects that store information and also provide the methods to work on this information.

For example, a document repository stores documents and methods of accessing and distributing them, as well as methods to create and store versions of the documents. A vote repository will collect and count votes and manage lists of people eligible to vote. A news repository will store news items and contain methods to collect and distribute the news items as well as registering readers of the news messages. People gain access to the repositories through communication services. Often, repositories are stored at a site known as the server and delivered to distant personal computers, known as clients. Such networks are now known as client-server networks and are described in the next chapter.

A platform can contain any number of repositories and often allows information to be moved between repositories. Some repositories may be connected to communication services while others may not. It is thus common to move information from a repository with no communication services to a repository on the same computer that has services to send information. Thus, information from a word processor repository may be moved to a message in an e-mail system for sending to another site.

7.3.3 Communication Services

Communication services primarily distribute information, but do not process it. A set of communication services is shown Figure 7.7. The services support both synchronous and asynchronous communications by individuals or groups.

News items	User registration News distribution Access methods News collection
Document	Version control Distribution rules Access methods Media support
Message collection	Message management access rules Receive message Send message
Reminder service	Reminder rules Reminder collection Sending reminders
Discussion	Organize statements Access rules Views on discussion
Project plan	Task assignment Task monitoring Task scheduling Schedule change
Vote repository	Vote collection Vote counting Voting lists Voting rules
Argument collection	Add arguments Organize arguments
Database	Access method Retrieval method Update methods
Ideas	Record idea Display idea Comment on idea

Figure 7.6 Repository services.

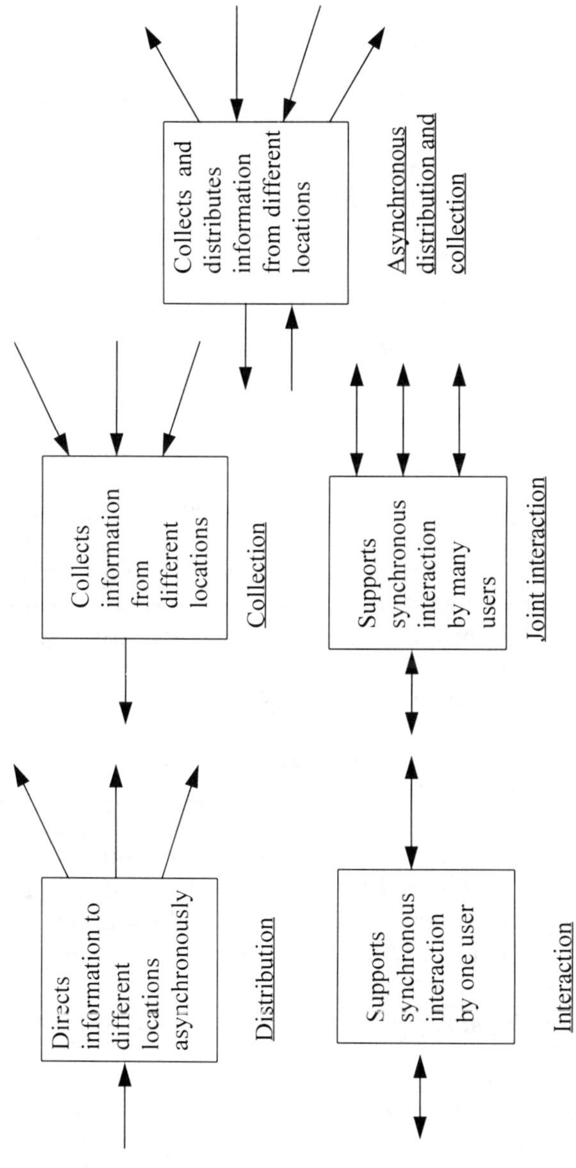

Figure 7.7 Communication services.

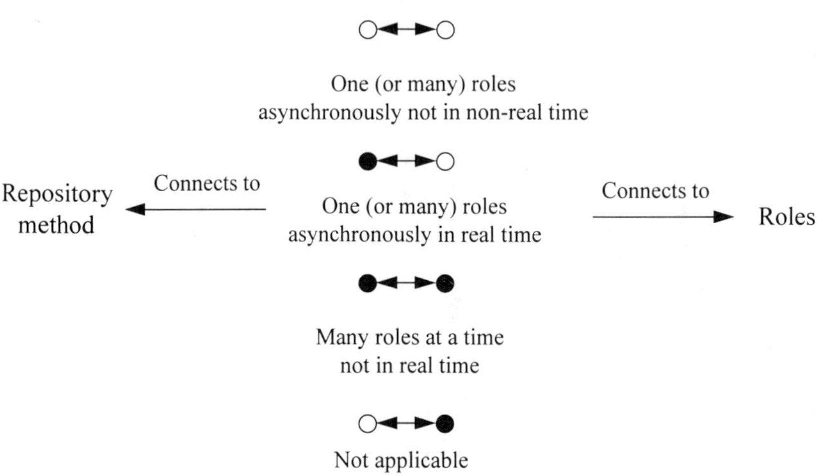

Figure 7.8 Notation for communication.

Thus distribution supports asynchronous distribution of messages, whereas collection gathers such messages. Both of these can also be combined into a service that supports both collection and distribution of messages. An interaction service allows a single user to interact synchronously with a repository, whereas joint interaction supports synchronous interaction by a number of users. A collaborative system can contain more than one messaging services.

This book uses a notation to logically describe communication services in platforms. This notation is briefly shown in Figure 7.8. It shows the direction of information flow by an arrow. The circles at the ends of the arrow end on repositories or roles. The diagram makes a distinction of whether the communication to repositories is in real time and whether more than one role can be connected to a repository at the same time. Thus, if the circle that ends on a repository method is filled in, then access is real time. If the circle that ends on roles is filled in then communication is synchronous. The circle that terminates on the role is labeled to identify the interaction that takes place using the communication service.

7.3.3.1 Examples Combining Repository and Communication Services

The notation shown in Figures 7.7 and 7.8 can be used to diagrammatically describe collaborative systems. One example is shown in Figure 7.9, which describes a simple messaging system. Here, there is a local message repository at the new where users can keep their messages. A user can interactively use the repository to read messages and compose new messages through interaction I1. The message repositories then exchange messages asynchronously through a message exchange,

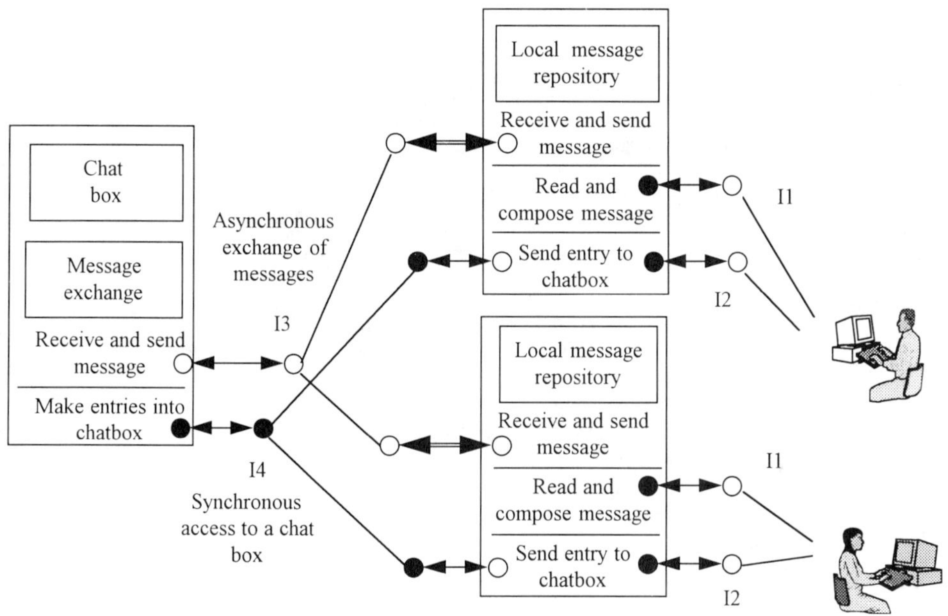

Figure 7.9 Messaging systems.

by the interaction labeled I3. The chat box, on the other hand, allows synchronous access to the chat box. Thus both users can initiate entries to the chat box through interaction I2. The entries are made synchronously and in real time through interaction I4 so that each user can see changes to the chat box simultaneously. However, chat box method rules ensure that at most one person is changing the chat box contents at one time.

Figure 7.10 illustrates another example where the *discussion* services are used to collect people's viewpoints. This is combined with a *votes* service that knows how to count votes to resolve any issues that might arise in the discussion. A facilitator role is supported to organize the discussion and maintain the database.

Both of the repositories are integrated with communication services. The discussion repository has an "add issue" method to collect issues asynchronously from discussion group members. It also has an "add comment" method to add people's comments on the issue, although now these comments can be collected through synchronous interaction I6. The votes service has a "vote collection" method that collects votes synchronously through interaction I9. It has a vote counting method that can be initiated in real time by the facilitator through interaction I7 to count votes when needed. Furthermore, the two services are linked to enable arguments to be moved to the votes service to collect people's votes on particular proposals. It is possible to add methods to each repository service. Thus for

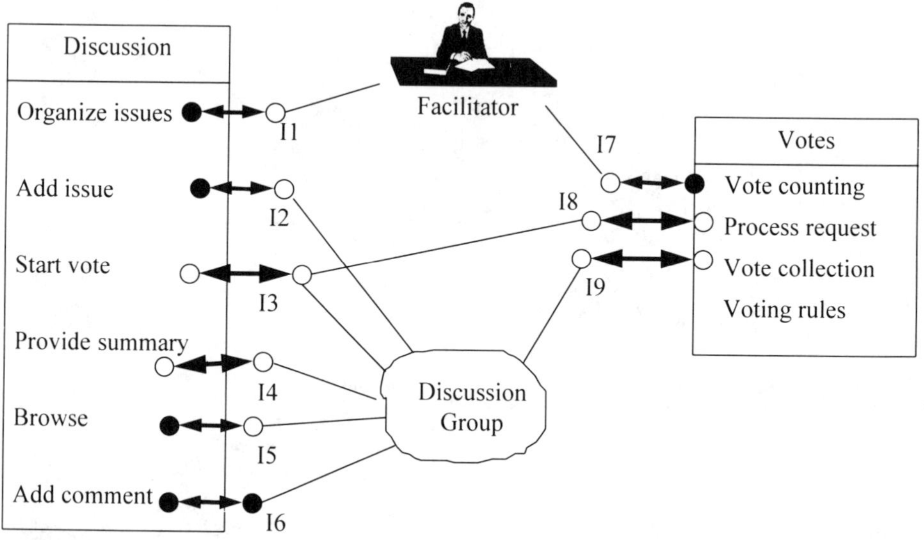

Figure 7.10 Another example.

example, the discussion service could include a "distribute notification" method to detect whenever a new argument appears for a selected issue. Selected roles could then be notified about the new argument.

7.3.4 Extending to Process Services

The combination of communication and repository services allows users to exchange information through the repositories. The process used to exchange the information is determined by the users. Process services provide assistance to support standard processes. They can be used to define flows and conversation rules and assign roles to process tasks. Some examples of process support can include

- *Negotiation processes,* where agreement is reached on arrangements made between a group of people;
- *Notification processes,* where people are informed of activities and changes taking place within their context;
- *Client-server processes,* where one person, the server, agrees to provide a service to another person, the client;
- *Decision processes,* where a group of people agree on a course of action.

One example of extension to include process services is shown in Figure 7.11.

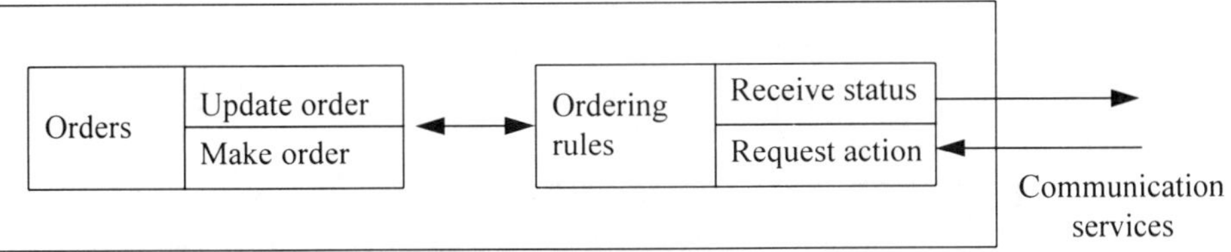

Figure 7.11 Including a process service.

The process service is similar to a repository service and stores the rules to be followed by the process. Thus, in Figure 7.11, the process service stores ordering rules whereas the repository contains the orders. The process service evaluates the status of other repositories and uses this status and its rules to notify users to take action. Thus, in Figure 7.11, the associated repository can be an order form. If the order status is found to be "new," the process service knows that it must be sent for approval. On return, the form has a new state that, again, is used to determine the next process step. Communication services are used by both the process service and the repository services.

7.3.4.1 Case Study: An Initial Service Selection for the Midvale Community Hospital

Figure 7.12 illustrates a first attempt at defining service requirements for Midvale Community Hospital. It uses the process definition in Figure 4.3 of Chapter 4 as a basis for choosing the services. Here, the major workflow centers around the patient record, both of which are in the same repository. The workflow process informs roles about the patient status. One of the ports from the workflow is to the examination process. It would be initiated once the patient is admitted. The examination process is a non-predefined process. It is a notification service that notifies doctors and other involved staff of new outcomes of examinations or tests. There are also ports from the notoification service process to various tests to initiate tests related to the patient. Each test centers around a test record and is embedded in a test workflow process.

7.3.5 The Time/Space Classification

The time/space classification corresponds to that defined by Johansen [2]. This classification is orthogonal to that in Figure 7.6. It determines the technology needed to implement the collaboration service. This technology is determined by the relative position of people in time and place, and the technical can be made independent of the logical level; that is, the system will posses the required knowledge of people's place and time constraints and provide the right support using

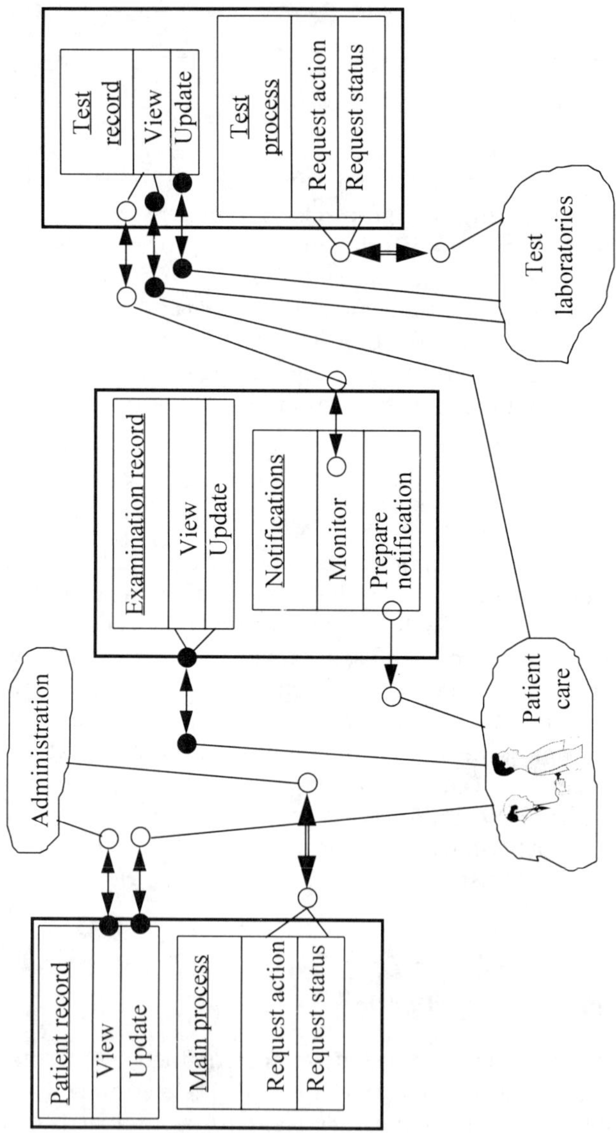

Figure 7.12 An initial selection for health services.

this knowledge. Thus a physical distribution of people, who need a shared view, would require technology that simultaneously displayed an artifact at a number of sites and supported discussion about the displayed artifact. Such access would be through ports.

7.4 PROVIDING PLATFORMS

An important goal is to simplify connection between services so that users can freely move from one service to another without using the underlying operating facilities, such as file managers or operating system commands. Integration achieved in this way is sometimes known as *seamless integration*, and relies heavily on developing a set of standards for repositories and communication services to support such connection.

7.4.1 Seamless Platforms

Seamless integration requires all services to be presented through one interface together with facilities to easily combine the services by easily moving information between them. The important factor, especially in Intranet design, is that users should be able to choose the best services for their collaboration, and furthermore easily adapt the service as their way of collaborating change. As shown in Figure 7.13, there are two ways to accomplish this. These are

- By using middleware, which provides an interface to each service and has the facilities to easily move information from one service to another;
- By one of the services itself providing interfaces to other services. We call this service the core technology.

In either case, the goal is to integrate services like e-mail, teleconferencing, or coauthoring into a single platform. Recently, the trend has been towards the second option, where platforms like LOTUS Notes now, for example, provide interfaces to many user services. LOTUS Notes will be described in more detail in Chapter 10.

7.4.1.1 Middleware

Middleware provides the methods that can be used to *integrate the collaborative services* into collaborative systems through ports and exchange objects. Thus, for example, it should be possible to use middleware to integrate a set of services that allow users to move a document from, say, an e-mail system to a word processor that is used in coauthoring, and integrate the coauthoring with a conferencing system that supports discussion about the document in real time. Users must be able to easily use the middleware to adapt the generic collaborative

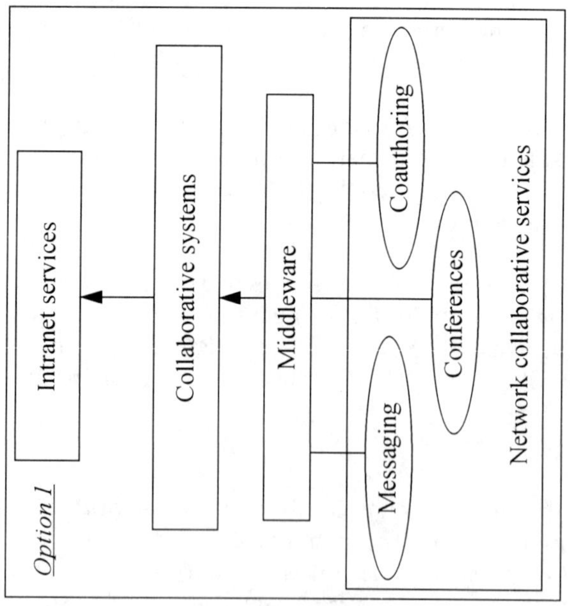

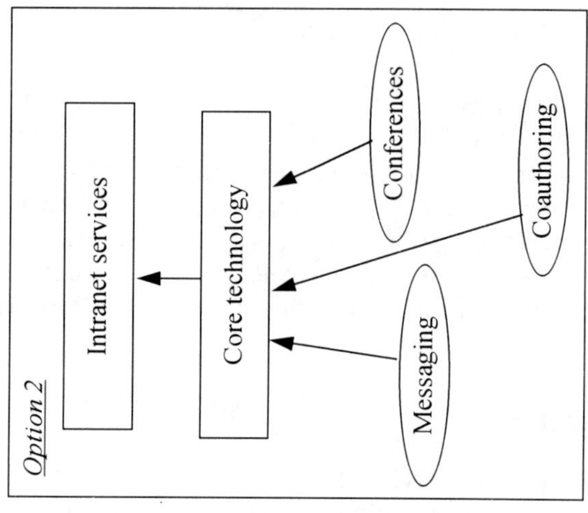

Figure 7.13 Building on platforms.

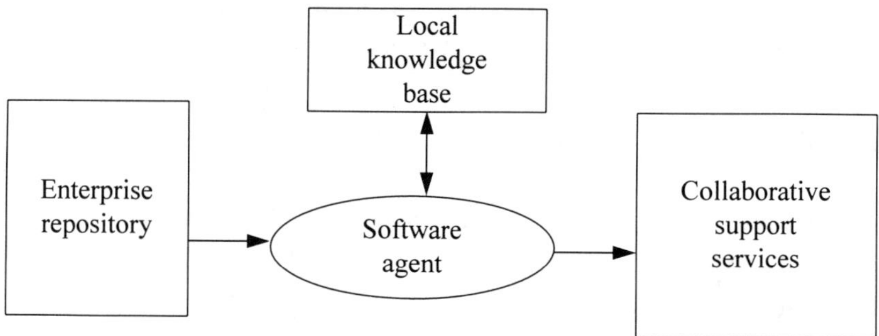

Figure 7.14 The role of agents. A wide range of such agents can be envisaged.

services to their working process and provide them through a platform, often supported by an Intranet.

Platform development is now in its early stages, and few commercial methods for constructing platforms using middleware exist. Those that do usually have limited services and facilities for their integration. However, often it is not easy to change services given current technology and suggestions have been made that special people, known as facilitators, be responsible for implementing change. Middleware that supports a seamless combination of services is also gradually becoming available. The suggestion has also been made that change be supported by intelligent software agents.

7.4.2 Software Agents

Recently, software agents have been proposed as a means of facilitating collaboration. A software agent would work in the way shown in Figure 7.14. The agent would have local knowledge about some particular collaboration task (for example, arranging a meeting). It could then access the enterprise database to find the location of people that will attend the meeting. Using this knowledge, the agent would set up the correct collaboration facilities for the meeting. If the roles of the meeting change or move, then their new location would be recorded in the enterprise database. The agent would use the new information to rearrange the collaborative services.

There could be agents to choose roles for a process, notify people about artifact changes, or even choose the best process, given knowledge about the people in a group.

7.5 SUMMARY

This chapter described the importance of platforms for supporting collaboration. It stressed that platforms must seamlessly integrate a number of services. It also defined a way of specifying collaborating platforms in terms of repository, communication, and process services.

7.6 DISCUSSION QUESTIONS

1. What do you understand by the term *platform of services*?
2. Why are seamless platforms important?
3. Can you name some additional repository services?
4. Describe some useful combinations of repository and communication services.
5. Define some useful process services.
6. Why are standards important when providing platforms?
7. What kind of software agents would be useful in communication?
8. What do you understand by the term *middleware*?
9. Define platform characteristics that make platforms useful for people other than computer professionals.

7.7 EXERCISES

1. Suggest a set of services for the construction industry.
2. How would you define process services to support conversations?
3. What services are required for keeping track of workflows?
4. Suggest a set of services for the Midvale Community Hospital.

References

[1] Rodden, T., and G. Blair, "CSCW and Distributed Systems: The Problem of Control," in Bannon, L., M. Robinson, and K. Schmidt, (Eds.), *Proc. of the Second European Conference on Computer-Supported Cooperative Work*, Amsterdam, Sept. 1991.

[2] Johansen, R., "Groupware: Future Directions and Wild Cards," *Journal of Organizational Computing*, (2)1, 1991, pp. 219–227.

Selected Bibliography

Busch, E., et al., "Issues and Obstacles in the Development of Team Support Systems," *Journal of Organizational Computing*, Vol. 2, No. 1, 1991, pp. 161–186.

Elbert, B. R., and Martyna, B., *"Client/server Computing: Architecture, Applications, and Distributed Systems Management,"* Norwood, MA: Artech House, 1994.

Greif, I., "Desktop Agents in Group Enabled Products," *Communications of the ACM*, Vol. 37, No. 7, July 1994, pp. 100–110.

Grudin, J., "Groupware and Social Dynamics," *Communications of the ACM*, Vol. 37, No. 1, Jan. 1994, pp. 92–105.

Hiltz, S. R., et al., "Distributed Group Support Systems," *Journal of Organizational Computing*, Vol. 2, No. 1, 1991.

Schmidt, K., and T. Rodden, "Putting it all together: Requirements of a CSCW Platform," *Proc. of the 12th Interdisciplinary Workshop on Informatics and Psychology*, Schaerding, Austria, 1993.

Reinhard, W., J. Schweizer, and G. Volksen, "CSCW Tools: Concepts and Architectures," *IEEE Computer*, May 1994, pp. 28–36.

Part B

The Technology

Chapter 8

Networks

LEARNING OBJECTIVES

- ❑ *Network structures*
- ❑ *Standards*
- ❑ *Network protocols*
- ❑ *Client-server networks*
- ❑ *E-mail and its limitations*
- ❑ *News services*
- ❑ *Bulletin boards*

8.1 INTRODUCTION

One of the critical factors of enabling people to work together is to provide them with the ability to communicate. This in turn requires computer systems that are connected together to support this communication. This chapter describes the facilities needed to connect computers into networks. These networks then support collaborative services.

Networks are very similar in concept to telephone communication, except they allow people to exchange information other than voice. There are a variety of networks. The most common is a local group supported by a local area network (LAN) within an organization or even within one office. There is often a file server, which is a computer that stores all the documents needed in the group. Each computer on the LAN can access these documents and the network provides the communication paths that enable people to easily distribute the documents and exchange messages. Most such networks support a number of repositories, which

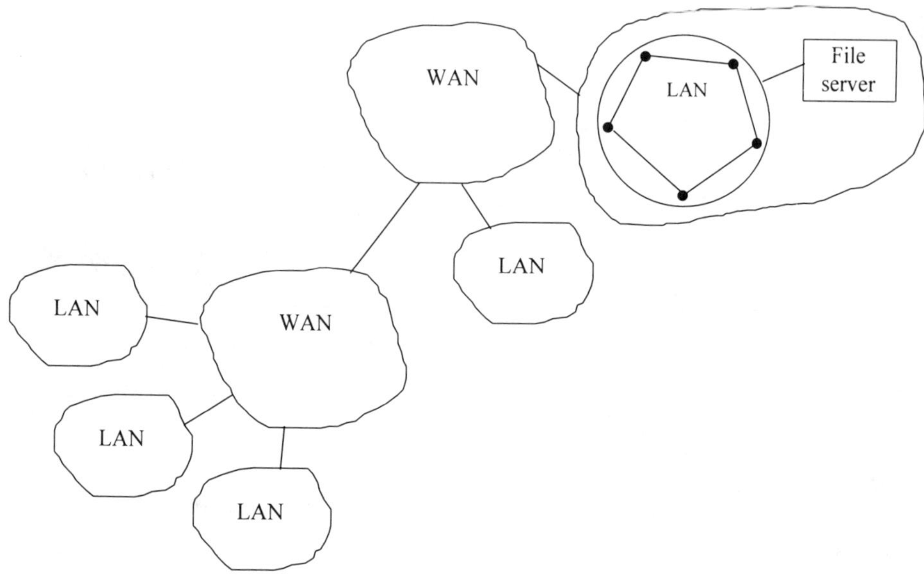

Figure 8.1 Networking.

contain information needed in the office, and provide communication services between these repositories.

Furthermore, it is also often necessary to connect people who may be connected to different LANs, which may be in different organizations. For this reason, it is now common to connect LANs into larger networks, often known as wide area networks (WANs) and shown in Figure 8.1. The WANs themselves may be connected, each with an increasing amount of users and services. Thus a local network can have a gateway to a WAN and access LANs through the WAN.

Constructing a network requires the connection and adaptation of a large number of software and hardware components.

8.2 NETWORK SOFTWARE AND HARDWARE

The components that make up a computer network include the communication lines themselves, the network computers and their software, as well as the personal computer, which is the main user contact with the network.

8.2.1 The Personal Computer

The personal computer or workstation is probably the most critical part of the network as it provides the interface to the user. The computer must also be configured

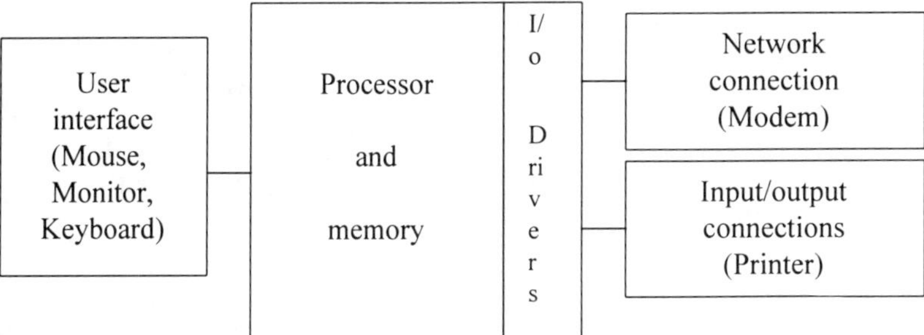

Figure 8.2 Setting up the workstation.

to the network. As shown in Figure 8.2, this usually means having a network card and modem to connect to the lines or having a connection to other I/O devices such as the printer.

The computer then needs software to drive the I/O devices. It must have sufficient memory to store the software and a sufficiently fast processor to quickly transfer information between the I/O devices and memory. This software becomes a part of the operating system. Most network operating systems now support *multitasking,* where I/O can be going on at the same time as other work by the user. Thus a user should be able to start a long file transfer, then carry on some other work while the file is being transmitted.

8.2.2 The Communication Channels

The hardware provides the communication channels for information transfer. The first networks relied on voice channels to transmit information. The voice channels are the same as those used for telephone conversations and are usually sufficient to transmit text or voice. However, the trend now is to go beyond text and include a variety of media in the communication, in particular digital images, video (which can support videoconferencing), as well as text and voice. Voice channels no longer have sufficient bandwidth to transmit this range of media. New kinds of communication channels have been developed for this purpose. They are known as ISDN (Integrated Data Network). Now, an even newer kind of channel is being designed known as ATM (asynchronous transfer mode) for multimedia transfer. Because of its importance, it is described in detail in a whole issue of the *Communications of the ACM* [1].

8.2.3 The Operating System

Network software manages the repositories and the communication protocols between them. It also supports the *protocols* that allow computers at both ends of a

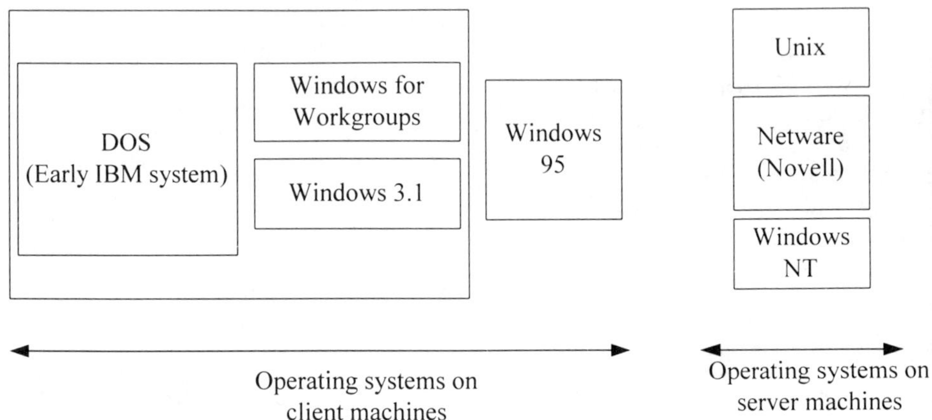

Figure 8.3 Operating systems.

link to communicate. If two organizations use different software, then it is not possible for their machines to be connected. What is needed is standard software at both ends that supports the same protocol.

The large variety of commonly used operating systems is illustrated in Figure 8.3. A distinction is made between systems that run on PCs, often known as client machines, and larger systems that store shared by a number of PCs, often known as server machines.

Historically, disk operating system (DOS) was one of the earliest systems primarily for standalone personal computers. It was then enhanced for communication. Many systems still use the basic DOS I/O components (known as BIOS). Thus Windows 3.1 and Windows for Workgroups are two Microsoft products based on DOS, using its input/output interface. Windows NT is one of the earliest systems developed for server machines on a PC platform. Windows 95 is a Microsoft operating system that has its own basic input/output system. Netware is Novell's operating system to support local area networks, which also has links to DOS and UNIX.

8.2.4 Putting it All Together

Setting up networks thus becomes a complex process. One way of seeing this construction is illustrated in Figure 8.4. Here, transmission links connect to the network hardware, which is controlled by the network protocol. The protocol, which is used to coordinate machines at both ends of a link (and is described in Section 8.3), is driven by the operating system that provides it with information to be transmitted. This information is obtained from the services or applications supported by the system. Operating systems at both ends must then be able to interpret the packets supplied by the protocols in the network layer into a form

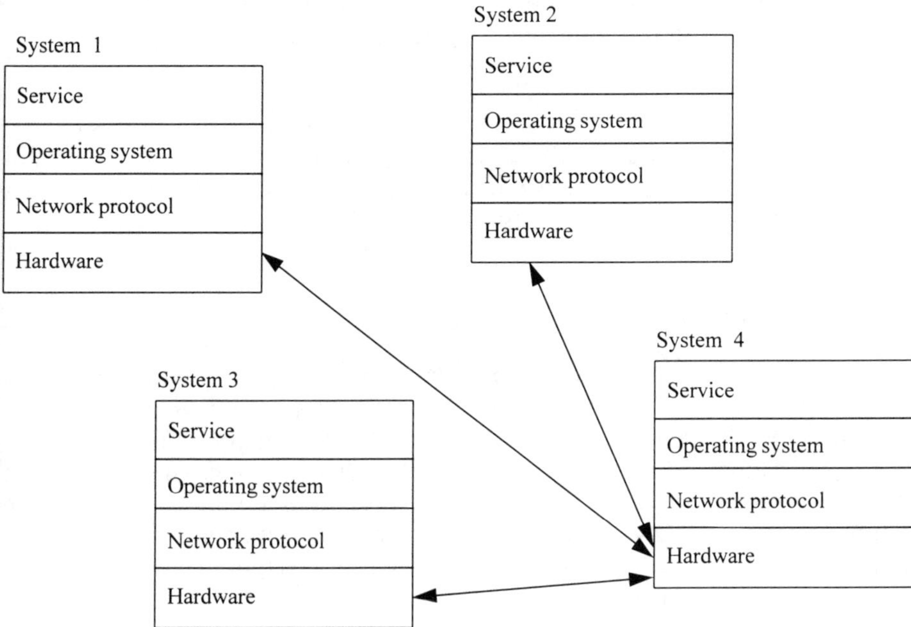

Figure 8.4 Forming networks.

acceptable by the services supported on the computer. Thus the services themselves must follow a standard so that information input at the service at one end is presented with the same meaning at the other end.

There are a large number of choices that can be made by both network designers and workstation users. These include

- Choosing the kinds of channels, in particular whether voice channels are sufficient or whether more advanced channels such as ISDN are needed;
- Choosing the network configuration or how computers are to be connected together;
- Choosing the kind of network protocol to be supported on the network and ensuring that the network can support the protocols, and making sure that each system supports the same protocol;
- Ensuring that the protocol is supported by the operating system and configuring the operating system to support the protocol;
- Ensuring that the operating system can support the chosen services;
- Choosing the computers to be connected to the network and ensuring that they have the software that matches the network protocol;

- Choosing the network cards to put into a personal computer;
- Choosing the protocols and adapting them to the network card.

At the same time, the designers and users must ensure that versions of software on different machines are the same, or that both ends use compatible channels. Anyone who has been involved in setting up networks knows the difficulties of ensuring that all these components match.

8.2.4.1 The Importance of Standards

Information interchange is only possible if the form of information transmitted by one computer can be used by another. This requires standards to be supported across the network. Network standards have now become an important industry issue and are defined within a reference model of Open Systems Interconnection (OSI). This reference model defines standards in terms of a number of levels. The definition in detail of these levels is outside the scope of this book, but a simple explanation can be given in terms of Figure 8.4, where each computer in a network is made up of the following levels:

- Collaborative services that are included in a platform and are presented as multimedia windows to users;
- Operating software, such as operating and window systems, that convert between information needed by a service and that needed for transmission using a network protocol;
- Network protocols, sometimes known as *transport protocols,* that transfer information provided by the operating system and the hardware;
- The hardware used to transmit information.

Computers that are to interchange information on the network must support the same network transport protocol. In Figure 8.4, information is interchanged through common network transport protocols. This in turn requires systems to be properly configured-including the workstation, operating system, and protocol.

The idea of configuration is illustrated in Figure 8.5. Here, there are input/output devices, such as a network connection that requires OS hardware adaptors to interface to them. These adaptors in turn interface to a protocol, which is configured both to the operating system and to the hardware adaptors. The operating system protocol is then adapted to communication services that provide interfaces to the user.

Perhaps the most critical issue is the network protocol.

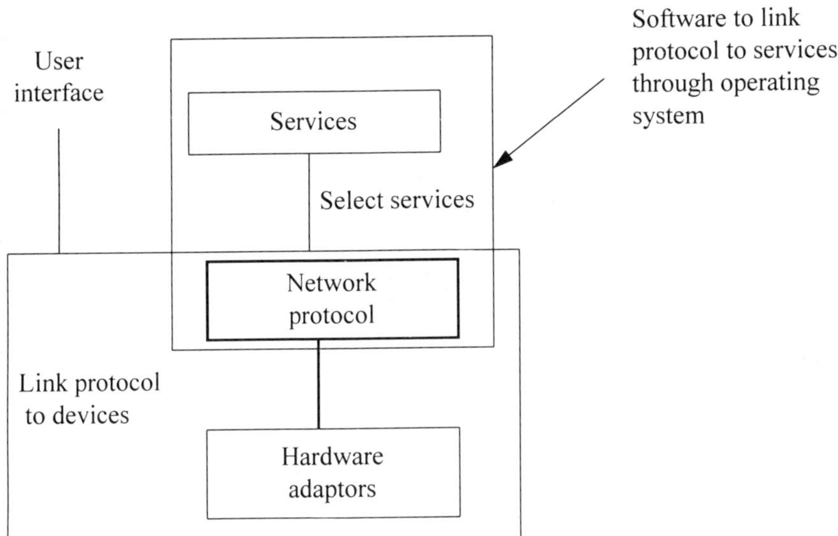

Figure 8.5 Configuring systems.

8.3 PROTOCOLS

The first standard is that of the network transport protocol that allows computers to exchange data packets. Perhaps the most widely used network standard at this level is known as TCP/IP, and has now been widely adopted as a standard for this purpose by many organizations, mainly because it is the standard used by Internet.

8.3.1 The TCP/IP Protocol

The TCP/IP protocol is one of the most important protocols because it is supported by the Internet. It now dominates the market, but there are other protocols such as SPX/IPX, IBM's SNA, DECNet, and ATM. It is illustrated in Figure 8.6. The application layers here are services supported by the network, such as e-mail or file transfer. The IP supports transfer of messages between sites, and each site in the Internet has its unique IP address. The IP layer is adapted to the communication hardware through the physical layer. The TCP packages application information into a format acceptable to the IP protocol.

One important characteristic of the TCP/IP protocol is that it is *routable*; that is, it is independent of the physical network and message paths can be chosen dynamically. Such paths are chosen by the IP level. Thus, even if the network connections can change, the protocol will eventually deliver the message.

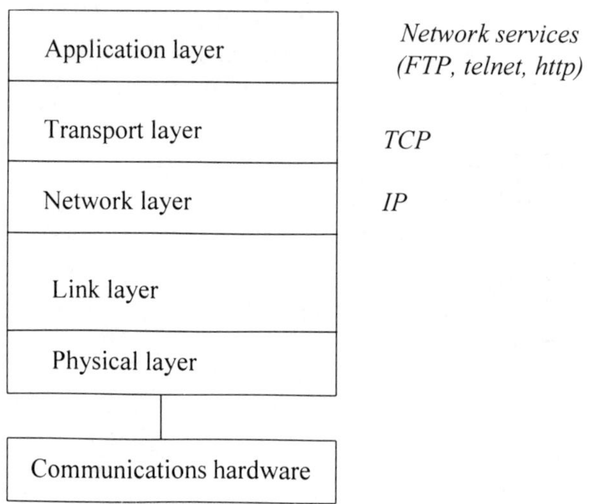

Figure 8.6 TCP/IP.

This differs from *nonroutable* protocols, which can only pass the messages on a predefined network. Many of the early protocols were nonroutable and are still applicable in some systems. To discuss some of these other protocols, one should go into history.

8.3.2 An Historical Perspective

One of the initial most widely used PC operating systems was IBM's DOS. This, of course, did not support any protocols initially, as it came before the days of networking. One of the earliest extensions to DOS was to include adaptors to its basic input/output system (BIOS) to create a system known as NetBIOS, which evolved over time to currently support a special protocol known as *NetBIOS frame protocol,* which is a nonroutable protocol. NetBIOS was later extended to NetBEUI, which has a more friendly interface and is supported by Microsoft, especially for LANs. NetBEUI also has its own protocol, here called the *NetBEUI protocol.* At the same time, additional systems were written to allow NetBIOS to also support the TCP/IP protocol. Perhaps the best known system here is PC/NFS, which interfaces a number of services to Windows. This evolution of protocols paralleled that of operating systems. Perhaps the most significant here is the development of Windows 3.1, which is one of the most common workstation operating systems. Windows 3.1 still uses most of the underlying DOS software, but can then be extended to use TCP/IP by using systems such as PC/NFS.

Operating systems often support different protocols, and it is usual for operating systems to support more than one protocol to improve its connectivity. For example:

- Windows for Workgroups uses NetBEUI to connect computers that run on Microsoft local area networks.
- Windows NT can support both TCP/IP and the NetBEUI local area protocol. Thus, NT can be used to interface Windows systems to UNIX systems or to Microsoft local area networks.
- IBM's LAN Server (version 4.0) can support TCP/IP, NetBIOS, NetBEUI, and Appletalk.
- Novell's NetWare (version 4.1) can support TCP/IP, NetBIOS, and Appleware.

It is important in designing networks to ensure that computers in a network all support the same transport protocol. Furthermore, the protocol that delivers the best performance for the network configuration must also be chosen. Thus, for example, Windows for Workgroups using NetBEUI is often used for peer-to-peer LANs. However, if one is to connect a UNIX and a Windows-based system, then one would use NT, which supports TCP/IP on the Windows system. Both use the TCP/IP standard and thus can be connected and interchange information, but may present this transmitted information in different formats to their respective operating systems.

However, networks are developing in an evolving environment. Many organizations are now moving from voice networks to ISDN to support multimedia data transfers, but here problems arise because of the number of options provided by these systems. Again, the choice of such options is outside of the scope of this book, but readers should be aware that considerable attention must be paid to choosing the right options if ISDN solutions are to be effective. One can expect that standards such as TCP/IP will change, or indeed new standards may be adopted in networks such as the Internet, to cater for the increasing demands for multimedia information transfer.

8.3.3 Performance

Performance is an issue that is especially important in any network operation. There are two aspects of performance in collaborative work. One is in synchronous collaboration, where it is important that communication between the two ends takes place with minimum delays. The other is a general requirement that communication is achieved with minimum movement of information across the network to avoid overload, which in turn leads to delays. One approach used to minimize overall traffic is to have a client site that contains all the repository services needed by one user. A server site is then used to collect all information from client sites that is to be shared and distribute it between client sites. Thus the client sites maintain information private to that site, whereas the server is the central repository of all information used by every client site, with copies of the commonly used information stored at each of the client sites. Changes made at one client site are replicated at the server and the other client sites. However, replication is controlled and occurs only at set intervals to reduce total traffic.

The choice of how services are distributed between the sites is often determined by *performance* and *cost* considerations. Performance must be such as to provide a good response to users. It is particularly difficult to achieve with video, where images captured at one end must be instantaneously transmitted to another. This calls for high-capacity channels, in turn raising the cost. Alternatively, still images can be transmitted at regular intervals, but these are not natural as they do not portray motion.

8.4 NETWORK CONFIGURATIONS

Computers may be configured in a variety of ways. The most common networks are LANs.

8.4.1 LANs

LANs usually connect people within an office environment. They can allow members in the group to exchange information, share documents, and make general announcements of interest to the whole group. A LAN usually has a server that enables communication between the workstations in the area and that is often a repository of shared documents. However, a local network does not allow members of the group to get access to information or people in other LANs. To do this, it is necessary to interconnect servers in these other LANs. Figure 8.7 shows one way in which local networks can be connected to other local networks so that information can be exchanged. Now, a LAN server is connected to a larger system, often known as a local host, which is the hub of a WAN. This host may have connections to other servers in the same organization and pass information between people connected to the different servers. It then becomes possible to connect hosts in different organizations through the WAN and allow exchange of information between people in those organizations.

Ultimately it should be possible not only for individuals to be interconnected using networks, but for information systems in two organizations to be also interconnected. This would enable the systems at both ends to send messages to each other, supporting any alliance made between the two organizations.

There are now a number of standard LAN systems in common use. A commonly used standard is Ethernet, which is based on an IEEE standard (IEEE 802.3). It is almost universally used with UNIX platforms, but is also becoming popular

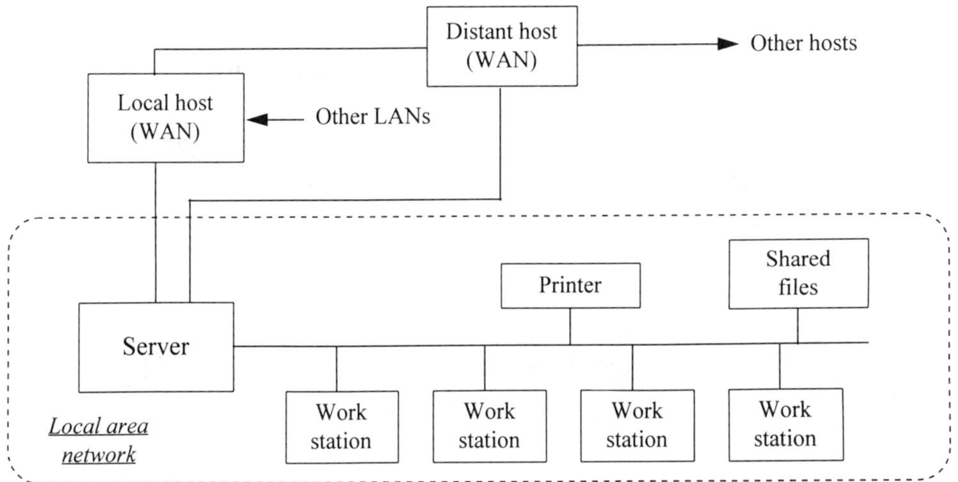

Figure 8.7 Connecting local networks.

with PC-based networks. Token ring is a topology that is frequently used, especially with IBM systems. It is possible to install LANs using standard commercial systems such as Windows for Workgroups, Netware, or Windows NT.

8.4.1.1 Windows for Workgroups

The Windows for Workgroups is a Microsoft product (see [2]) that provides a platform for supporting coordination between workgroup members in a local area network. The idea of Windows for Workgroups is illustrated in Figure 8.8.

It supports peer-to-peer communication and provides a number of generic collaborative services including:

- E-mail that allows a number of workgroups to be set up;
- A chat box;
- A scheduler for appointments;
- A large number of utilities including a word processor, card file, calendar, chatbox, clock and many others.

A number of utilities are also provided but these are not mail-enabled. For example, there is a clipbook that can be used to share information between workgroup members. This provides a degree of seamlessness by allowing information to be moved between documents. Users can move objects into clipbook pages and make these pages accessible by other selected users. They can also move this information into a mail-enabled document and send it to a distant site.

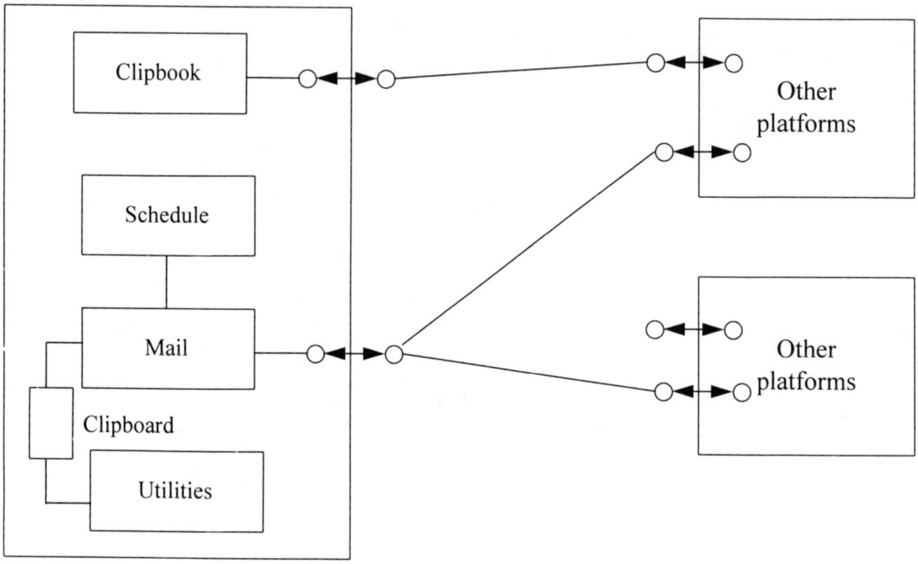

Figure 8.8 Windows for Workgroups.

8.4.2 Client-Server Systems

Client-server systems provide a variety of options for distributing services or applications across the network. The way work is distributed is described in three components: the presentation component, which describes the interface presented to the user, the data management component, which describes the data; and the processing component, which are the actual computations. Perhaps the simplest distribution is shown in Figure 8.9. Here, only the presentation is distributed to the client, with all the processing carried out on the server. Distributing presentation is important given the trend to more powerful, interactive multimedia interfaces. Now a lot of the work associated with providing graphical interfaces is carried out on the client, whereas the more routine arithmetic computations and file operations are carried out on the server. There are, of course, other ways of distributing components on a client-server network. For example, some application processing could be placed on the client, or even some data could be stored or duplicated on the client.

Messages between clients are sent through the server, which has all knowledge about the network, including the location of people, their passwords, and other profiles. It receives messages from clients and uses its knowledge to distribute these messages to other clients. The client-server approach offers considerable flexibility. Computers with personal databases can be added to the network at any

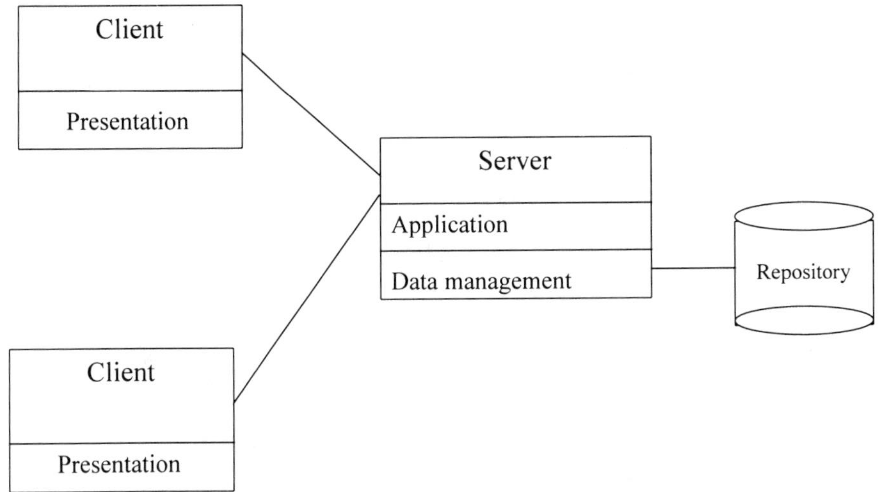

Figure 8.9 A client-server architecture.

time by connecting them to a server, where they become clients of the server. Server tables can be updated to include the new clients, thus connecting them into the network.

8.4.3 Peer-to-Peer Connections

The client-server approach is not the only way to form local area networks. The other kind of configuration is known as the *peer-to-peer* configuration and is shown in Figure 8.10. DECNet or SNA are typical vendor systems that support such configurations. Here, each machine supports one user and that user determines how data is to be shared. The user also initiates all data transfers. Generally, it is found that peer-to-peer networks can generate more traffic for the same number of users than in the more controlled client-server configuration. Consequently, peer-to-peer networks are often limited to LANs with up to about 25 users whereas client-server configurations are preferred for larger groups.

A typical office system would provide each workstation with standard services such as a word processor and a spreadsheet. It often also provides a diagramming tool such as, for example, Powerpoint. Often, the repositories are local to each workstation, although workstation operators can move files from one station to another. Chapter 10 will cover such platforms in more detail and describe ways of interconnecting services on them.

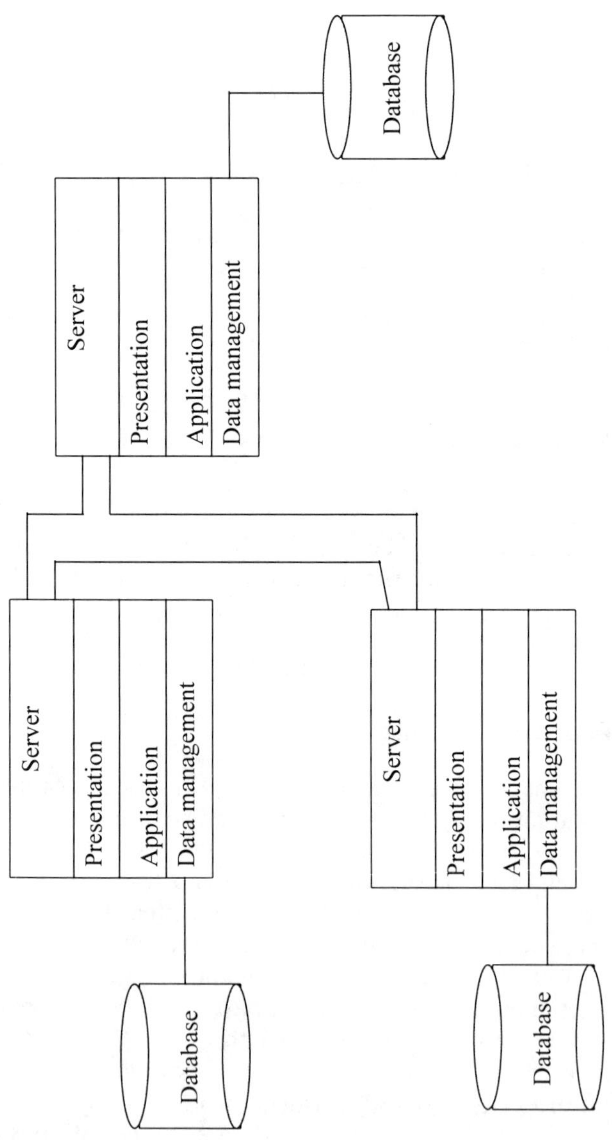

Figure 8.10 Peer-to-peer networks.

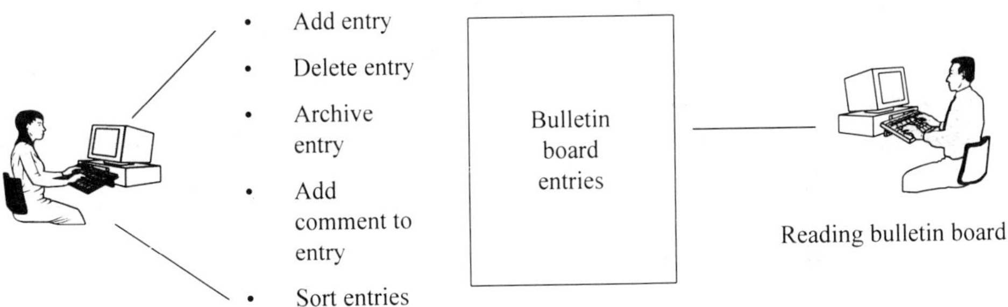

Figure 8.11 Bulletin board.

8.5 BASIC MESSAGING SERVICES

A variety of services exist to support interchange of messages between people connected to the network. Most local networks include services such as:

- Electronic mail (e-mail) to asynchronously exchange messages;
- Bulletin boards to asynchronously share information;
- Chat boxes to synchronously exchange written messages.

8.5.1 Bulletin Boards

A bulletin board is one of the simplest services provided by a network. As shown in Figure 8.11, a bulletin board is a repository that collects messages. The repository, which is usually stored in the server, can be accessed by all group members. It is usually supported by two communication services: one to collect information from a number of users, and the other to allow users to access that information. The bulletin board can serve as a group memory for the group.

The most common use of bulletin boards is for informal sharing of information. In this case, they are totally passive and rely on their users to initiate both input and retrieval of information. However, they can be made more active by enhancing them with a variety of methods to support

- Assigning a special member to enter relevant information onto the bulletin board;
- Including criteria to classify information and notify selected participants of information of particular interest to them;
- Issuing regular reports about information usage;
- Distributing a regular list or index of newly arrived information.

Bulletin boards can result in better awareness as notes of general interest now become available to all people in the system. If used in this way, they can become the repository for organizational memory, particularly if posting messages of significant events becomes mandatory in the organization. This will also improve awareness in a distributed group by making information available to all group members.

8.5.1.1 Interactive Bulletin Boards

It is becoming increasingly common to allow registered bulletin board users to interact with the bulletin board by making comments or observations about bulletin board entries. This is one of the services now commonly made available to support interpersonal relationships, where people can register their viewpoints on issues registered on the bulletin board. Later, in Chapters 10 and 11, the book will describe discussion databases that can also be used for this purpose.

8.5.2 Electronic Mail (E-Mail)

E-mail can connect any number of workstations and allow users of these workstations to communicate. It can be simply a local network in an organization or it can be within a growing number of networks.

E-mail is most suited as a communication service that supports three basic communication protocols, as shown in Figure 8.12. Its basic function is to send messages. It does not keep track of the relationship between any two messages. Most e-mail systems provide simple ways of replying to a message, recording a message, or sending the message on to any group member. E-mail can be made more effective by providing a variety of task support services at the user end.

8.5.2.1 Support for E-Mail

Workstations connected to e-mail usually have a message repository for storing all the workstation messages. It is now common to provide a variety of support services for e-mail users through methods associated with these repositories. Such methods support users by keeping track of messages in the workstation message repository. Typical repository methods include the following:

- *Attachment of documents* to messages to allow users to send existing documents, prepared using a word processing package;
- *Message management services* to allow users to manage messages by allowing them to sort messages into folders and thus keep better track of particular conversations;
- *Media support* to allow information other than text to be transmitted;

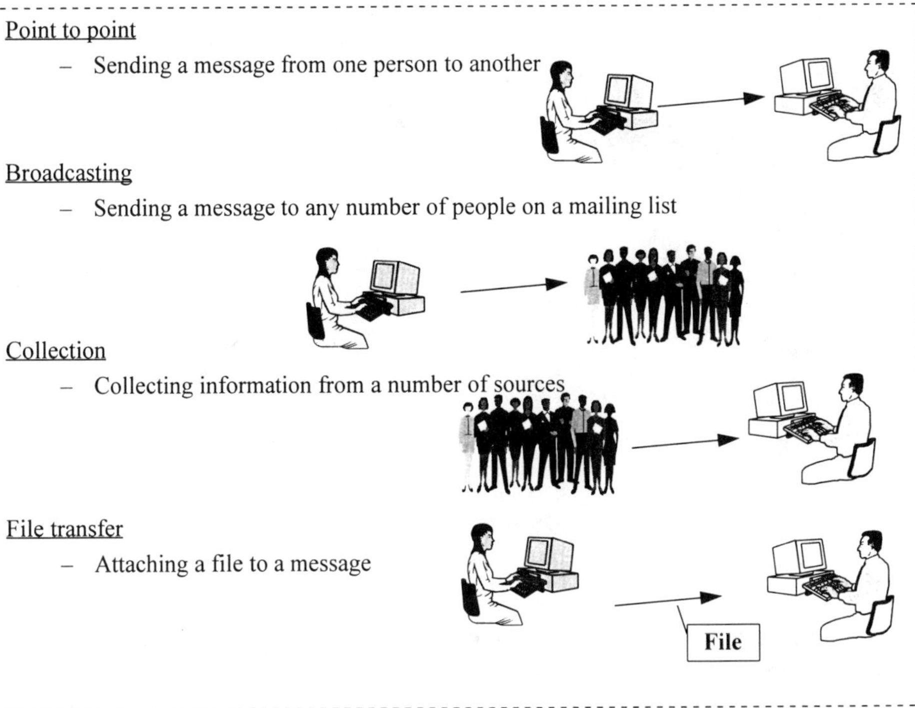

Point to point
– Sending a message from one person to another
Broadcasting
– Sending a message to any number of people on a mailing list
Collection
– Collecting information from a number of sources
File transfer
– Attaching a file to a message
File

Figure 8.12 Basic messaging.

- *Message redirection* to allow users to redirect messages to other users and send copies to more than one user.

8.5.2.2 Extending E-Mail With Process Support

One limitation of e-mail is that it does not provide any process support. Each message is considered as an independent message and users must keep track of any action that this message is to initiate or any relationships between messages, such as, for example, linking a reply to a previous outgoing message. It is, of course, always possible to add extensions, and in this way to eventually evolve to more structured collaboration patterns. Extensions usually concern extending the communication process with richer protocols.

E-mail users must keep track of any conversations that involve the exchange of more than one message. Examples of such conversations are shown in Figure 8.13. Here, one person broadcasts a message to a group of people asking for a convenient meeting time. Following the collection of responses, that person finds a common time and broadcasts this to the group.

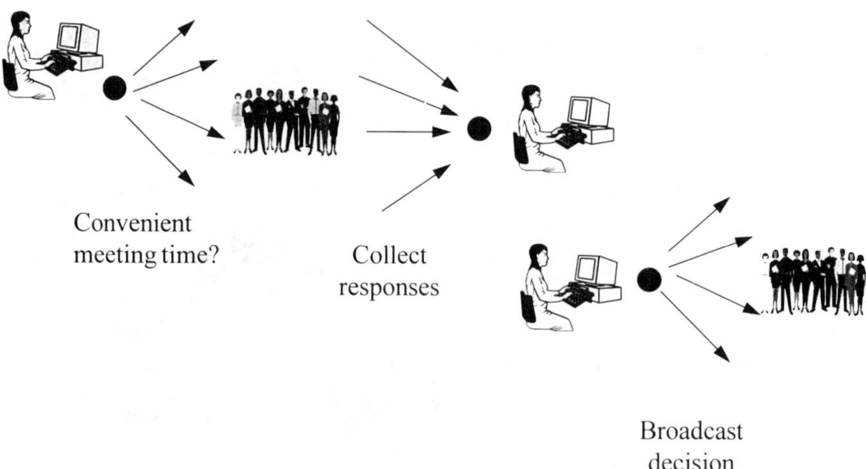

Figure 8.13 Processes with e-mail.

Any more complex kinds of collaboration will require establishing an understanding of the communication protocols to be used between users, and managing these protocols outside the e-mail system. For example, consider how e-mail can be used by groups to work on documents as shown in Figure 8.14.

With two people, communication is effective. One person can make a change and request a reply from the second. This reply can then be used to change the document. To extend this idea to three people, the document would first be broadcast by one person to the other two. However, now it is possible to receive replies that are in conflict. Additional messages must then be sent to resolve these conflicts. As the number of people grows, so does the problem of resolving such conflicts.

It then becomes necessary to assign roles to people and maintain these roles through understandings external to e-mail. Such manual support for processes can become quite complex and impractical using e-mail, and recourse must often be made to synchronous coordination or other asynchronous coordination methods. Such coordination is described in more detail in Chapter 11.

8.6 REPOSITORY SERVICES FOR GROUPS

So far, this chapter has described how networks are used to interchange information between network users. Apart from exchange of messages between users, networks can also provide a variety of other network services. Such services usually center around repositories that take a more active role in sharing information. One

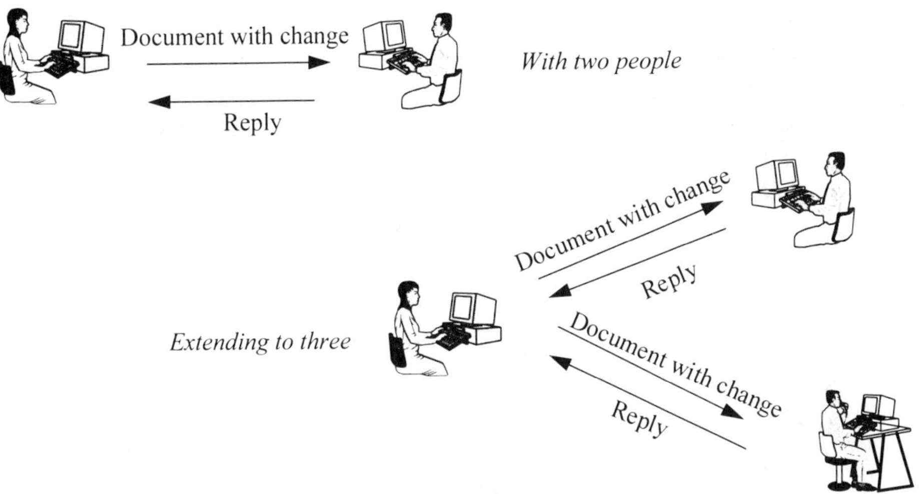

Figure 8.14 An example of using processes with e-mail.

common repository service is news services for groups of users. A news service is usually a repository that contains messages or documents of interest to special groups. For example, subscribers to Internet can get access to a large variety of newsgroups ranging from social items such as sports results to technical topics such as CSCW. News services can vary both in the type of information that they supply and the controls imposed on access to this information.

8.6.1 Setting up Support for Special Interest Groups

One example of a news service is the services provided to special interest groups. One can now "register" with a group and be a recipient of information in that group. Basically, as shown in Figure 8.14, this is a combination of a repository with its methods and communication services to collect and distribute information between users and the repository. Special interest repositories include methods to register new users, discontinue subscriptions, and collect and distribute information. The usual process is to announce the formation of a special interest group and describe its proposed services often to news groups already on a net. This results in what is sometimes known as the formation of *social fields*; that is, people with a common interest. The communication path here is shown in Figure 8.15 and corresponds to collection and distribution, which is so adequately supported in e-mail. It differs from the bulletin board because collected information is sent out on a regular basis to registered users.

The primary methods included in news repositories are as follows:

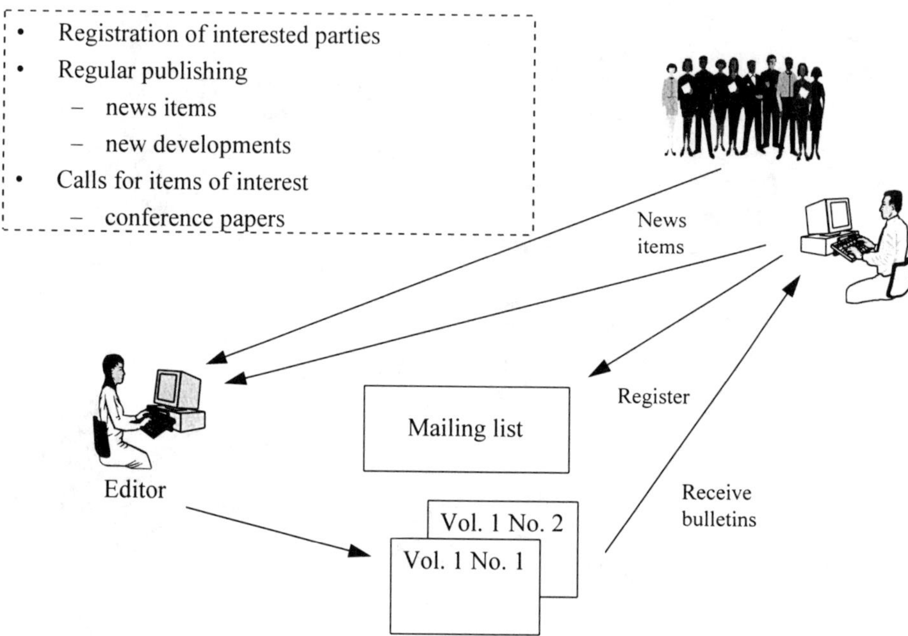

Figure 8.15 News services.

- Registration services to allow new users to register within the group;
  Sorting and organizing repository information;
- Various editing services;
- Collection and distribution of messages.

A number of these functions are carried out by the maintainer or facilitator of the group. This person will collect the news items, edit them, and arrange regular distribution.

8.6.2 Combining News and Bulletin Boards

Bulletin boards and news groups can be combined to get the benefits of both. Thus, as shown in Figure 8.16, the news items are now also recorded on a bulletin board so that members can both get the news bulletins and also read the bulletin board. They may in some cases also record their comments on the board. The facilitator now has an additional responsibility-removing old items from the board.

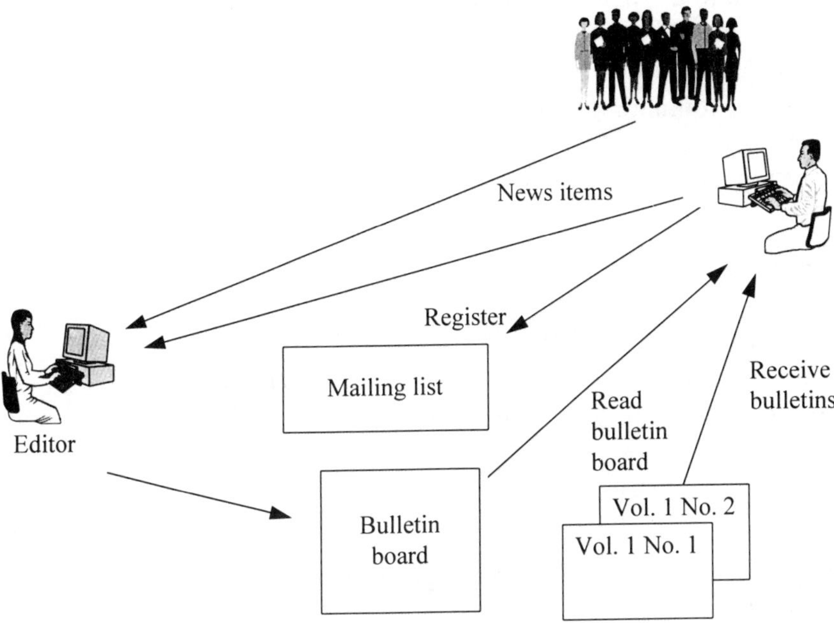

Figure 8.16 Combining news service and bulletin board.

8.7 SUMMARY

This chapter described ways in which computer networks are constructed. It stressed the importance of standards in integrating operating systems, protocols, and various user services. The chapter then described a number of user services, concentrating on those used for e-mail and information distribution. It described bulletin boards and news services, and described them in terms of repositories with links to users through communication services.

8.8 DISCUSSION QUESTIONS

1. What is the difference between a network and a message service?
2. What choices must be made when setting up a network?
3. What is the difference between client-server and peer-to-peer network configurations?
4. Describe some situations where a bulletin board would be useful in business.

5. What criteria would have to be met before a bulletin board becomes an effective group memory?
6. What do you understand by the term *message service*?
7. What kind of communication services do you think are essential in a network?
8. Describe some protocols that you use that could be supported by e-mail.

8.9 EXERCISES

1. Can you suggest additional services for global planning?
2. How can messaging services be used in the learning environment?
3. Would they be of value in software development?

References

[1] Vetter, R. J., (Guest Ed.), "ATM Concepts, Architectures, and Protocols," *Communications of the ACM*, Feb. 1995.

[2] Cheryl Currid and Company, *Networking with Windows for Workgroup's SYBEX*, San Francisco, CA, 1993.

Selected Bibliography

Bersons, A., *Client-Server Architectures*, New York, NY: McGraw-Hill, 1996.

Comer, D. E., and D. L. Stevens, *Internetworking with TCP/IP Vol. 2: Design, Implementation, and Internals*, Englewood Cliffs, NJ: Prentice-Hall International Inc., (2nd. Ed.), 1991.

Dern, D., *The Internet Guide to New Users*, New York, NY: McGraw-Hill, 1994.

Elbert, B. R., and B. Martyna, *Client/Server Computing: Architecture, Applications, and Distributed Systems Management*, Norwood, MA: Artech House, 1994.

Goldberg, Y., M. Safran, and E. Shapiro, "Active Mail - A framework for Implementing Groupware," *Proc. of the CSCW92 Conference*, Toronto, 1992, pp. 75–83.

Hunt, C., *Networking Personal Computers*, Sebastopol, CA: O'Reilly and Associates, 1995.

Chapter 9

Commerce on the Internet

LEARNING OBJECTIVES

- *Network structures*
- *Internet*
- *World Wide Web*
- *Publishing and advice*
- *Trading on the Web*
- *Security issues*
- *Intranets*

9.1 INTRODUCTION

One of the critical factors of enabling people to work together is to provide them with the ability to communicate. A large number of networks that support message interchange have been built over the last number of years. Many of these networks are interconnected into national and international grids that share information between thousands of people. In fact, the largest of such networks, the Internet, has over 30 million users and is growing by 1.5 million a year.

The Internet itself is a system that has evolved over a number of years. As shown in Figure 9.1, communication between computers began around 1970, using early ARPA networks funded by the U.S. Government. At that stage, it was still highly specialized and expensive. The second generation of messaging systems began with the introduction of the UNIX operating system and the connection of many academic institutions to the network, including Berkeley whose UNIX version incorporated TCP/IP. UNIX is still seen by many people as a system built for programmers and requires considerable expertise and knowledge of operating sys-

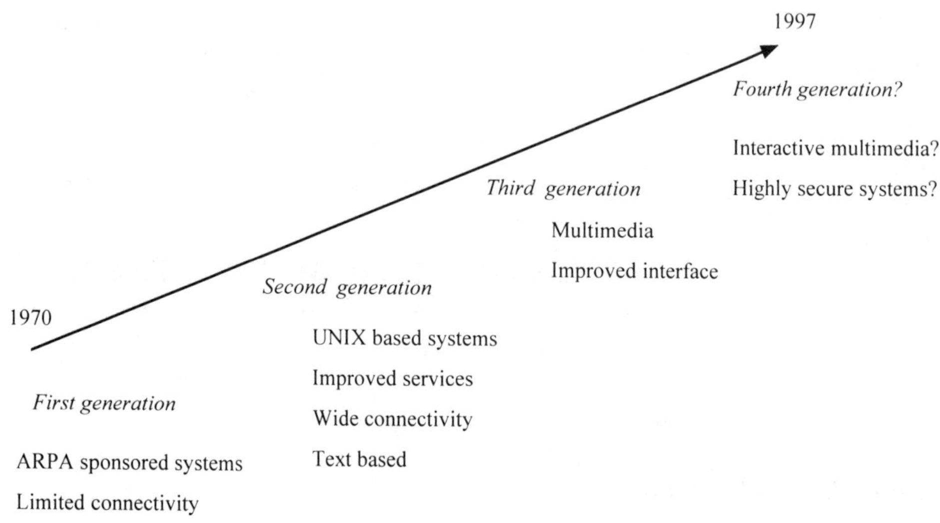

Figure 9.1 Network generations.

tem features to use properly. There are, however, applications such as Telnet or the X Windows System that provide a friendlier interface. Recently the World Wide Web (WWW) has provided an improved interface to the Internet and its has begun to take off. Further developments will ensure that such growth continues to penetrate more and more new classes of customers.

The Internet, of course, is not the only computer communication network. There are many others, both public and private. Another widely used network is CompuServe, while Microsoft is developing its network (known as MSN). Networks can be distinguished by the services that they provide. Services are important factors in network provision as it is no good simply to have a network that enables us to connect people to a computer and then give no other support. It is thus necessary for that network to provide services to support repositories and communication between people on the network. Some of these services can be quite simple, like making a connection using a person's address. Others tend to be more complex and to support a variety of communication patterns. This chapter describes common services provided by many networks. The book, however, does not describe the networks in detail. Such details can be found in the references given at the end of the chapter.

9.2 INTERNET

The Internet is one of the most widely known networks and has been instrumental in connecting academic and research institutions across the globe. The idea be-

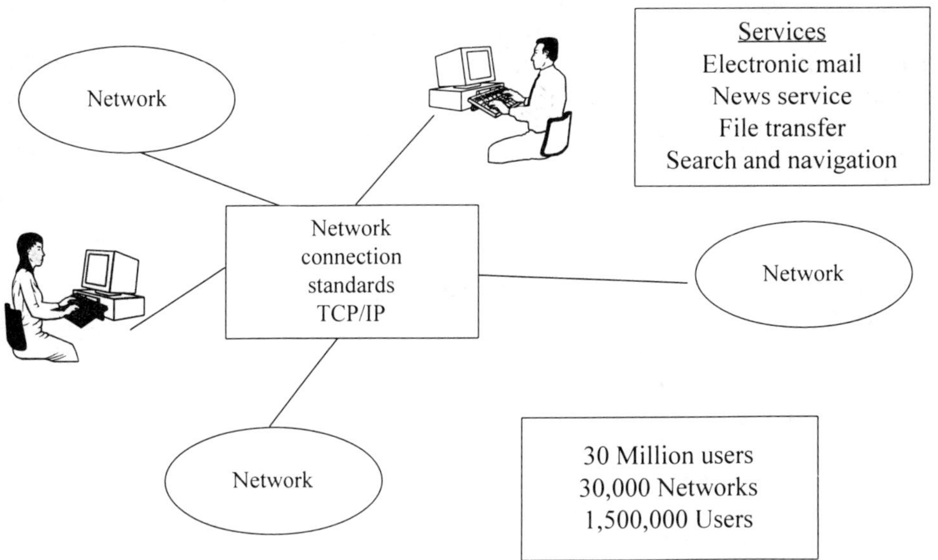

Figure 9.2 The Internet.

hind the Internet is shown in Figure 9.2. In its simplest concept, it is a loose connection of computers through a standard network. In addition, it provides a large number of repository and communication services for its users.

A detailed description of the services provided by this network can be found in the book by Comer and Stevens [1](see references). The Internet has perhaps become synonymous with the term information highway. Internet is an international network of networks. So far, it has mainly been used to connect researchers, but its use in business is rapidly spreading. It now has over 30 million users and is growing at a rate of 1.5 million new users a year. The Internet provides a large variety of services, which are available to any organization. It is almost the ultimate repository, one that contains all possible information and provides the services that enable users to select information needed by them. The primary service of the Internet is the ability for two people to send messages to each other. Each user on the Internet has a unique address that takes the form

<local-user-id>@<computer-name>

The computer name is usually made up of a number of parts that are used to locate the user's server. The major parts are the local computer name, the domain name, and the country. For example the author's e-mail address is igorh@socs.uts.edu.au, where "au" identifies Australia and "edu" is the domain name educational network. Other common domain names are "com" for commercial business, "gov" for government agency, "mil" for military, and "org" for organization. The author's local computer name is made up of "uts," which is the author's university and "socs," which defines the server within the university. The "igorh" part is the author's address on this server.

Messages themselves also follow a standard. This standard allows people connected to networks other than Internet to exchange messages with Internet users. The Internet standard, known as the simple mail transport protocol (SMTP) message takes the form:

From: Igor Hawryszkiewycz <igorh@socs.uts.edu.au>
To: Mary Smith <mary@xyz.com>
Cc: lenny@abc.gov.aqu>
Subject: agenda - next meeting
Date: 21 Nov. 1995

This defines the sender and receiver as well as any people that are to receive a copy (following cc:). The subject of the message and its date are also part of the header. The text of the message follows the header.

Other standards have also been proposed here. The X.400 is an ISO standard used to support interoperability of e-mail services on different networks. This standard encompasses fax messages as well as e-mail messages. A standard for handling multimedia messages for the Internet has also been proposed and is known as the multipurpose Internet mail extension (MIME).

9.2.1 Internet Services

Some examples of services provided by the Internet are the following:

- *Electronic mail:* to allow one user to send a message to another selected user or a group of users;
- *News services for groups:* where groups can subscribe to receive information on some topic;
- *File exchange protocol:* known as anonymous FTP, where files available at one host can be copied by users at another host;
- *Browsers:* to assist users to find information on the Internet.

Details of these and other services can be found in the references at the end of this chapter [2] or many other books on this topic.

9.2.2 Using the Internet

Increasing demand for access to network services has been combined with calls for better interfaces to access the services. Early Internet systems were based on UNIX operating system and required their users to be familiar with this operating system. It required users to be familiar with commands that initiate message transfers, store messages in files, retrieve them, and so on. Recent developments have been towards more user-friendly menu-driven systems. Use of the Internet can be considerably simplified by interfaces such as, for example, Eudora, which contains facilities such as:

- A repository where messages are divided into folders, where each folder stores a subset of messages related to a particular user or task;
- A menu that allows a message to be sent forward to other people or a reply to be formulated to the sender;
- Ability to move text between messages and word processing documents;
- Ability to attach documents to a message.

9.2.2.1 Searching for Information

As the number of network users grows, so does the amount of information stored on networks. Consequently, services are needed to help users search through this information. The Internet includes services such as Gopher, Veronica, or Jughead to assist users to search the Internet for information. Again details of these can be found in the literature [2].

There are also services to find other Internet users. A service known as Finger can be used to do this.

9.2.3 Repository Services for Groups

So far, this chapter described how networks are used to interchange information between network users. Apart from exchange of messages between users, e-mail is also used to provide a variety of other network services. Such services usually center around repositories that take a more active role in sharing information. One common repository service includes news services for groups of users. A news service is usually a repository that contains messages or documents of interest to special groups. For example, subscribers to the Internet can get access to a large variety of news groups, ranging from social items such as sports results to technical topics such, as for example, CSCW. News services can vary both in the type of information that they supply and in the controls imposed on access to this information.

9.2.3.1 Setting Up Support for Special Interest Groups

News service special interest groups, as shown in Figure 9.3, provide a repository with its methods to collect and distribute information between users and the repository. They include the methods described earlier in Section 8.6.1.

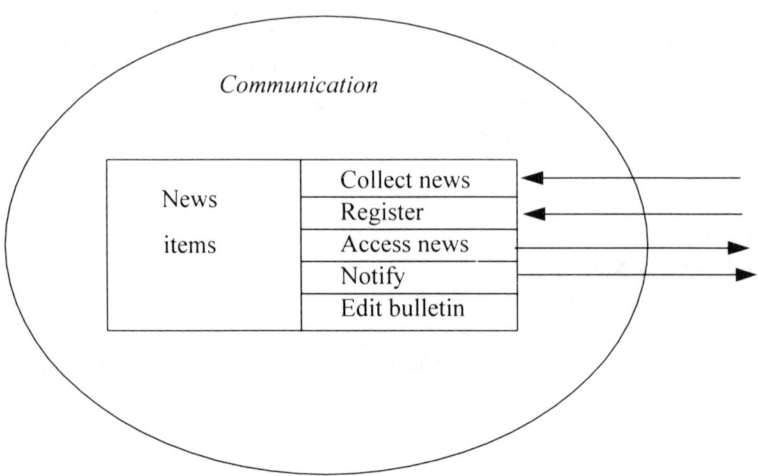

Figure 9.3 News services.

9.2.4 Networks Other Than the Internet

Apart from the Internet, there are other networks available for commerce. In contrast to the Internet, which evolved from a general open academic community, the other networks were designed more specifically for commercial applications. Perhaps the best known of these is CompuServe, which was primarily set up to cater to the commercial user. It was established in 1979, and now has over 2.5 million subscribers. Prodigy, a network established primarily for family-oriented material, was established in 1990 and now has over 1.2 million subscribers. Microsoft is also developing a network, known as MSN, based on its Windows95 system.

These networks offer many of the services available on the Internet, but because of their restricted access, claim to provide better security. However, the popularity and widespread penetration of the Internet almost makes it mandatory for any network to provide Internet access.

9.3 WORLD WIDE WEB

The Internet is basically a text messaging service. Messages appear as free text (although it is possible to attach files to the messages). The Internet has introduced many of the basic services needed in messaging. However, a number of shortfalls became obvious with its growth. One was that the larger and larger volumes of information required more and more time and effort to be spent in browsing to find this information. Secondly, the Internet was primarily text-based and command-driven and thus unable to support multimedia artifacts in its messaging services. World Wide Web has recently become available for this purpose, and has immediately found wide acceptance. It firstly supports the storage of multimedia documents, but also simplifies searches for information by providing navigation paths through the information.

The World Wide Web uses the idea of hypertext to support navigation. The idea is that one retrieves what is known as a *home page* for some entity, which most often is an organization (often a university). The home page is like an index that includes buttons, or key words, to find more detailed information. Each such page can include additional information and so on. For example, Figure 9.4, shows a typical structure that one can find for a university.

Here, the home page is a general description of a university. This includes two key words, "courses" and "research." Clicking on a key word selects a page that provides information on that key word. Thus clicking on "courses" will display a page that describes the university's courses.

World Wide Web extends the Internet by providing services such as:

- Storage of multimedia information in hypertext format;
- Browsing through hypertext documents;
- Ability to collect comments and information from distant users.

9.3.1 Setting Up a World Wide Web Site

The question now is to how to set up the World Wide Web pages. The most common way is shown in Figure 9.5.

A WWW site is set up on a server and accessed by users on a client. Each of the multimedia pages is set up on the server in what is known as a marked-up format. This means that the page is not stored in its actual form, but is stored using a set of mark-up commands that can be used by browsers at the client to set up the actual page. A language known as HTML is used to create such pages. To access these pages, it is necessary for the client to have a World Wide Web browser, the two most common being Mosaic or Netscape. The browser allows people to open new pages and uses the marked-up language to present a multimedia page to the client user. The browser can then be used to follow links provided in these pages.

The process used in Figure 9.5 thus begins with the browser sending a request for a page through a protocol known as HTTP. The request is received by the server, which then sends an HTML page to the browser. The browser then "decodes" the HTML page and displays it. Wide interoperability is only possible because of standards set by HTTP and the mark-up format. These standards are in a constant state of evolution and maintaining them is one of the most critical issues in maintaining wide WWW use.

The HTTP protocol first establishes a connection to the server and then sends the client request to the server. If the request can be met, the server transmits the

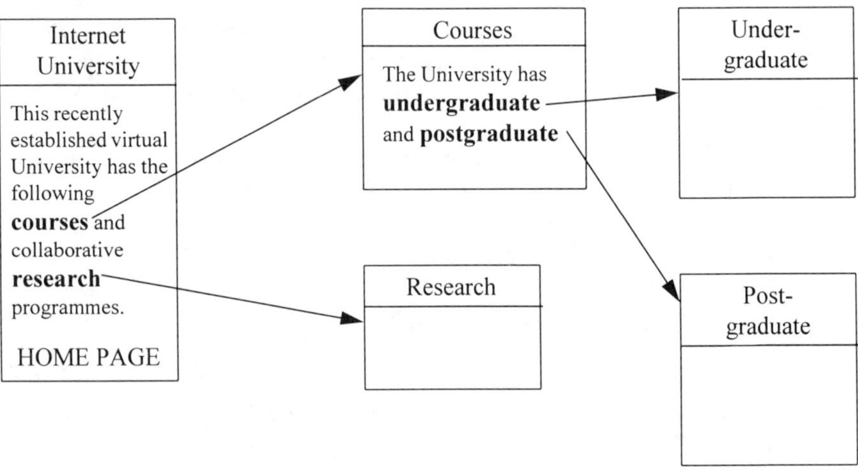

Figure 9.4 A hypertext structure.

requested information and closes the connection. The server then loads the received data and displays it on the screen. The received data can include both text data and graphic images (known as GIF files), as well as video and audio files.

9.3.1.2 Extending to Multimedia

The operation shown in Figure 9.5 is suitable for displaying static information, but will not be effective for displaying multimedia material stored on the server because of transmission line limitations between the client and server. For example, display of videos or voice would require many images to be sent each second, something that cannot be done without having the most expensive communication lines. An alternative to this exists, using a new language known as Java. Java uses the alternative to the method shown in Figure 9.5. Now, instead of the server sending pages to the browser one by one, a whole set of pages, or a video file, together with the program is sent to the browser. The program, known as the Java applet, then executes on the client machine. The browser on the client machine has to be designed in a way that can accommodate the reception of the Java applet together with its pages. This calls for changes in the mark-up format, and not all WWW browsers have the ability to interpret these changed mark-up formats.

The rapid growth of interest in Java simply illustrates the rapidly changing nature of WWW publishing. There is an ever-growing demand for more information to be available on the WWW, and for better and better ways of publishing it. This in itself results in many new products coming into the field and the need to rapidly change standards. The importance of standards cannot be overstressed, as

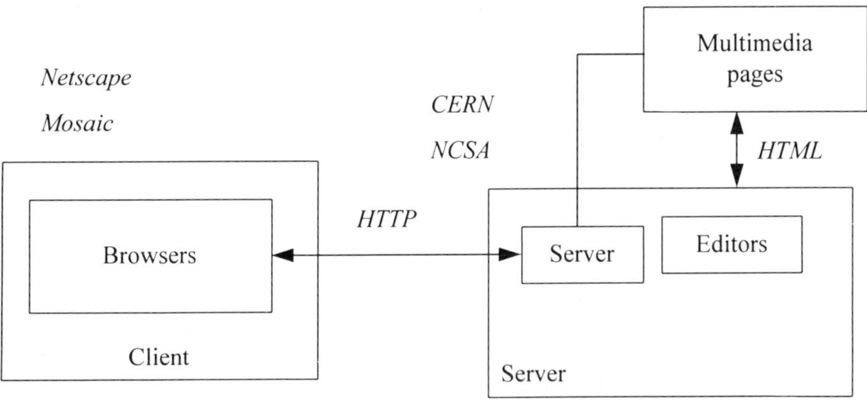

Figure 9.5 Setting up the WWW site.

it is only with standards that wide connectivity can be maintained. Some divergence of standards is already apparent, with alternatives to Java no doubt possible and different browsers adopting different alternatives. But then there is yet another possible direction- rather than embedding JAVA applets in a browser, they can be embedded in an operating system, doing away with the need of browsers.

9.3.2 Supporting Interaction on the Web

Apart from its ability to present information, the WWW has also supported interaction. It is possible to create pages that look like forms using the HTML. Users can make entries into the form fields. The input is then sent to the server, which replies with a response. Forms can be used to do things like allowing users to contact people in the enterprise directly through the form or to place orders or other requests without having to resort to another service for this purpose. This ability to interact through a friendly interface goes a long way towards providing the services needed by networked organizations.

9.3.3 Integration With Databases

It is also possible to integrate WWW pages with underlying databases to support retrieval of information from corporate databases. Users can then make entries through the WWW page that retrieve and display the data in the database. The creation of such links requires an extension to the way the WWW site is set up. Now, in addition to the facilities shown in Figure 9.5, an interface known as the common gateway interface (CGI) is added, as shown in Figure 9.6. This interface can be written in a programming language such as C++ or some of the higher languages now becoming available for this purpose. The programs can include commands that access corporate databases, with the retrieved information returned to the client.

9.4 BUSINESS USE OF THE WWW

Chapter 5 described a wide variety of business uses now appearing on the WWW. Most business started with the WWW as a publishing medium advertising their products and services, but growth now is into the consumer market. The variety of sites and their structure is growing almost by the day. Most of the large organizations now have a World Wide Web home page, which leads to other pages that describe the organization in more detail. In addition, indexing services are now available, one being the commercial index page that allows a user to locate the home page of many commercial organizations. The address of this page, incidentally, is http://www.directory.net/.

To give a better idea, the kinds of pages that can be found on a typical business site are listed in Figure 9.7. The home page is usually an introductory page that introduces the enterprise and its main activities. It usually provides the guidelines to other pages and suggestions on how to use them. The usual structure contains pages to provide the following services:

- Personal support services to support contacts and allow its customers to contact people for advice and service, and maintain interpersonal relationships through interactive bulletin boards;
- Information services that provide clients with advice on the enterprise's current role, policies and structure, and services. It also includes information on the business products and services as well as advice on how to use them and where to get them. In addition, news pages are often provided to publish any recent interesting events in the organization;
- Work process support services for processes such as ordering or purchasing.

Figure 9.7 is relatively static and shows only the kinds of pages found on a site. However, a good site design must go beyond simply a set of pages that describe the structure of an enterprise. A good design must be interesting and help the user focus on their interest rather than going through a set of dreary pages about its structure. It must also provide an easy to use interface that allows the general public easy access and navigational paths to quickly focus on enterprise services. What most organizations have now realized is that the WWW site is increasingly being used as a marketing medium. The WWW pages are an enterprise window to the world and should whenever possible be used to attract, and what is more important, *keep*, clients at the site. It is often said that a site should use the same approach to marketing as occurs through personal contact. It is thus important to use the site to *establish credibility* by publishing interesting information

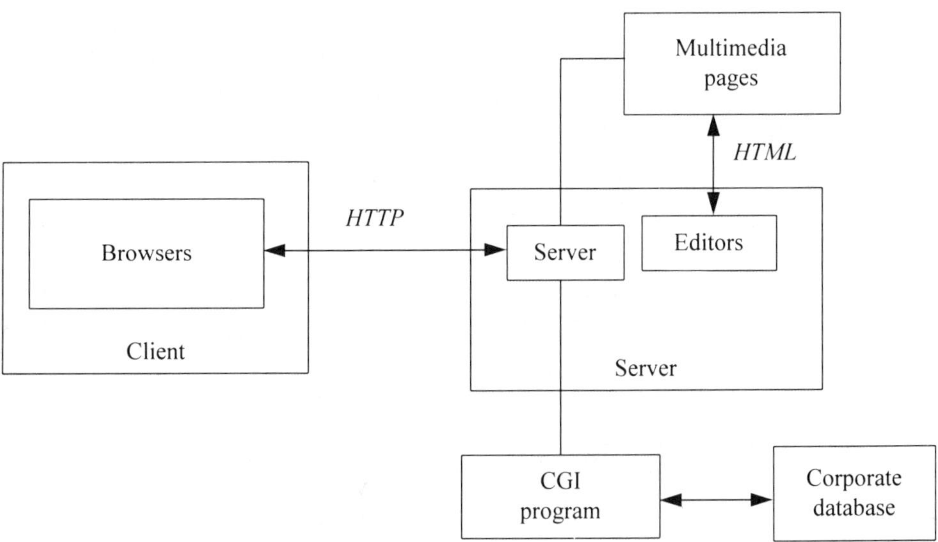

Figure 9.6 Extending to corporate gateways.

about products and services, especially stressing their potential benefit to the client. It should allow the client to select and read further information if they need it, and then, if their interest rises, to place orders. The rest of chapter will describe how some sites achieve this, and readers are encouraged to browse through these and other sites to better understand site dynamics.

9.4.1 Publishing on the Web

The growth of commercial WWW sites has been phenomenal since the WWW was first introduced. It is now very common for almost every organization to have a WWW site, and most provide the kind of information shown in Figure 9.7. Many early sites were really descriptions of an organization and facts about them without involving the reader in any significant way. This has changed very quickly to being an increasingly marketing oriented and more interesting front. The design of World Wide Web pages will become even more important in the future, as they become an increasingly more important window to the world.

9.4.1.1 An Example of Interface Design

One example of such a site that includes these three parts is the Goodyear site (http://www.goodyear.com). What is interesting about this site is its graphical highway, shown in Figure 9.8, which on a single page shows pictorially all the

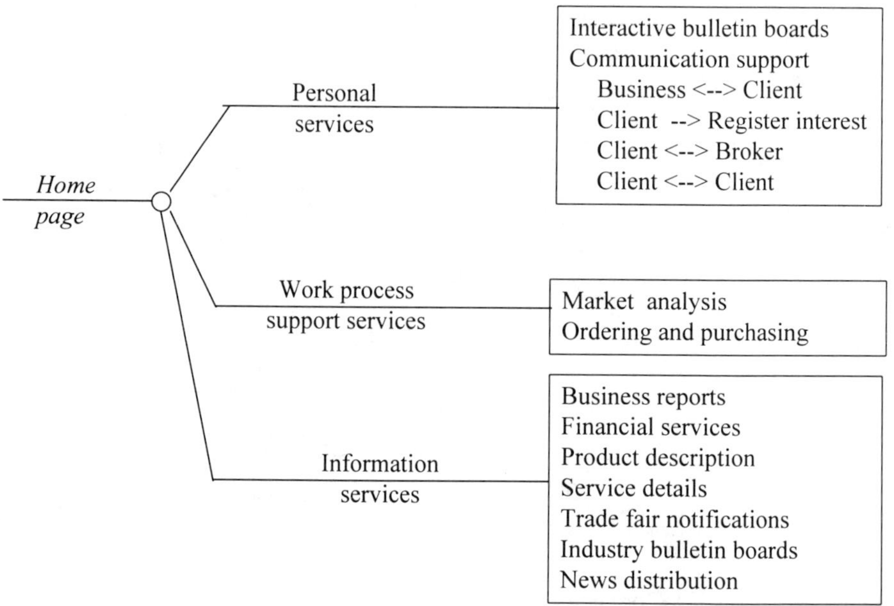

Figure 9.7 Typical pages on a site.

kinds of information in the site and provides easy access to it by clicking on any of the images on the highway.

This highway presents a total view of the pages supported on this site. It includes a number of pages of information pages on tires and how to choose the most appropriate tire for a given vehicle and driving characteristics. For example, selecting TREAD gives advice about what various tire grades mean, while another page describes the effects of underinflation. This volume of information gives a reader more and more confidence in the information in the site. Furthermore, at each point there are further links to more detailed information and its relevance to Goodyear products. The site also provides services to find a nearby store and place an order with that store.

9.4.2 Providing Advice

Another major use, especially with government, is provision of advice. A large number of sites are available on the net for this purpose. Links to a number of federal agency pages can be found from the address http://www.fedworld.gov. One interesting site, for example, is the Small Business Administration (SBA) site that provides support for small businesses. Its home page states that its main purpose is to "aid, counsel, and protect the interests of the Nation's small business community." It then goes on to say how it does this. It includes pages

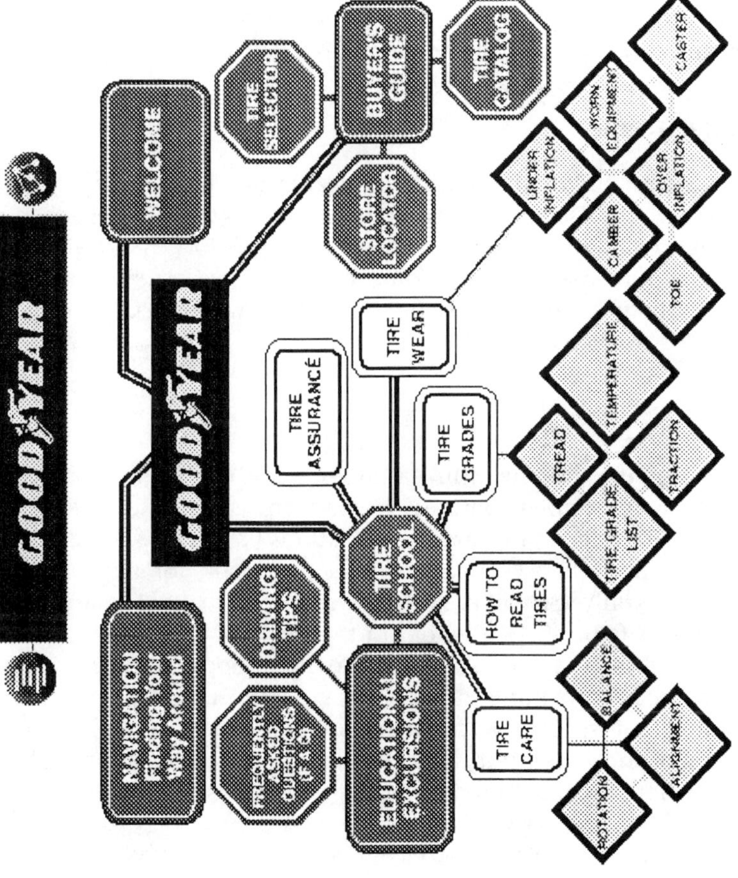

Figure 9.8 The Goodyear graphical highway.

such as, for example, starting a business, expanding a business, or financing a business. Selecting any of these gives further advice. For example, selecting the financial assistance part provides information about the kind of assistance provided by SBA to small businesses.

Many commercial organizations also provide advice and access to services (for example, airlines or banks), although very few provide direct access to their services (for example, making an airline reservation).

9.4.3 Consumer Trade

Many sites provide the ability to trade with an organization by directly placing orders using the WWW. The use of the Internet for this purpose has only been minor to date. According to the Price Waterhouse forecast, only 5% of U.S. retail, wholesale, and mail-order purchases were electronic in 1994. However, this figure is expected to triple by the year 2000. A number of different ways of supporting such consumer trade are now evolving.

9.4.3.1 Electronic Purchasing

One natural extension is to use the form features of the WWW to initiate business transactions. One of the simplest business transactions is to make an electronic purchase by placing an order. Many large organizations have now established sites that provide this facility. There is a growth of such support for smaller traders, through what is commonly known as electronic shopping malls. The idea here is that a shopping mall site provides a "shopfront" to a number of clients, who are most often small traders. The shopping mall provides a focus for such stores or small businesses, perhaps providing the store owners with services such as setting up and maintaining their sites. However, many shopping malls go beyond this to also provide many of the community services found in physical shopping malls, creating a "virtual shopping mall" environment to attract visitors. One site here is Downtown anywhere (http://www.awa.com/), which presents a virtual street to the browser. This site provides a set of services that allows small traders to set up a shopfront on the site. A shopper or site visitor can get a client listing of these traders and access their pages. Apart from this, the site, like other sites of this kind, provides a set of "community" services to visitors to the site. These include a library and newsstand where access is obtained to a variety of electronic newspapers, museums and galleries, financial information, and so on. The goal is to provide the visitor with the same kind of information as is found in most shopping centers.

Another interesting site , with perhaps a more community-oriented background, is that established by BizNet Technologies, whose home page (http://www.biznet.com.blacksburg.va.us/) is shown in Figure 9.9. It also provides what is almost a marketing service, offering various packages to help businesses establish an Internet presence. These start with simple listings and extend to enhanced services that include a number of WWW pages. Each of the clients can provide information about its products to the consumer using the site. For the browser, there are four main selections. Client listings give a list of clients, which can often be followed to the client's home page. There are also pages that inform potential clients of services provided by the site. But what is unique here is a group of community services, known as the Blacksburg Electronic Village (BEV), that support general information and communication for its community, Blacksburg, Virginia, served by the site. Apart from a directory of local resources, the site also supports community discussion based on bulletin boards that keep track of news items of interest to the community.

9.4.3.2 Business Services

Interaction can go beyond simple transaction and into the provision of personalized services. One example is the Dun and Bradstreet site (http://www.dbisna.com/) to provide advice on a variety of business problems. The home page includes selections on news, some tips or business how-to's, and a product catalog, among others. The business how-to's page can be used to get some points about various aspects of running a business-again establishing the credibility of the site. An interested person can then view the products catalog, which includes areas such as credit solutions or marketing solutions. The site visitor can select one of these. On selection, typical solution strategies are described, and contact information provided if the visitor requires advice on a particular need. The marketing component is again very prevalent here. If one follows the site, there is considerable interesting, high-quality information in the area of the organization's services, but always giving the opportunity of getting individual advice. Another example of a site for providing more focused advice (http://www.corporate.com/) is that established by Corporate Agents to provide advice on ways for expanding businesses. The site again has a number of pages of advice, but then provides the browser with an "Incorporate Online" form that is sent to the company, thus making initial contact.

9.4.3.3 Integrating the Consumer

There is now also a trend to integrate sites closer with an organization's clients. Such sites should enable clients to keep track of services requested by them. One example is the FedEx site.

Like all other sites, the FedEx home page (http://www.fedex.com/) includes links to information about their services and their availability. But what is more important is that once you place a package for delivery with them, you can actually track its progress using their site. Each package gets an airbill number, and on entry to a tracking page you can find out its current progress.

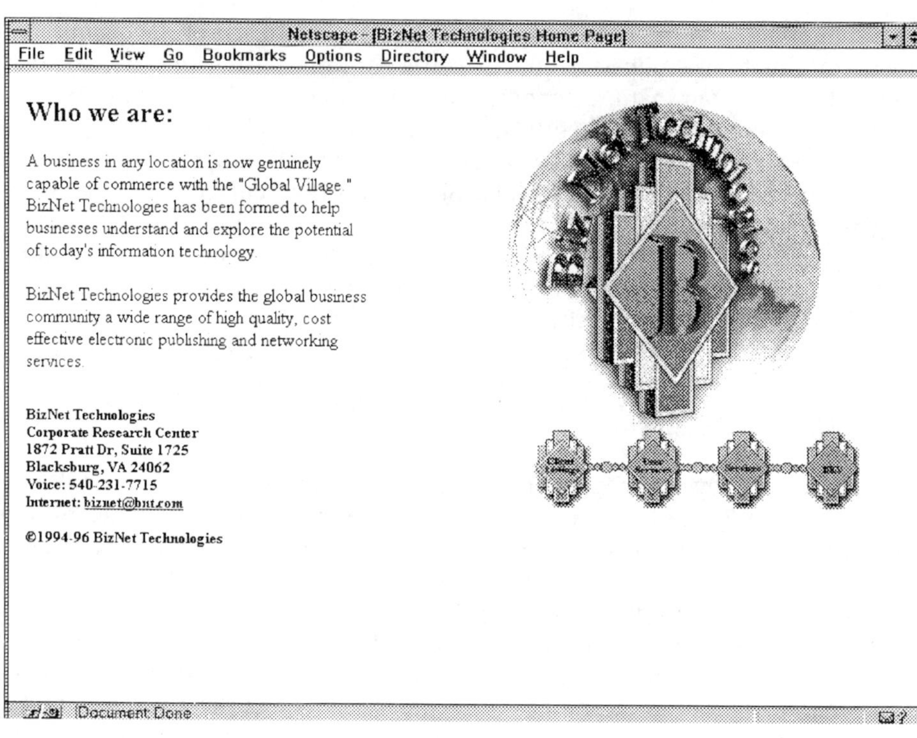

Figure 9.9 BizNet Technologies.

9.4.4 Security on the Web

The issue of security has always been an important consideration in using the Web for electronic commerce, especially consumer trade. Security and confidentiality often come into question when choosing what to place on public computer networks. This becomes particularly important with commercially confidential information. What is needed is a method to identify possible threats, a way to assess the risks to the site of these threats, and a way to determine the kind of preventative action to take to eliminate the threats. The kinds of possible threats are

- Impersonation to get information;
- Unauthorized use of information, as for example, costs quoted for a contract;
- Unauthorized disclosure;
- Theft and fraud;
- Denial of service;
- Alteration of message;
- Repudiation actions.

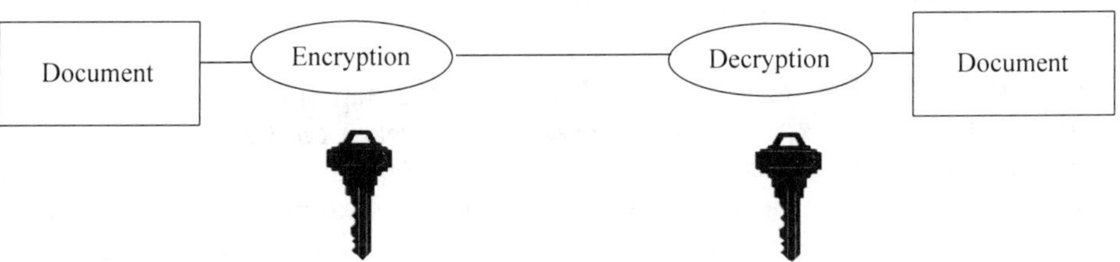

Figure 9.10 Encryption.

In particular, risk analysis involves the assessment of damage if a threat succeeds; that is, the consequences of data being changed, lost, or simply read. Often, a risk assessment is made to determine the abilities of commercial competitors to get unauthorized access, as well as the consequences of unauthorized people getting such access. The risk, if considered serious, can then be removed. One way to do so is to use private networks for confidential data with special gateways from these networks to public networks. There are also possibilities of reducing the consequences of unauthorized access by using key encryption techniques on public networks. Cryptography is a complex science, but from an implementation viewpoint, what is needed is an algorithm and a key known to users at either end. The idea is illustrated in Figure 9.10. A document is encrypted using a private key at the sending site and decrypted using the same key at the receiving site. The question is how to get the key across from one site to the other without the risk of someone intercepting it.

One approach is to enter the key at the time of access and change it regularly, thus considerably reducing the risk of access by all but the most sophisticated of intruders. Another is to work through a trusted intermediary. This intermediary maintains the private keys of all its clients. It can receive a message from one site, decrypt it with that site's key and encrypt it with the receiving site's key for onward transmission. Procedural approaches to maintaining confidentiality are also becoming better understood as a means of providing security. For example, perhaps we should not have to give our credit card number to each merchant with whom we trade. Another approach is to simply authorize the credit provider to pay the merchant. In that case, a secure link is only needed between the credit providers, their customers, and the merchants but not between customers and merchants.

A further important security issue that must be solved before the consumer market can take off is to ensure that the public can make payments for goods electronically without the potential of fraud. Major credit card providers are developing trusted sites to ensure payment security and the idea of using tokens rather than cash in transactions.

9.5 BUSINESS-TO-BUSINESS LINKS

The support provided for business-to-business links depends very much on the kind of alliance between the businesses and the level of connectivity supported. The large variety of alliances described in Chapter 5 means that there is no such thing as a standard business to business link and most support must therefore be customized to need. Depending on the network, links can be work process oriented to support prearranged activities, such as ongoing joint links in manufacturing pyramids. More support for interpersonal communication is needed where there is a high-level discussion (for example, in joint sales promotion). Or information exchange may be sufficient (for example, where arrangements are in place for catalog distribution). But what often adds to the complexity is that for many alliances, different kinds of connectivity may be needed at different stages of networking. This in turn calls for network support that can seamlessly move between many services.

9.5.1 Case Study: Business Network Support

Like any design, support for business networking depends on the process followed. Often, as described in Chapter 5, different kinds of support are needed at different stages of networking. The World Wide Web has been suggested as a way of supporting the early phases of business networking. The proposal is for the World Wide Web service to provide easy access to government support services and brokers in the service, and to facilitate information exchange and contacts. The goal is to develop a site where a communities working to the same goal can share information to mutual benefit. Figure 9.11 illustrates some initial services provided for this purpose (see also http://www.socs.uts. edu.au/research/bnp).

Every box on this guide map represents a group of pages in the system. It includes

- A group of general information pages about networks and the advantages of networking, together with links to current networks in the system;
- A group of pages "about industry assistance" that allows people to find out about the particular programs supported by this site, including pages that give access to approved government programs;
- A group of contact pages to people associated with the program, each with the ability to initiate a contact directly from the page;
- An opportunities bulletin board that supports direct contact to the opportunity owner;
- Links to a number of marketing databases;
- A variety of news items with many pages to keep people aware of latest developments or tips and leads that can be followed.

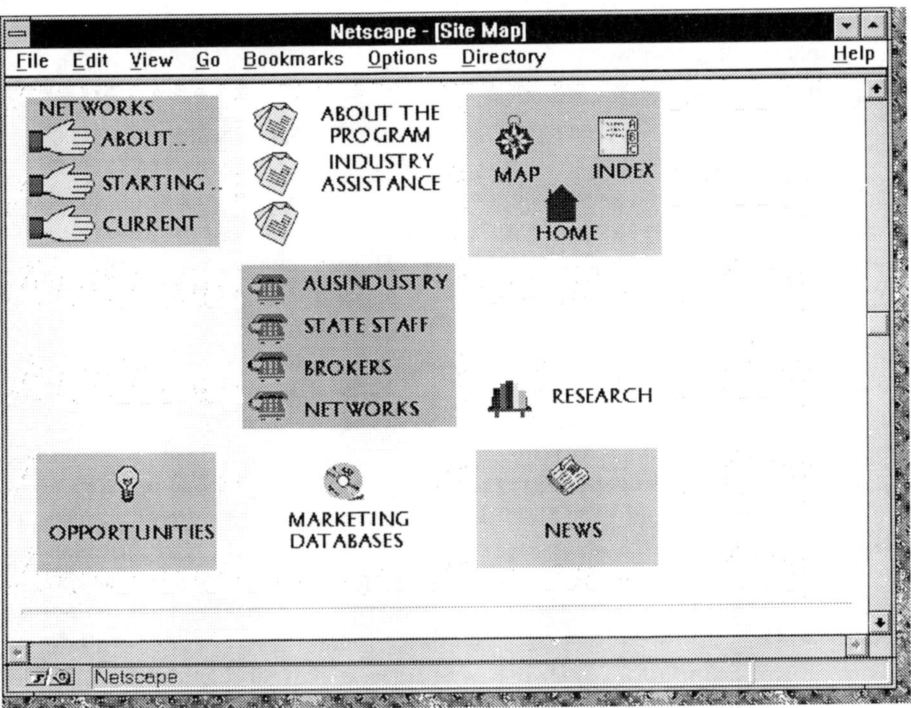

Figure 9.11 Supporting business networks.

The site can be used to distribute new information on the networking program using the news service. New announcements can be easily posted to brokers connected to the service. People can also follow up on opportunities on the opportunities bulletin board. It is also possible for a member of the public to make an enquiry about the BNP and then decide whether there is some advantage in participating. Such users are provided with an advisory section on the network formation process and the steps followed in network formation.

But what is even more important is that sites developed in different countries can be linked to extend small business networking across the globe. Thus USNet is also developing a site (http://www.technet.org/usnet/index.html) that provides a valuable set of references to publications and a newsletter, and development of a set of network connections to share their experiences. This site is also being linked with related sites across the globe.

The development of sites like those in the preceding example perhaps illustrates another important purpose of the WWW: to establish communities that have common practices and allow them to easily establish contacts and share information and experiences.

9.6 THE INTRANETS—USING THE WWW WITHIN THE ENTERPRISE

So far, we have considered the WWW as a public network. There is no reason why the technologies used for the public Internet cannot be adopted for use within individual enterprises. It is possible to set up WWW sites together with browsers and servers on a network restricted to one enterprise, and thus restrict access to only that enterprise. The opportunities provided by CGI scripts and Java applets to integrate WWW pages with corporate databases give the enterprise the opportunity to provide its employees with easy access to corporate data-not only traditional records, but also documents, directories, and so on. It also provides opportunities to support work processes in the enterprise.

The definition and overall structure of intranets is still not generally agreed upon. For example, should it be an enterprise-wide monolith or a combination of networks, each supporting one enterprise unit, with perhaps some enterprise-wide support. A possible structure of an intranet site that has separate components for each enterprise unit is illustrated in Figure 9.12. Each unit site can have pages that are quite similar to those shown for a business in Figure 9.7. It provides the three kinds of network support for each unit, but often places more emphasis on work process support and coordination together with interactive bulletin boards for interpersonal support. Integration is achieved through directories that can be used to navigate to other units or enterprise support. It is perhaps one of the challenges of intranet design to build systems that integrate all three connectivity levels—information exchange, personal communication, and work process—into an integrated seamless system.

9.6.1 Developing Intranet Sites

Development of intranets is expected to grow in the future. The advantage is that once a WWW network exists, very little additional infrastructure is needed to establish an intranet that provides basic connectivity. However, the real challenge is to go beyond information exchange and integrate the intranet into the work practices of the enterprise-to make it ubiquitous to the user. Furthermore, process support should include all the desirable features described in earlier chapters, in particular it should improve awareness and keep users notified of events important to their work. New processes should be integrated into desktop systems on personal computers, adhering to the principles described in Chapter 3. In this way, ease of access is simplified and the intranet is integrated into people's everyday work without causing major disruption. Thus, for example, discussion databases may be added to the desktop as an add-on service to allow people to asynchronously keep track of their work and pick up discussions interactively at any time,

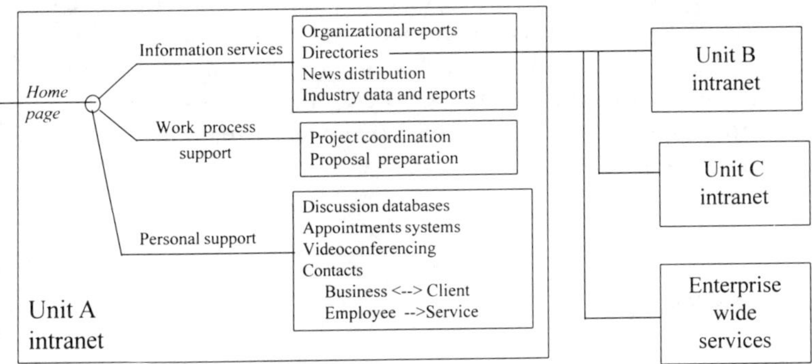

Figure 9.12 A typical intranet site.

without necessarily having to change systems. Similarly, direct video connections can then be added. Adding process support sometimes gets more difficult, as it often requires a clearer definition of processes, which is often followed by clearer definition of roles and provision of services that support these roles. Chapters 15 through 17 describe some methods used to identify such services.

As more and more services become available, the possibilities of intranet design are almost limitless and only bounded by the designers imagination-ranging from transmitting the chief executive's speech to all desktops to the idea of a paperless office, with all information provided to all employees courtesy of intelligent agents.

9.6.1.1 Case Study: An Intranet for Worldlink Consultants

Worldlink Consultants is planning to use Netscape software as the enabling software for its intranet. The intranet will support the many consultants on assignment or preparing proposals to gain access to the most up to date information on their areas of expertise. The intranet known as Expertise at your Desktop will facilitate exchange of ideas between their employees, with corporate desktops and mobile laptops using the Netscape Navigator as the front end. The interface provides access to knowledge repositories. It is made up of major applications, namely,

- Information services to give access to a variety of topics, filed by industry and geography;
- An expert skill directory, which is a corporate directory of the company's personnel and their skills;
- A system to support interpersonal communication for client teams, including bulletin boards and discussion databases;
- Access to legacy systems to existing corporate information.

Just like with any system, it is becoming obvious that support will have to go beyond these early stages that primarily concern access exchange and access to information together with some support for interpersonal communication. The next step is to extend the network to provide work process support, especially by providing the services to support the proposal preparation process, and to improve both the quality of documents and reduce the preparation time.

9.6.2 Development Tools

Development of intranets will call for better and better development tools. Effective sites will need to integrate databases into the Web and provide support for workflows that operate on these databases. Support for interaction through the World Wide Web will also become important so that people can interact with the databases and each other. HTML provides a standard language for use in developing such sites. It then becomes possible to integrate the network with corporate databases using CGI scripts, or to go even further and add videoconferencing and realize the concept of the virtual office. It is extremely useful for publishing, especially where graphic images are to be included with text. However, there are now a large number of publishing systems becoming available to simplify application development on the WWW, and Java is perhaps the best known of these, but more and more new languages can be expected for this purpose.

9.7 SUMMARY

This chapter described the kinds of services provided through networks. The chapter concentrated on the Internet, and especially on the World Wide Web. The Internet was primarily based around the exchange of text documents. The World Wide Web is an extension to the Internet and supports multimedia information exchange. The chapter described ways in which the WWW can be used in electronic commerce between enterprises, and then led to a discussion of using it as an intranet to support networking within the one enterprise. The WWW is only one of the emerging platforms for networking enterprises and is now proposed as one way for building enterprise intranets. Other platforms for intranet design will be described in the next chapter.

9.8 DISCUSSION QUESTIONS

1. What do you understand by the term *Internet*?
2. Describe the components of a good World Wide Web design.

3. What is the mark-up language?
4. What is the difference between HTML and HTTP?
5. Why are standards important in public networks?
6. What additional facilities are provided by World Wide Web when compared to Internet?
7. How would you link the WWW to the corporate database?
8. What is the difference between the Internet and intranets?

9.9 EXERCISES

1. How can messaging services be used in the learning environment?
2. Would they be of value in software development ?

References

[1] Comer, D. E., and D. L. Stevens, *Internetworking with TCP/IP Vol. 2: Design, Implementation, and Internals*, Englewood Cliffs, NJ: Prentice-Hall International Inc., (2nd. Ed.), 1991.

[2] Hahn, H., and R. Stout, *The Internet Complete Reference*, Berkeley, CA: Osborne McGraw-Hill, 1994.

Selected Bibliography

Berners-Lee, T., et al., "The World Wide Web," *Communications of the ACM*, Vol. 37, No. 8, 1994, pp. 76–82.

Bieber, M., and T. Isakowitz, (Guest Eds.), "Designing Hypermedia Applications," *Special Issue ACM Communications*, Aug. 1995.

Cronic, M. J., *The Internet Strategy Handbook*, Harvard, MA: Harvard Business School Press, 1996.

December, J. A., and N. Randall, *The World wide Web Unleashed*, Indianapolis, IN: Sams Net, 1996.

December, J. A., and M. Ginsburg, *HTML and CGI Unleashed*, Indianapolis, IN: Sams/Net/Macmillan, 1995.

Denning, D. E., and D. K. Branstad, "A taxonomy for Key Escrow Encryption Systems" *Communications of the ACM*, Vol. 39, No. 4, March, 1996, pp. 34–40.

Dern, D., *The Internet Guide to New Users*, New York, NY: McGraw-Hill, 1994.

Ellsworth, J. H., and M. V. Ellsworth, *The New Internet Business Book*, New York, NY: John Wiley & Sons, 1996.

Emery, V., *How to grow your Business on the Internet*, Scottsdale, AZ: Coriolis Group Books, 1996.

Fraase, M. *The PC Internet Tour Guide*, Chapel Hill, NC: Ventura Press, 1994.

Gilster, P., *The INTERNET Navigator*, New York, NY: John Wiley & Sons, 1993.

Hawryszkiewycz, I. T., "Support Services For Business Networking," in *Proceedings IFIP96*, Canberra, E. Altman, and N. Terashima, (Eds.), Chapman and Hall, London, ISBN 0-412-75560-2, 1996.

Judson, B., *Net Marketing*, Wolff New Media LLC, New York, NY: Random House, 1996.

Pyle, R., guest editor, "Electronic Commerce and the Internet," special issue of *Communications of the ACM,* Vol. 39, No. 6, June 1996.

Siyan, K., and C. Hare, *Internet firewalls and Network Security*, Indianapolis, IN: New Radio Publishing, 1996.

Walker, S. T., et al., "Commercial Key Recovery," *Communication of the ACM*, Vol. 39, No. 3, March, 1996, pp. 41–47.

Wood, L., and P. Blankenhorn, *Bulletin Board Systems for Business*, New York, NY: John Wiley & Sons, 1992.

Chapter 10

Platforms for Collaborative Systems

LEARNING OBJECTIVES

- ❑ *Platform requirements*
- ❑ *Combining generic services*
- ❑ *Platforms for information exchange*
- ❑ *Commercial platforms*
- ❑ *Object exchanges*
- ❑ *Commercial platforms*
- ❑ *Discussion databases*

10.1 INTRODUCTION

Chapter 7 described the importance of combining networking services into platforms. Platform design is growing in importance, with most platforms constructed from commercial network services. Chapter 7 described a number of such services and pointed out that there is almost a limitless number of ways to combine them. These services include messaging services, document services for joint artifact production, and workflow services. It also identified one important property of successful platforms: seamless transitions between services.

Most platforms are built by using commercial products. Consequently, design is constrained by the available commercial services, which the user or designer must integrate into a platform. At the moment, there are no universally accepted standards for combining collaborative network services into platforms. Indeed, commercial developments are constantly and rapidly evolving, making the development of such permanent standards difficult. Nevertheless, some trends are becoming apparent. One is to use middleware, but the other and more prevalent trend is to base platforms on one enabling technology. The WWW is increasingly proposed as one such technology, although LOTUS Notes is a product that is also widely used to provide collaborative platforms.

This chapter describes platform requirements and then goes on to describe LOTUS Notes as a platform technology together with other products. The chapter cannot, obviously, provide details about all these products, but only gives an indication of some of their important features. Readers should refer to the latest manuals, or World Wide Web sites, to find current product details.

10.2 IDENTIFYING REQUIREMENTS

Chapter 8 described that platforms are supported by networks made up of four broad levels: collaborative services, operating software, network protocols, and hardware. This chapter concentrates only on the services level. It particularly looks at the way that the collaborative services, many of which were described earlier in Chapter 7, can be integrated into seamless platforms. Typical collaborative services supported by platforms include

- E-mail, often allowing transmission of multimedia images;
- Document management with annotations and version control;
- A conferencing facility that supports communication between users using different media.

They can also include repository services such as:

- An appointments system to arrange meetings;
- A system to support voting;
- Authoring tools for preparing documents;
- Brainstorming facilities;
- Methods for keeping private notes;
- Customized discussion databases;
- Tools for cooperative design that support shared screens (for example, a sketchpad).

10.2.1 Integrating Services

The usual approach to seamlessness is to simplify the movement of information between services. The idea is illustrated in Figure 10.1. Here, a user can open any of a number of services by selecting the button on the left (for example, document system). Any information can be moved from the service into a workspace. Any number of actions can be carried out on the workspace and then

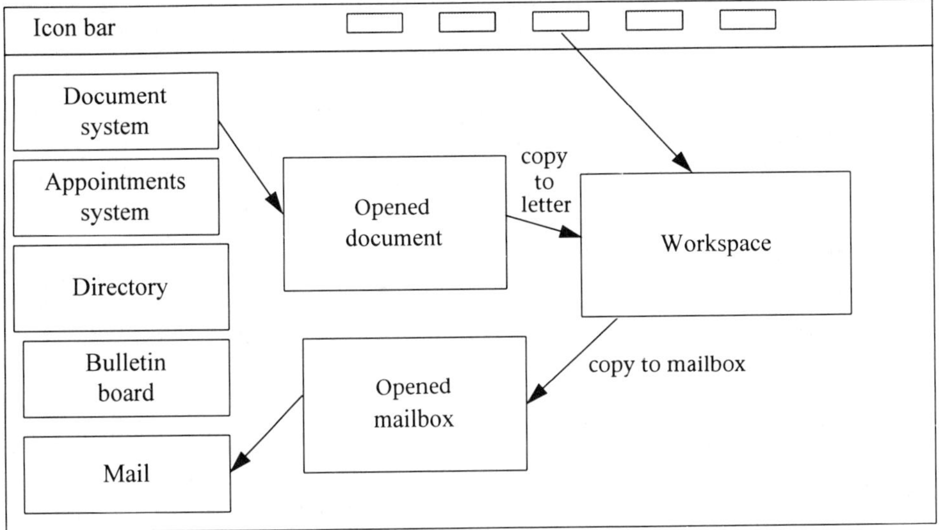

Figure 10.1 Interface integration.

the information can be moved back into another service (for example, mail). The idea can be extended to include enterprise applications (for example, ordering or accounts) as one of the services.

10.2.2 Platform Requirements

Platforms must meet a number of requirements. One of the important requirements, especially when developing work process support on intranets, is seamless integration. There are, however, many other requirements, derived from both social and technical needs, that must also be considered. Some of the important requirements are

- *Services provided by the platform*—the kind of collaborative services provided by the system, in particular the type of repositories supported and the communication services that are supported.
- *The type of repositories supported*—the kind of questions important here are
    - Does the system support only one type of repository or is a range of artifact structures supported?
    - Are standard templates for repositories available?
    - Can the corporate database be easily integrated as a repository?
    - Are repositories mail enabled?

- *Communication services that are supported*—the range of services including synchronous and asynchronous methods of communication. Important aspects here are
  - Are these available for automatically informing users about changes in the context and reminding them of what they should do?
  - Is a work group participant fully aware of the status of the activity of other workgroup members?
- *Process services that are supported—important questions here are*
  - Can workflows be easily specified?
  - Can the service be easily adapted by its users to support changes to work processes?
- *Service integration—can the system seamlessly* integrate a number of services to construct a collaborative application?
- *Platform adaptability*—can a platform be adapted by users during operational use, and in particular can they be introduced in a gradual manner?
- *Service interfaces*—are the interfaces easily adapted to user requirements?
- *Platform extensibility*—can new services be easily added to a platform or can existing services easily be adapted to new uses?

10.2.3 Integrating Collaborative Services

Integration poses different kinds of problems for synchronous and asynchronous collaboration. Synchronous communication requires the simultaneous use of many communication services by more than one person. Thus a person can be talking on a video line while using a shared screen. In some cases, even stronger integration may be needed. For example, it may be useful to insert a drawing created using a shared screen into a document on the same screen. It is often very difficult to write middleware to support such parallel integration of any chosen products. Asynchronous integration does not need such simultaneous use, but must allow users to easily move from one network service to another, often through the interchange of messages.

10.2.4 Middleware or Enabling Technology

An attractive approach to platform development is to use middleware products that can interface with many services. This means that middleware software can read information from any service that conforms to the interface. It then converts the information into an internal standard form. This form can be read by any other services through the standard interface. However, such middleware is difficult to write because most collaborative network services do not conform to standard interfaces. Furthermore, it is difficult to maintain because of the rapid evolution of such services. However, standards are slowly evolving here, and standards such as CORBA (common object request broker architecture) are beginning to improve the feasibility of a middleware solution.

The alternative approach is to use one core or enabling technology, which provides a number of internal services. These services are already bundled with middleware internal to the technology, allowing the services to exchange information. Even in this case, however, there are always demands for new external services to be quickly added to the enabling technology. This demand can be met in two ways. One is for the vendors of the external service to provide an interface to the core technology. The other, and more common approach, is for the core technology to provide a high-level language that can be used to easily develop interfaces to external services.

The major two candidates for the core technology are the World Wide Web and LOTUS Notes. Sometimes these two technologies are viewed as complementary. Thus the WWW, in essence, was initially built to exchange information. It has recently grown to provide other services. LOTUS Notes grew out of document management and workflow support. Each of these two technologies thus needs additional features to satisfy all the major platform requirements. They can do this by either growing in their own right, or through interfaces with each other, or with other external products. Network designers must thus be ever vigilant to identify products that are particularly suited to their specific needs, and integrate them with the core technology.

10.2.5 The WWW as the Enabling Technology

There is no doubt that WWW technologies can be easily adapted for information exchange. They can also be used to maintain interpersonal relationships, although mostly relying on asynchronous collaboration, such as, for example, bulletin boards. Integration to corporate databases can also be supported through the use of a high-level languages such as Java. The question is how can the WWW be used to support work processes? One way is to develop forms using the HTML language and make the forms available at client interfaces. The client users can record the progress of their tasks on the form. Specially written CGI programs will use this information to notify persons of any actions required in related tasks. Such programming can become quite involved, especially where there are lengthy workflows.

Adaptability to new work practices can then only be achieved with difficulty, as changes require a change to CGI code. Thus adaptability depends on the ingenuity of the person writing the CGI code in making it adaptable to change. However, Netscape, the vendor of one of the most popular browsers on the WWW, has entered into an agreement with Collabra Technologies, a supplier of workflow products, to combine these products with the WWW and provide work process support.

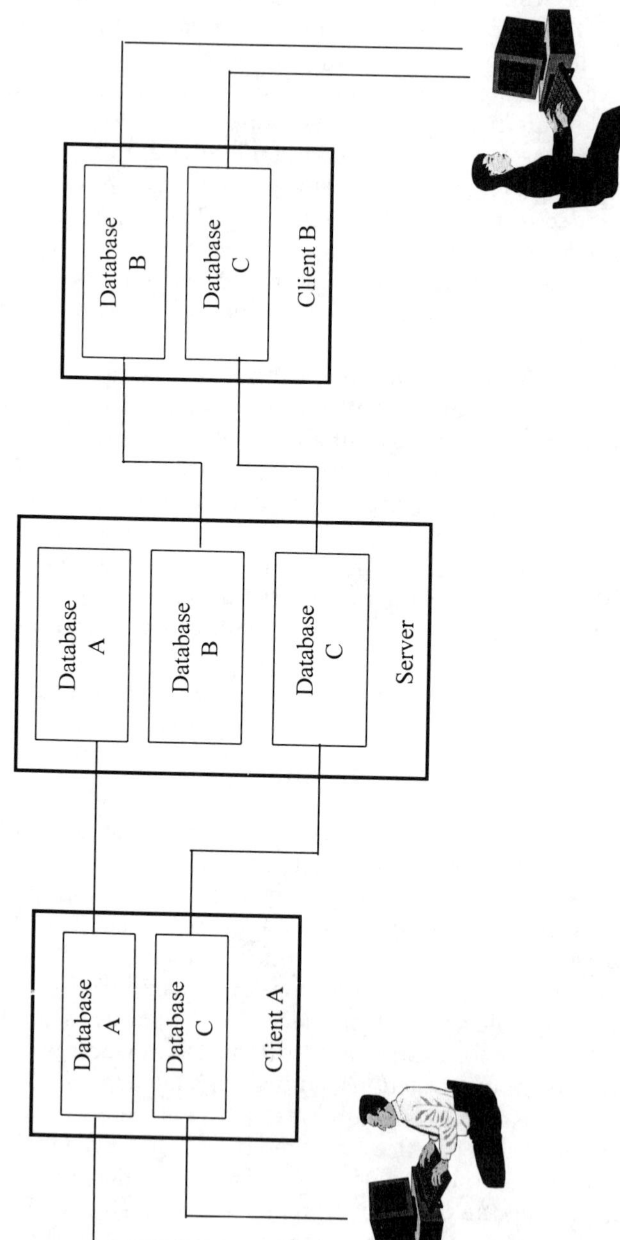

Figure 10.2 LOTUS Notes.

The WWW has another advantage: it be easily placed on desktops on existing LANs. Hence, it can be easily introduced into existing networks without excessive infrastructure costs. The result is that intranets based on the WWW will soon pose competition to these commercial platforms, and have the potential to become the dominant technology for internal enterprise networks.

10.3 LOTUS Notes

LOTUS Notes (see [1]) is a system that first achieved popularity in the early 1990s to support predefined document flows within enterprises. It obtained almost instantaneous popularity because of its ability to be quickly installed and programmed to support document flows. It is based on the client-server architecture. The idea behind LOTUS Notes is shown in Figure 10.2. The server is a repository of information held as databases. Clients can access these databases. The usual approach is that the databases are held on the server, with copies stored (or replicated) on the client machines. Thus in Figure 10.2, the server has three databases. Two of these, A and C, are *replicated* on client A and another two, B and C, on client B. The copying of databases is known as replication. Changes made by one client are first automatically copied or *replicated* to the server. They are then replicated in other client machines. The idea of replication is important here as it can improve system performance. When a user changes database contents at one client site, the system detects this change and initiates a transfer, first to the server and then to the other client sites. Users at these clients then have direct access to the database on their own machine. In fact, this machine can be a laptop, and thus the database can be used while the user is traveling. This is one substantial difference from the WWW, where data is always stored at one location unless explicitly copied by the user.

It is possible to build networks that have any number of servers, as illustrated in Figure 10.3. Systems that include more than one server can improve performance by replicating databases on servers and connecting clients to the nearest server. To do this, designers should ensure that there is a server close to most client sites. If there is a server site always close to a client site, then performance is further improved.

LOTUS Notes is supported by a number of operating systems, including Windows and UNIX.

10.3.1 Databases, Forms, Documents, and Views

LOTUS Notes centers around documents that are stored in databases. Databases can be defined in two different ways. One is to use an interactive forms definition facility to define document structures in terms of *forms*. The form defines a number of fields, some of which can be defined to be computed fields. A number of buttons can also

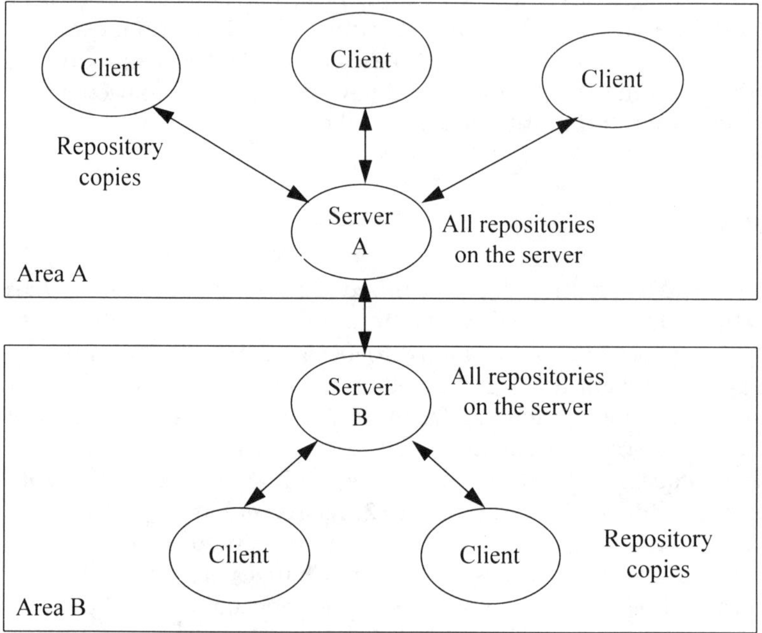

Figure 10.3 A LOTUS network.

be defined for a form. A program using LOTUS script can be associated with each button. This program is activated when the button is selected and can be used to change documents in LOTUS Notes databases. New documents can be created or composed (in terms of LOTUS Notes terminology), using the defined form structure and stored in a selected database. A single form can be used to compose any number of documents. It is also possible to create what are known as responses to documents, and responses to these responses. Thus a document can flow through a process with comments, recorded as responses and attached to the document.

The second way to define documents is to use templates, which are predefined forms. These templates include the following:

- A discussion database;
- A conference database;
- A correspondence database;
- A customer tracking database;
- A document library;
- A message repository;
- A meeting tracking repository;
- A reminder repository.

Document structures have many other features that are outside the scope of this book, but are described in detail in the references. For example, documents can contain rich fields that can include information other than text (for example, graphic images). Forms can also contain subforms. LOTUS Notes also allows a response to be made to each document. This response is then stored and displayed together with the document.

Documents are used to create and store information. The complementary feature are views that can be used to display the stored documents. Views specify conditions that must be satisfied by a document for display by a selected user. A different view can be provided to each user, depending on their role in the system, thus enabling workflows to be easily specified. A document can proceed through a number of states (say, "new," "checked," and "mailed"), where state is defined by a document field. A different user operates on the document in each of the states. Thus documents in state "new" can appear in the view of a user who checks documents. When checking is complete, the state is changed to "checked," in which case the document appears in the view of the user who mails it. It then enters the state "mailed."

The latest version of LOTUS Notes also has agents, which correspond to the software agents described in Chapter 7. These are written using a language known as LOTUS Script and can be used to do things like scan a set of documents to see if there are any conditions that require attention or carry out database searches using selected conditions

10.3.2 Creating Systems Using LOTUS Notes

Creating applications with LOTUS Notes centers around capturing information through documents, distributing these documents through replication, and providing access to them through views. Accessed documents can be changed, if needed. A system is usually defined by first identifying the information needed by the different roles in the system. Forms are then used to define the structure of the documents that will store the information. Buttons are included in the documents to initiate changes to the databases or to change the states of documents. These changes will be replicated on other client sites, and the documents displayed in views of those clients if they satisfy the right conditions. Additional changes can then be made to the documents at those sites.

There are, however, many other features that were described in earlier chapters and should be provided to create good processes. One is to provide the ability to *maintain awareness* by monitoring how work is progressing and the status of various documents. To do this, it may be necessary to provide a view that displays the status of all the documents in a particular process. Alternatively, views that only display documents relevant to one or a subset of users can also be developed. Another feature is to provide some form of a reminder or *notification service* that informs the various users of what they have to do. In addition, statistics are also very useful, providing data on times taken to carry out particular tasks, thus providing useful information for future designers. Similarly, records of encountered problems can also be useful for such designers.

10.3.2.1 Case Study: Setting Up Tasks for Workflow Enterprises

Workflow Enterprises decided that LOTUS Notes can be used to improve coordination in software engineering projects by better management of workflows between the system development tasks. Their goal is to use LOTUS Notes not only to keep records of tasks, but also to provide all of the features found useful in expediting task progress. This includes notifications and maintaining awareness through monitoring and informal communication services. The following databases were first created:

- A database of tasks that defines the work to be carried out in each task;
- The definition of the relationships between the tasks;
- A discussion database that allows group memory about the project to be maintained and supports personal communication.

The task form is central to the system as it defines what the user must do; its structure is shown in Figure 10.4.

This task form was defined using the Forms menu in LOTUS Notes. Each task has one task owner, who is the person responsible for carrying out the task. It also has one or more successor tasks, which are initiated once this task is completed. The form also includes information about its completion status, the amount of work on the task, and other information about the task (such as its starting and completion dates).

The task form was designed to follow a task work process that is common in any project and includes buttons to initiate parts of this cycle. For example, it includes a button to initiate a review and also one to request approval of task completion. When the review button is selected, a reviewer will be notified and asked to review the task output; when the approval button is selected, the project manager will be notified and asked to approve the task as being complete. Comments made during reviews and approval are stored as response documents to the task. Thus all the documents associated with one task are grouped together. The files button on the task form is used to access the documents that are needed by the task.

Tasks thus pass through a number of status states, starting with "ready," then "in progress," then "under review" or "under approval," and finally "complete" when approved. Comments made at each state are attached as response documents to the task document. There was one other extension to the idea of tasks. It is found that there are many different tasks carried out in software development, each needing slightly different actions. Thus there are development, testing,

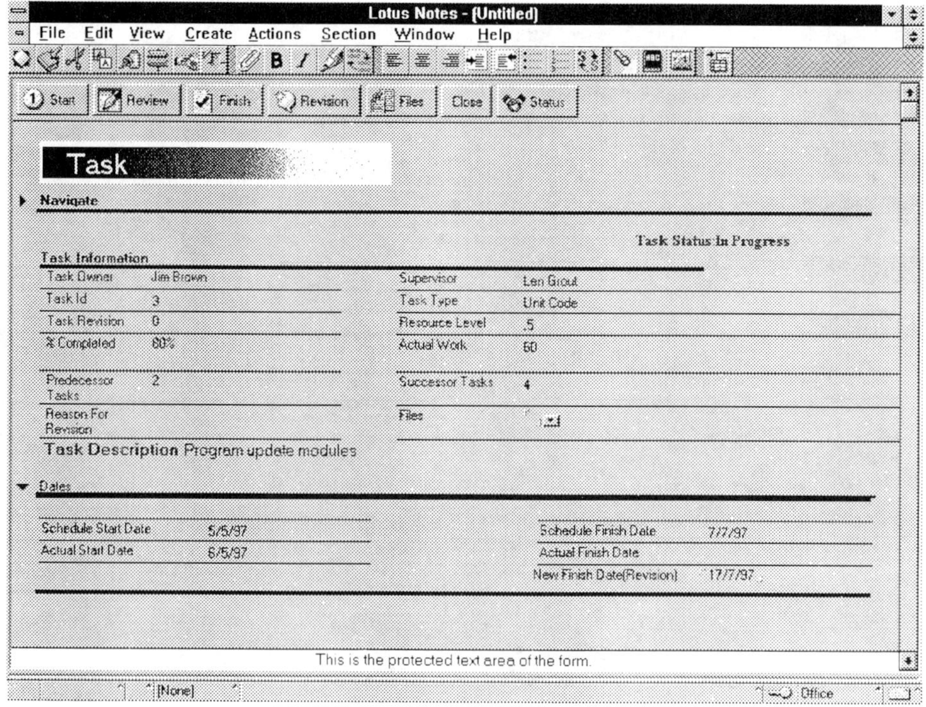

Figure 10.4 A task form.

coordination, and specification tasks. They all have much common data such as ownership and scheduling, but may differ in the actions allowed by the owner-for example, a testing task needs no review. Rather than using one generic form, which can be confusing, it was decided to provide customized forms for each different kind of task. Thus there is one form for development, another for design, another for testing, and so on. This was readily done using the forms definition facility of LOTUS Notes.

10.3.3 Using Views for Notification

The standard way of distributing documents is by automatic replication. Thus when a button is selected, it will either create a new document or change an existing document. This document will then be replicated at other client sites, thus distributing information. There is also another way to distribute information: by mail-enabling selected documents. In this case, a document is dispatched to particular users, following a change or creation of the document. The documents are

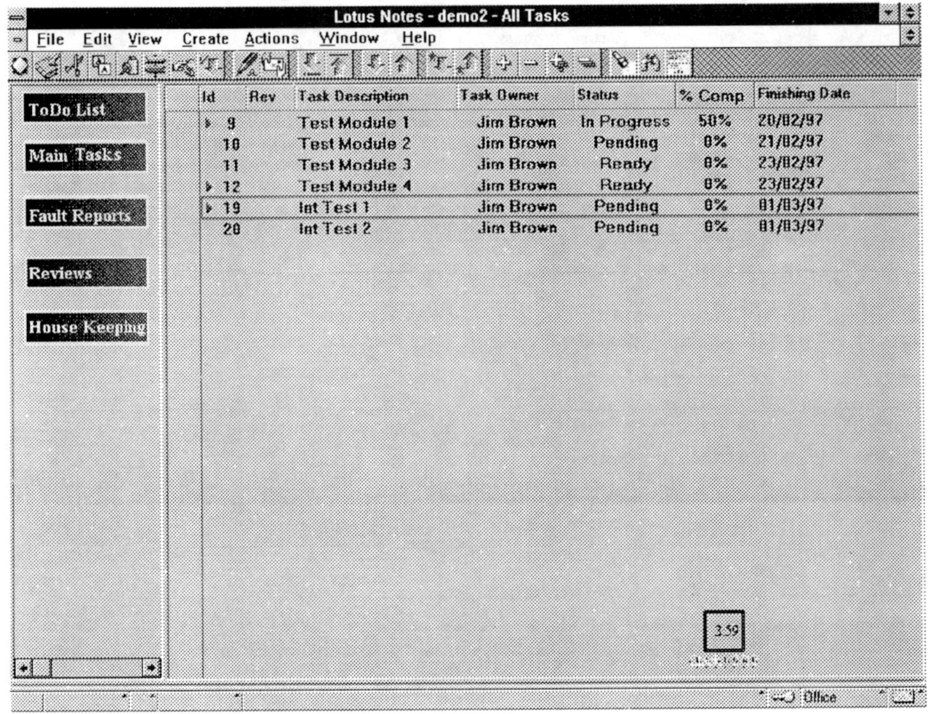

Figure 10.5 To-Do list.

sent through the mail system, informing the distant user of its arrival. What is also important in good process design is to notify people when work is ready for them. Views become very important here as they can be used to define conditions under which users become notified about documents waiting for their attention.

10.3.3.1 Case Study: Notifying People About Their Tasks in Software Engineering

Workflow Enterprises decided that the best way to keep team members informed about work waiting for them is through a To-Do list. This To-Do list is shown in Figure 10.5. It shows a number of testing tasks that are to be carried out by a team member.

The To-Do list is created by selecting those tasks that are "Ready," "Pending," or "In Progress" and owned by the team member on the client machine. This is done using the view facility. The view lists all the tasks that are ready for processing by the user at the site. The team member can then select one of the tasks, in which case the task document for that task is displayed on the screen. Team members are expected to consult the To-Do list at regular intervals to see what tasks are waiting for them.

Figure 10.5 also illustrates the general appearance of a view in LOTUS Notes. Each view can be made up of a Navigation menu on the left, a list of documents, and a preview panel that shows part of a form. The Navigation menu can be used to navigate from one view to another or to initiate document updates, provided the view user has the access ability to do this.

10.3.4 Using Views to Maintain Awareness

Another important part of any application is to monitor the current activities in that application. Views can be developed to maintain awareness in very specific ways. One way is to have a general view that displays the status of all activities in the application. In this way, all team members can be kept aware of what everyone else is doing and the status of individual tasks. Alternatively, views can be more selective-displaying all documents in a given state (for example, "complete"), or those with one particular user, or those that satisfy some computational condition such as some value falling within a specified range. A view selects the documents that satisfy the view and displays them in a predefined sequence. A user can then select a particular document by clicking on it from the view list. What is even more important is that a view can be easily interactively defined, allowing new conditions to be determined during the course of operation and documents satisfying these conditions to be quickly displayed.

10.3.4.1 Case Study: Monitoring Software Engineering Projects

Workflow Enterprises has decided to provide a number of standard views to support project monitoring. One is a personal view that display all the tasks owned by one team member. The other is an "all tasks" view that displays all the tasks in the project. This second view is illustrated in Figure 10.6 and has been created using the LOTUS Notes view facility. It shows the tasks necessary to develop a software system made up of four modules. Each module is designed, coded, and tested, including two integration tests.

10.3.5 Satisfying Platform Requirements

LOTUS Notes is viewed by many as primarily providing document management and process services. However, over time it has been extended to include facilities to improve interpersonal communication and to integrate these services with its document management services. Thus it also includes e-mail and discussion databases, where documents can be included in both.

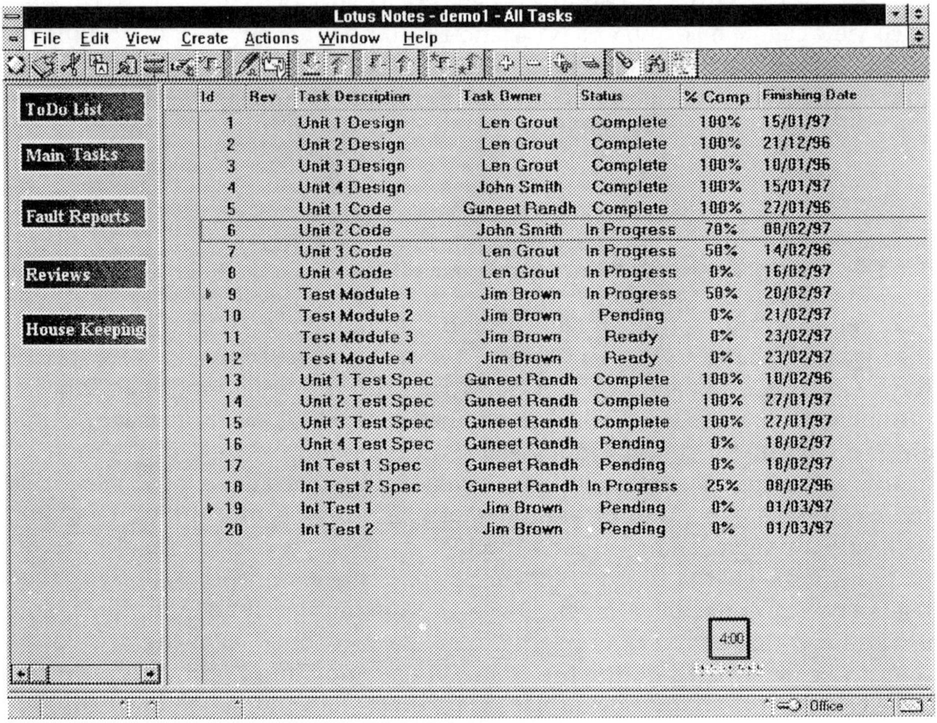

Lotus Notes - demo1 - All Tasks

File Edit View Create Actions Window Help

ToDo List

Main Tasks

Fault Reports

Reviews

House Keeping

Id	Rev	Task Description	Task Owner	Status	% Comp	Finishing Date
1		Unit 1 Design	Len Grout	Complete	100%	15/01/97
2		Unit 2 Design	Len Grout	Complete	100%	21/12/96
3		Unit 3 Design	Len Grout	Complete	100%	10/01/96
4		Unit 4 Design	John Smith	Complete	100%	15/01/97
5		Unit 1 Code	Guneet Randh	Complete	100%	27/01/96
6		Unit 2 Code	John Smith	In Progress	70%	08/02/97
7		Unit 3 Code	Len Grout	In Progress	50%	14/02/96
8		Unit 4 Code	Len Grout	In Progress	0%	16/02/97
9		Test Module 1	Jim Brown	In Progress	50%	20/02/97
10		Test Module 2	Jim Brown	Pending	0%	21/02/97
11		Test Module 3	Jim Brown	Ready	0%	23/02/97
12		Test Module 4	Jim Brown	Ready	0%	23/02/97
13		Unit 1 Test Spec	Guneet Randh	Complete	100%	10/02/96
14		Unit 2 Test Spec	Guneet Randh	Complete	100%	27/01/97
15		Unit 3 Test Spec	Guneet Randh	Complete	100%	27/01/97
16		Unit 4 Test Spec	Guneet Randh	Pending	0%	18/02/97
17		Int Test 1 Spec	Guneet Randh	Pending	0%	18/02/97
18		Int Test 2 Spec	Guneet Randh	In Progress	25%	08/02/96
19		Int Test 1	Jim Brown	Pending	0%	01/03/97
20		Int Test 2	Jim Brown	Pending	0%	01/03/97

Office

Figure 10.6 A view of tasks.

Thus LOTUS Notes can support all three kinds of communication and is increasingly suggested as a core technology for intranets. Documents can be used to exchange information, while personal relationships can be maintained through e-mail or discussion databases. It is possible to integrate a workflow system into the desktop platform to support workflows in the way described above.

When it comes to integration, LOTUS Notes does not contain links to the corporate database. However, there has been a tremendous improvement to facilities to construct such links using the scripting language in version 4. This can be used to construct links to corporate databases through an SQL interface. There are also developments towards providing a wider range of communication ports, including audio and video, thus placing more emphasis on support for interpersonal relationships. A number of telephone companies are now developing value-added services based on LOTUS Notes to encourage business to business communication using the networks.

Other important features of LOTUS Notes that make it suitable as an intranet platform are that it provides a high-level scripting language that can be used to develop support for a large variety of work processes. This also makes it possible to link some databases to third-party products, thus allowing legacy systems to be integrated into an application.

10.3.6 Extending With Agents and Third-Party Products

LOTUS Notes primarily started as a standalone document transfer system. However, its popularity has gradually led to interest in extending it to a wider range of applications by integrating it with other external products. In addition, products developed by other vendors now come with an interface that allows them to be integrated with LOTUS Notes. One set of programs that are often integrated with LOTUS Notes are workflow products that allow users to set up workflows using graphical interfaces.

Another important interface for workflow systems is to project management systems, such as for example, Microsoft Project. Thus the project plan can be defined using a project management system and the project plan automatically converted to tasks on a LOTUS Notes database. This can be achieved with the interfaces supported through software known as HiTest.

Version 4 of LOTUS Notes now provides a scripting language very similar to Visual Basic. In this way, it becomes possible to integrate LOTUS Notes with any other system that can be called from Visual Basic. A link with Internet is now being provided by the Internotes facility that allows WWW pages to be read as LOTUS Notes documents. It also allows what is known as a DOMINO server to be set up using LOTUS Notes and for documents to be published through this server. The server can thus access a page of the World Wide Web, send it to a client for editing, and then replace it on World Wide Web.

10.3.7 Discussion Databases

Interpersonal communication across distance can be supported by both the WWW and LOTUS Notes. LOTUS Notes provides a template of discussion databases that can be quickly set up to gather comments on issues.

Perhaps the simplest kind of database to support interpersonal communication across distance is the discussion database. A discussion database is usually a set of statements organized under a number of headings. Often, responses to statements can also be recorded in the discussion database and records maintained of people who made those responses. LOTUS Notes, for example, supports discussion databases where statements are organized by category and issue. Responses can be made to be recorded with the original statements; responses to responses are also supported. The discussion can then be produced as a list of statements followed by their responses.

One such discussion database is illustrated in Figure 10.7. Here, an initial statement is made raising the issue of, "What is important to our intranet?" Others

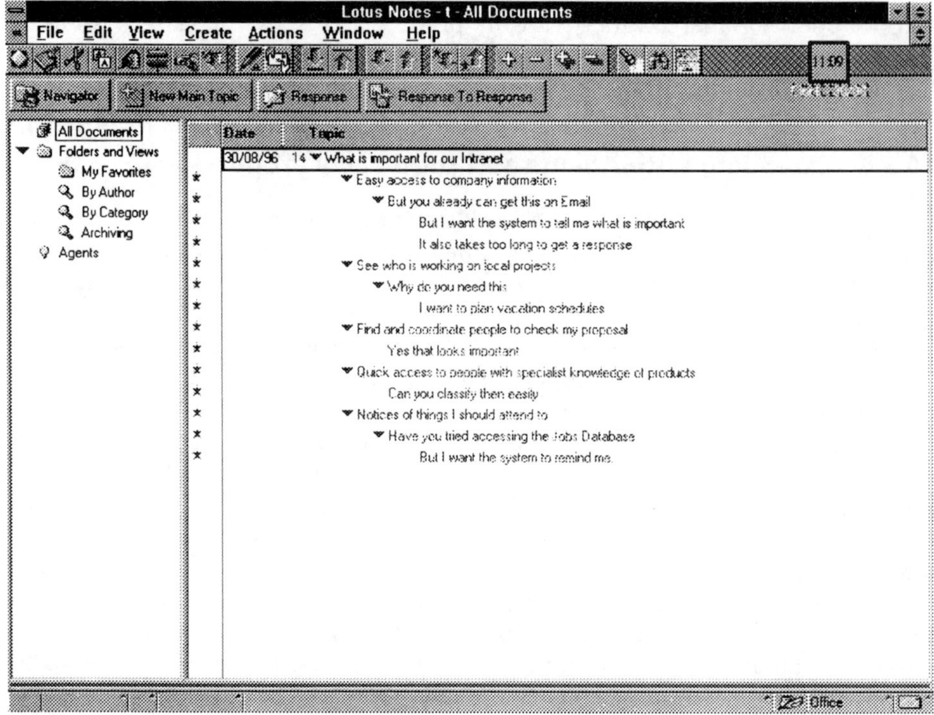

Figure 10.7 A discussion database on LOTUS Notes.

can then make their responses. The first of these is, "Easy access to company information." Someone who disagrees with this then states, "But you already can do this on e-mail." However, further responses say that e-mail takes too long and that what is needed is the system to initiate any information transfers. The discussion database thus provides a very organized way of keeping track of a discussion. They are superior to e-mail for this purpose-imagine how much more difficult this would be if all of the messages in Figure 10.7 were on a linear input e-mail list interspersed with other mail information.

10.4 OTHER PLATFORMS

Although the chapter concentrated on LOTUS Notes, there are other organizations that are beginning to provide collaborative service (or workgroup) platforms. One of a number of examples is Oracle, which is adding collaboration services to its database systems. Another example is the Microsoft Object Ex-

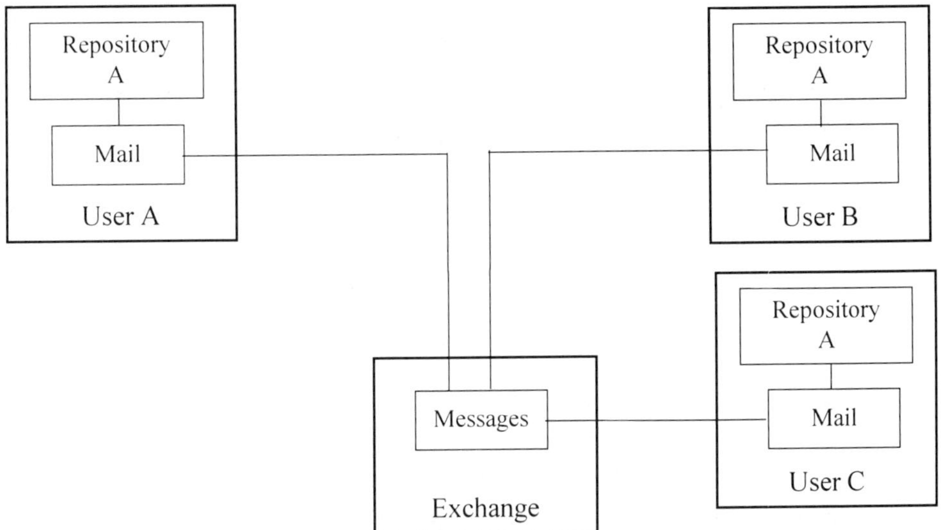

Figure 10.8 Object exchanges.

change and Sun Systems, who are providing a set of products for developing intranet infrastructures.

10.4.1 Object Exchanges

In an object exchange, transfers of information are initiated from the repositories. Repository users determine the sites to which their information is to be sent and initiate transfers to these sites. One can envisage such a system in the way shown in Figure 10.8, where each repository is directly connected to a mail system.

The result is that e-mail effectively becomes the middleware of the system and is used to integrate the services through an information exchange. Thus a number of repository services can be interfaced to an e-mail service that can interchange messages with other sites. Information can be moved between the repositories in a number of ways. One is to attach repository parts to mail messages, another is to copy them into the message. Repository methods can initiate transfer of documents to the message repository. The communication ports, whenever necessary, carry out format conversions between the different services.

The important point to note in the object exchange is that it supports the value-added notion described in Chapter 3. Existing users are not required to rewrite applications specifically for collaboration. They simply get a mail option in their application and need not use it if they don't want to.

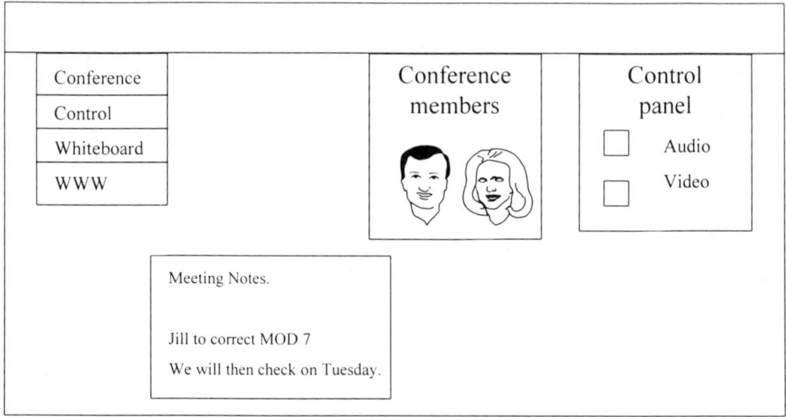

Figure 10.9 Integrating at the operating system level.

10.4.2 Toolkits

There has been continuing research into integrating services for synchronous coordination that require a number of seamlessly connected services to support easy discussion between collaborators. This may include a videoconferencing service, a word processor, a connection to World Wide Web, and a whiteboard. There is also a design tool that provides design drawing. It should be easy to view all of these at the same time at both ends and move information from one to another.

It is therefore not surprising that a number of experimental toolkits have been developed to construct such seamless platforms. Many of these are still at the research stage and most such development is within research environments. One example is GroupKit [2], being developed at the University of Calgary in Canada. It is publicly available through the World Wide Web. GroupKit runs on UNIX workstations under the X11 Window system and is primarily used for prototyping platforms. It provides users with a special language to construct groupware applications. Applications center around *conferences*, which are selected teams of people.

Applications here center on a conference interface designed using the special language. Thus, for example, it is possible to set up an interface like that shown in Figure 10.9. It provides a control panel with the conference participants together with services to support them. The control panel can allow users to choose any of a number of media such as video or sound for communication. Other services on this platform concern information. Such services can display a document or provide a sketchpad or whiteboard that is synchronously displayed at all locations. It can also allow users to refer to a page from the World Wide Web or produce a record of their discussion on the word processor. Conference participants can discuss documents and make notes on the sketchpad, defining any future actions.

10.5 SETTING UP DISCUSSION DATABASES

Earlier, this chapter introduced discussion databases as a medium for interpersonal communication in asynchronous environments, and Figure 10.7 illustrated one such database. Discussion databases are increasingly proposed for use in asynchronous personal communication. Discussion databases can be supported by both LOTUS Notes and the WWW. LOTUS Notes has a template that can be used to define a standard discussion structure. This allows any one team member to make a statement and other members to respond to it. Then, responses could be made to these responses. There are, of course, other structures that can be built on either LOTUS Notes or the WWW.

10.5.1 Discussion Databases on the WWW

The WWW is increasingly used to support discussion forums. These are similar to the LOTUS Notes discussion database and allow people authorized to use a discussion site to enter issues onto the site. Others can respond to these statements. As an example, the Center for Information Systems Management at the University of Texas has developed a discussion forum (http://cism.bus.utexas.edu/) to promote discussion on information systems research. It works in the same way as any discussion database. People can enter new issues in this database while others can respond to these issues, and still others browse and add to the discussion.

10.5.2 Discussion Database Structures

So far, this chapter has described discussion databases that allow people to raise issues and gather comments on these issues from others. It is also possible to design more elaborate structures for discussion databases to store the various comments in ways that reflect the kind of argumentation that may take place in decision making.

10.5.2.1 Databases for Argumentation

Figure 10.10 shows one way for managing such discussion. Here, a number of people have proposed alternate structured ways of maintaining design rationale. Important issues are clear statements of criteria used in a change and alternatives considered. The idea here is to organize statements in a way that enables a decision to be made. The structure must therefore follow some argumentation or design rationale.

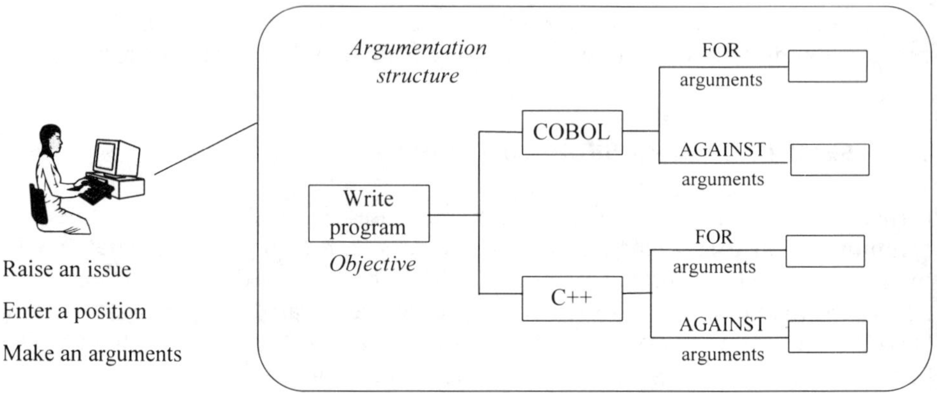

Figure 10.10 Argumentation databases.

Users of this argumentation database can raise new issues, take positions on the issue, and argue for or against each position. Listings of issues together with their arguments can be produced and used in meetings to make any decisions. For example, in Figure 10.10 a decision has to be made whether to use COBOL or C++ to develop a program. For and Against statements are entered for each alternative. Eventually, these statements will have to be consolidated and a decision made based on them. Often, arguments are collected asynchronously, but a decision is made in a face-to-face meeting.

10.5.2.2 Databases for Design Rationale

Keeping group rationale adds to the workload as it requires people to actually state the reasons for any change. This often is unacceptable, and people simply forget to do so because of time pressures. It then becomes necessary to look for ways of reducing this overhead. Multimedia systems produce one alternative by recording rationale in voice. A person can verbally state changes made to the documents as they are made, and thus it does not require additional time to make written rationale entries.

Another way to support rationale is to provide a structure that simplifies methods for entering rationale. An example on such a structure is described by Lee and Lai [3]. The structure provided here is to:

- *First identify the issue*—an example may be to design a database;
- *Identify an alternative approach to the issue*—may be to use a relational system or develop a special file;
- *Specify the evaluation*—for example, poor, good, excellent, and so on;
- *Clearly define the criteria*—such as easier retrieval;

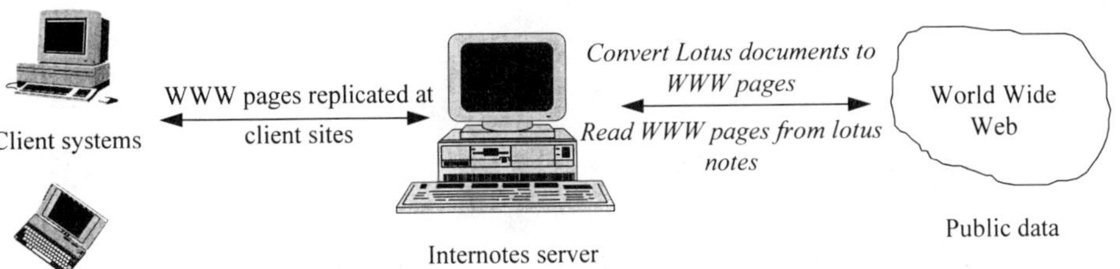

Figure 10.11 Combining LOTUS and WWW.

All arguments or reasons for taking an action will then be based on this structure.

A statement in the rationale may them simply look like, "I chose a relational database because retrieval facilities are good."

10.6 COMBINING TECHNOLOGIES

It is now increasingly recognized that one way to gain market share will be to simplify interfaces between technologies to create ever more powerful platforms. Thus, for example, there is considerable advantage to combining platforms with complementary sets of services. Why not combine the workflow capability of LOTUS Notes with the publishing and information exchange services of the Internet and the WWW. This is exactly the goal of the LOTUS Notes DOMINO servers shown in Figure 10.11 (for details, see site http://www.lotus.com/inotes/). The details of such a combination can itself take a whole book (there is one such book quoted in the references). It allows users to set up a DOMINO server and exchange WWW pages with a LOTUS Notes database. These pages can then be replicated in any number of (possibly portable) clients.

10.6.1 Possible Strategies

The range of core technologies and their combinations leads to almost an infinite set of possible network possibilities. It is not possible to outline all possibilities in a book, but an example is given in Figure 10.12. Here, LOTUS Notes is used as the core technology that provides support for work processes in two enterprise units, which are part of the enterprise intranet. The operations can be integrated through a WWW distribution service. Thus task C in unit A and task X in unit B communicate through the WWW, where the primary goal of the intranet is to support workflows within the enterprise. There is also a corporate WWW component that provides consumer support as well as directory support and policy information for internal use.

This figure illustrates one of the many possibilities. Figure 9.12 of Chapter 9 illustrated another possibility based on the WWW only. There are many others

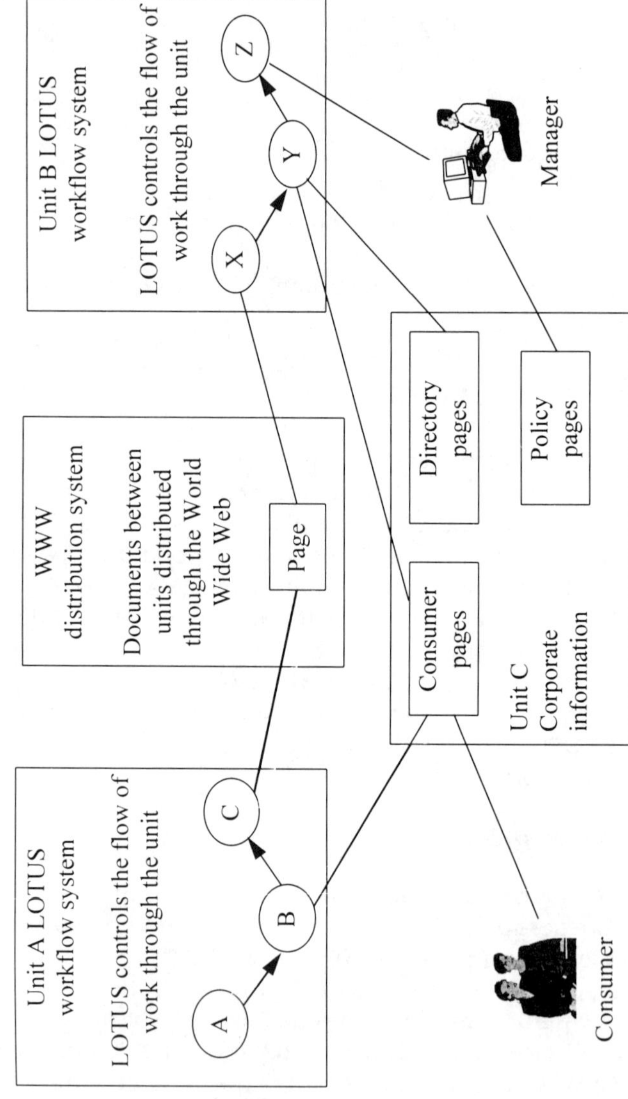

Figure 10.12 One of the unlimited possibilities.

using a mix of these or supporting a different distribution of responsibilities between units.

10.7 SUMMARY

This chapter described platforms that are used to build networking systems. The chapter began by describing basic platforms that can be assembled from generic applications. Basic platforms have the disadvantage of being seamless. Consequently, there are now a number of commercial systems that provide platforms with various degrees of seamlessness. Two architectures to building networks were described, namely, using an object exchange or using a client-server architecture. The chapter then described LOTUS Notes as a core technology for such networks and outlined possibilities for building intranets based on this technology and integrating it with the WWW.

10.8 DISCUSSION QUESTIONS

1. Are there any advantages of using middleware instead of a core technology in platform design?
2. What is the difference between client-server and object exchange platforms?
3. Why is seamlessness needed in platforms?
4. What is the difference between forms and documents in LOTUS Notes?
5. What are alternate ways of maintaining awareness using LOTUS Notes?
6. How would you maintain awareness with the WWW?
7. Would it be easier to use LOTUS Notes or the WWW to develop notification methods?
8. How would you design a work-oriented context described in Figure 2.7 using LOTUS Notes or the WWW?
9. Why can the WWW become an important competitor for developing intranets?
10. Compare LOTUS Notes and the WWW as intranet platforms.

References

[1] Shulman, M., *QUE Guide to Lotus Notes*, NH: New HQUE Press, 1993.

[2] Roseman, M., "GroupKit: A groupware Toolkit for building Real-Time Conferencing Applications," *Proc. of the Conference on Computer Supported Collaborative Work, CSCW92*, Toronto, 1992.

[3] Lee, J., and Y-K. Lai, "What´s in Design Rationale," Human-Computer Interaction, Vol. 6, 1991, pp. 251–280.

Selected Bibliography

Bernstein, P. A., "Middleware: A Model for Distributed Systems Services," *Communications of the ACM,* Vol. 39, No. 2, Feb. 1996, pp. 86–98.

Borland, R., *Microsoft Exchange in Business,* Redmond, WA: Microsoft Press, 1996.

Hawryszkiewycz, I. T., I. Gorton, and L. Fung, "Putting Software Development on the Information Highway: Going Beyond e-mail," *The American Programmer,* Aug. 1995, pp. 8–14.

Object Management Group and X Open, *The Common Object Request Broker: Architecture and Specification,* Document Number 91.12.1, Revision 1.1, 1991.

Pyle, L., *Creating Lotus Notes Applications,* NH: QUE Press, 1994.

Tamura, R. A., et al., *LOTUS 4 Notes Unleashed,* Indianapolis, IN: Sams Publishing, 1996.

Chapter 11

Repositories and Coauthoring

LEARNING OBJECTIVES

- *Document configurations*
- *Versions and version control*
- *Processes for document evolution*
- *Traceability*

11.1 INTRODUCTION

The previous chapters described platforms for creating and distributing information in the networked enterprise. This information is mostly in the form of documents, but is increasingly graphic images, technical designs, or even videos or audio tracks. The general term *artifact* is often used to describe items of such information. Furthermore, the information, rather than being static, often evolves over time. People in the enterprise usually work on these artifacts-designing them, storing them, reviewing and studying them, and discussing them with their fellow workers. Document or artifact management must often consider documents that are outside the computer system (such as, for example, complex drawings) as well as electronically stored documents. Often, computer systems store references to these documents but find it quite difficult to keep track of them, especially if they are frequently changed.

Document development itself can be seen as a design process, as it always involves changing a document to meet some goal. Two kinds of support are needed to create and change artifacts in networked enterprises. One is task support, which is needed to actually create and amend the document, for example, a word processor for text editing or a tool for making drawings in industrial design. One important aspect of support is to keep track of the changes and ensure consistency between documents. The term *configuration management* is often used to describe systems that manage jointly owned related documents in an environment of change. The other kind of support is process support, which is needed to support the steps needed by people to maintain document consistency. Process support looks at ways of passing documents between people or allowing people to work together on the same object at the same time. This chapter begins by introducing document configurations and then continues to describe processes that are needed to maintain consistency when people are jointly working on documents in different places and times. It then looks at the kind of commercial systems available to support document management.

11.2 REPOSITORY STRUCTURES

Most people are used to the idea of databases that contain records mostly of the same type. Thus there can be a personnel database that contains records of people's names, addresses, positions, and so on, or an accounts database that contains transaction records of credits and debits to accounts. Furthermore, all documents in a transaction database often have the same structure. Design of such databases is a topic on its own and many books have been written about it. Early database systems were built around the idea of hierarchical and network structures, whereas the trend recently has been to flat files mostly supported by relational database management systems.

Repositories that are used in cooperative systems are often more complex and contain many different kinds of documents, each with its own structure. Thus, for example, the repository that is needed for software development will include documents that specify user requirements as well as a variety of design and test specifications. It will also include the software modules themselves. In addition, more structured information about individual tasks and their progress will also need to be kept. Not only does such a repository contain many different records and documents, but such records can be related and made up of information displayed using different media as well as entries from other repositories. Furthermore, rather than organizing documents as sequential pages, more complex links between documents can be established.

Such multimedia documents now almost always are made up of many components, with each component presented using a different medium. Thus a personnel record could contain the picture of the person whose details are recorded in the record. Similarly, there could be a design document that is made up of a design drawing together with text that describes specifications or production details. It is also possible to have documents that contain video clips or statements using voice. Thus, for example, a help screen could include a voice explanation of some interface process based on menus together with an ongoing video that illustrates this process.

Furthermore, documents in a repository are often related and document management must maintain references between related documents. A help file can be provided for this purpose, but often links between documents help people to use repositories. Hypertext, which was introduced in Chapter 9 in the context of the World Wide Web, is often suggested for linking documents. Each document in a hypertext structure has pointers to related documents. Hypertext systems have the additional advantage in that they can support multimedia documents.

11.2.1 Keeping Track of Documents

Repositories are increasingly required to keep track of document evolution. This is often done using the idea of annotations to record reasons for document changes and versioning to make it easy to track document progress.

11.2.1.1 Recording Changes

In a typical development process, documents evolve over time. There may be a number of people who continually change or make additions to a document. For example, a specification may be continually changed as client requirements change. These changes then affect related documents such as designs. Configuration management systems must include the facilities to keep track of such changes and their effects.

Figure 11.1 illustrates one way to record document changes. Each person in a group makes a small change and records the reason for the change in a note. In this way, rationale behind each version is defined as it passes from coauthor to coauthor. This rationale can be a note that is simply a written comment or it can take on a more formal structure like that described in the earlier section.

11.2.1.2 Versioning

The events that lead to a document change usually result from changes to other documents. Thus in software engineering, if system requirements change, then the requirements document changes. This in turn requires a change to the design documents as well as the test documents, followed by changes to the actual programs. One way to proceed through such changes is to create new document versions. Each change results in a new version, with the reasons for the changes attached to the document.

The idea of document states and versions is to keep track of changes. Documents go through a number of states, and a document is assigned new version numbers as it passes through the states. The *states* ways of numbering versions can be chosen to suit a particular need. As an example, Figure 11.2 shows a document going through three states, with the first being a *working* state, where it is being created or changed, perhaps by a number of people in a group. This corresponds to

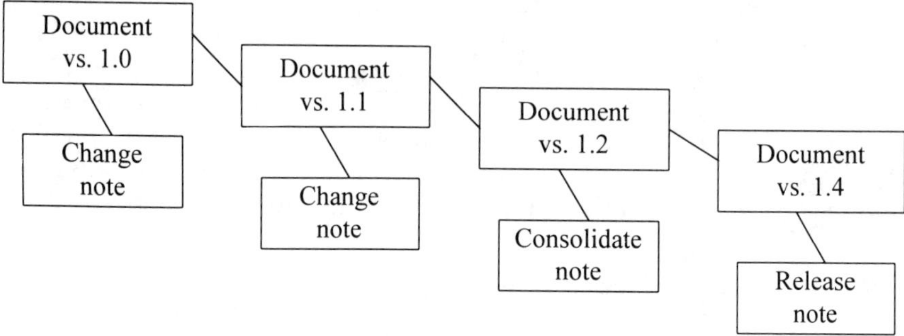

Figure 11.1 Design in an asynchronous environment.

the way the enterprise works in practice-first documents are known as working documents, then there may be an interim release of documents, and, finally, once everybody is happy, the document may be released. Initially, any number of working versions (1.1 and 1.2 initially in Figure 11.2) may be created with each one assigned to each group member. At some point in time agreement is made to consolidate these working versions into one version, which in this case becomes version 1.3. This version is then sent for approval and enters a *pending approval* state. In Figure 11.2, approval creates an interim release (version 1.4), with the document sent to a successor task. Comments received from a successor task may lead to a new working version (version 1.4), which is subsequently approved and released. It is, of course, possible to design systems with different states than those shown in Figure 11.2 (for example, draft, issued and so on).

Version numbering must be closely related to the process followed in creating documents. Many processes can be used. One process is for a version to pass from one group member to the next, with each making an amendment in sequence. Another is for each person in the group to work on their own version and examine each others' versions. Then, all group members will have to come together to decide what the final version is to be. This becomes the pending document, which contains material acceptable to all group members, perhaps with some comments on some alternatives. The pending version must be approved by management and becomes the released version after approval. There may also be customizations of a released document. For example, a manual for released software may be customized to particular hardware.

11.2.1.3 Verification and Traceability

Keeping track of change is often important to keep people aware of what is happening to a document. They describe the history of a document and reasons why certain actions were taken. *Traceability* is a term that describes how the evolution

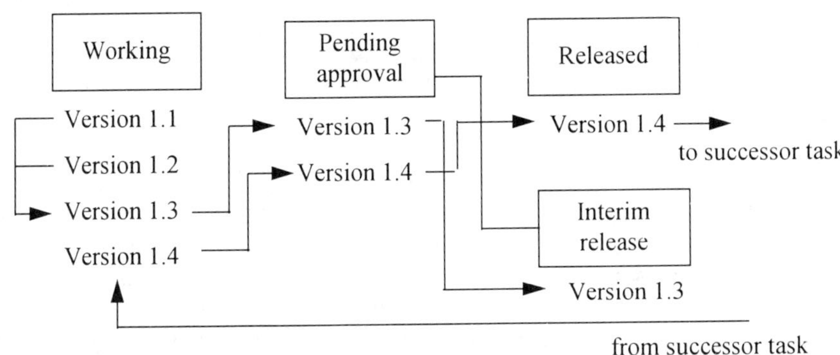

Figure 11.2 Versions and states.

of a document can be traced. *Verification* describes checking to find the source of change in a particular document. Often, as in the previous case, verification is used to check whether one or more documents can be traced to the same source.

11.2.2 Configurations

Rather than dealing with a single document, it is more common to work in environments with a number of related documents-often called configurations. The collection of documents and their relationship to each other is often called the document configuration.

11.2.2.1 Case Study: Workflow Enterprises—Document Configuration

One challenge in supporting distributed software development is to keep track of the large number of related documents. Figure 11.3 gives a very broad idea of the kinds of documents found in a typical software engineering project. First there is the requirements specification, which is used to develop system and test specifications. The system specification leads to a number of module designs, with program modules then constructed to satisfy these designs. There may be a large number of module design documents, one for each system module. Similarly, test specifications are used to eventually develop test cases, which are then used to test the modules. Test specifications are defined in parallel to system development and are used to test the program modules. The test specifications are made up of:

- A test design specification, which defines a set of tests related to a requirement specification and used to ensure that a particular group of modules correctly realize this part of the specification;

- Test cases that define how particular requirements features are tested. Test cases are related to a design specification and to parts of the requirement specification.

Tests using test cases are made whenever a particular set of modules are ready for testing. A test creates a test log, which is used to amend the modules if necessary. The test log is related to a test case. A test failure will require an evaluation of the reasons for failure. When no document numbering system exists, one question that always arises is whether the failure occurred because the tests did not match the latest document version, or whether there is a fault in the developed system. This itself can take up valuable time and the goal is to smooth the testing process and ensure that any test failures result from system faults rather than being the result of using the wrong documents. To do this, it should be possible to easily trace a test log document to a set of modules and trace the module to a requirement specification. The test case can also be traced back to see if it originated from the same set of specifications and requirements.

The system becomes much more complicated when one remembers that each of the boxes in Figure 11.3 can itself go through a number of versions. Thus a change to requirements may lead to new versions of specifications. Similarly, a test may require the revision of a program. As a result, the number of documents encountered in a large project can be quite large-often in the order of thousands-and it can become difficult to keep track of all document changes. Traceability is also important because often it is necessary to trace back a reason why a program module has been developed in a certain way, or the effects of a change of one document on all others. There is a gradual realization that what is needed is a formal document policy and numbering scheme to gain better control of the document management process.

11.3 MANAGING DOCUMENT DEVELOPMENT

Systems that include many documents require a configuration management system to manage the documents. Configuration management includes both maintaining the documents and keeping track of them as they are changed. There are many aspects to configuration management. These include the following.

- Identifying documents together with the various versions of each document;
- Identifying relationships between documents;
- Assigning responsibilities for documents or document parts;
- Defining a process for document preparation;

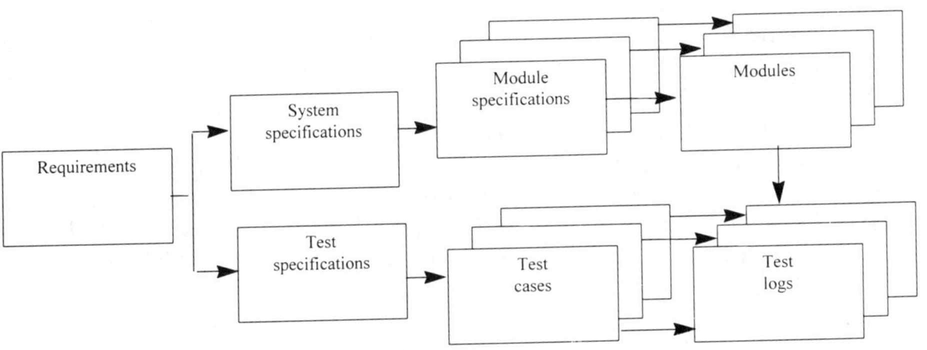

Figure 11.3 A document repository for software engineering.

- Keeping track of changes to documents;
- Maintaining awareness of document status through the whole team;
- Monitoring progress in the development of a document;
- Maintaining a history of changes to documents and the reasoning behind the changes.

The first two requirements are satisfied by giving each document a unique name. This, together with a version number, is usually sufficient to keep track of changes. Relationships are maintained by including cross references between the documents using these unique names and version numbers.

Document location is also an important issue. Where documents are not close to a person, then that person may find it difficult to get hold of the latest copy. On the other hand, keeping more than one copy means that the copies are inconsistent as changes may not as yet have reached all copies.

Both task and process support must be provided to do this. Such support should allow documents to flow in an orderly manner with all developers fully aware of document progress.

11.3.1 Task Support

Task support concerns the way that documents are created and maintained during the development process and the annotations that are allowed with the documents. This will often depend on the type of document being developed and is most often provided by the software used to develop the document (for example, Word, when developing a text document). The task support tools here will be the facilities needed to produce a document, annotate changes, and so on. They are usually based on a word processor, but allow more than one person to work on the document at the same time. They also include many of the facilities described in the earlier parts of this chapter- versioning, annotation, and so on.

For visual documents, the task support tool is usually a drawing tool. In a synchronous environment, such a tool will allow a person to draw an artifact using that tool, and this drawing appears at the other end. Some discussion can follow, with sketches added by persons involved in the collaboration. Additional task support facilities can include things like a sketch pad with *overlays*, which are similar to document versions. Such overlays make it possible to have a basic design and electronically overlay it with design suggestions. Such suggestions can be stored with comments, or design rationales, for future reference and/or decisions on adopting a final design.

11.3.2 Process Support

Process support defines how a document flows between the coauthors and depends on group structure. A highly structured group will follow a different pattern of document production than an unstructured group. Consequently, the development process supported for different kinds of groups must be different. However, in all cases, processes have common characteristics. They must give people the ability to create new documents, to discuss them, and eventually to reach some agreement on the final document form. The development process can differ in synchronous and asynchronous environments. In synchronous environments, agreements are reached by discussion, whereas asynchronous environments require the use of some voting method.

11.4 COAUTHORING PROCESSES

Coauthoring is defined here as a number of people developing a jointly owned document, *including* the discussion and argumentation used resolve any conflicts about document content. In coauthoring, the participants are often provided with document goals and the important issues to be considered. Coauthoring of documents should:

- Include the ability of each person to suggest and make contributions to the documents;
- Provide support to resolve any conflicts and ensure that eventual consensus is reached on any final document form;
- Maintain a level of awareness where all participants know about the current document state;

- Maintain privacy to documents that may be confidential;
- Ensure that all participants are working with consistent data.

Coauthoring requires both task support and process support. Task support is provided by the software used to change the document, record changes, and so on. Process support defines the steps followed to make such changes and is defined by the development process, the roles of the group members, and the tasks to be carried out by each member.

11.4.1 Synchronous Environments

The important aspect of support in synchronous environments is for people to be able to make changes during discussion. Thus a typical synchronous system should provide an interface that contains the people at distant sites and the document under discussion. It should allow people to manipulate the document while explaining their changes verbally.

The process can be relatively straightforward. All that is needed is to arrange mutually convenient meeting times to discuss issues related to documents.

11.4.2 Asynchronous Environments

Process support here differs substantially from that provided for synchronous groups. Asynchronous collaboration is often supported by defining specific roles and document structures, and assigning specific responsibilities to the roles. The following roles are commonly found in document creation:

- Originator—responsible for the documents;
- Coordinator—keeps track of document changes;
- Coauthor—works on the document or parts of the document;
- Observer—person who is affected by the documents and may comment on them;
- Editor—checks the documents and returns them.

Document structures can include objects such as comments, suggested revisions, public comments, private messages, and so on, and roles are given the ability to work or create specific kinds of objects. The process then defines flows between the different roles and the abilities of each role with respect to the documents. It is also becoming more important to keep track of changes and the rationale behind them, especially for large groups, to ensure that earlier mistakes are not repeated.

Document changes now proceed in steps with longer time intervals than those in synchronous work. For this reason, asynchronous design is sometimes seen to be unsuitable for creative work. Some people claim that asynchronous work reduces creativity because it is difficult to quickly build upon each other's ideas or come up with the kinds of ideas that seem to emerge in fast spontaneous discussion.

On the other hand, asynchronous design allows people to make contribution at any time. This is claimed to be important by some writers, who often see good ideas that come on the spur of the moment forgotten when they get involved in spontaneous discussion. Hence, different approaches may suit different people, again requiring the chosen system to match the group culture.

It is important in asynchronous work is to maintain awareness by all people involved in joint work. Notification schemes become very useful here. A change to one document can affect other documents. Persons responsible for those other documents should be informed as early as possible, to either comment on the change or make changes to their documents. There is also now a suggestion that software agents be provided to keep track of changes and notify people about them.

11.4.3 Mixing Synchronous and Asynchronous Design

In most cases, a mix of the two approaches will satisfy many requirements. Synchronous meetings can create new ideas that may perhaps have to be evaluated later. Following the evaluation, another meeting may be called. Such meetings are also often needed to resolve all outstanding issues, as asynchronous resolution of controversial issues may be impossible or take an unnecessarily long time.

Another criterion for combining synchronous and asynchronous work is for processes that contain a mix of deterministic and nondeterministic workflow. It often happens that the nondeterministic work is carried out in a small group, perhaps working on one small component of an artifact. The work of such groups is then combined into a deterministic process. In such cases, the small groups can work synchronously, whereas the main workflow proceeds asynchronously.

At the same time there may be different ways of managing document flow. There three most obvious ways are as follows:

- Keep a central copy and allow people to check copies out into their private workspaces as needed.
- Circulate a document in some predetermined sequence.
- Allow a number of people to concurrently work on their copies and then resolve any conflicts through a synchronous meeting.

11.5 DEVELOPMENT PROCESSES IN ASYNCHRONOUS ENVIRONMENTS

Often, changes made to one document must be followed by changes to related documents. Documents thus evolve through processes followed by roles involved in document production. Any support system must ensure that all documents in a given configuration are always consistent with each other. Coauthoring requires a definition of what is meant by a correct structure and the specification of processes that must be followed to realize correct structures. Correctness and its definition will depend on the group structure and ownership of documents. As an example, three structures of increasing complexity are where:

- There is one owner of each document in a related set of documents.
- There is one document that is jointly owned by a number of people.
- There are a number of related documents, each jointly owned by a number of people.

The next sections describe these alternatives.

11.5.1 One Owner per Document

The simplest process here is to create a separate originator role for each document and pass changes from one role to the next. Thus, as shown in Figure 11.4, one originator role changes their document and passes the change to the originator role of a related document. This role makes a change to a related document and passes on this change. Thus a requirements change may be passed to the architectural designer, who changes the system specification, which is then passed to the module designer. The module designer creates a module design document and passes it to both the programmer, who uses it to build the module, and to the test designer, who creates the test case. The tester then applies the test case to the module and reports on any faults in the test log document. This is then passed to the designer to correct the program.

Asynchronous changes are easily supported in this environment. Thus there is nothing to stop the designer from amending the design while all other people are working on their documents. However, such changes must be passed on to the people who own related documents. This is where version control comes in. Each time an amendment is made, the document is given a new version number. Persons working on related documents must ensure that the document versions match. Thus, the first amended version of a design specification may be version 1. There may be a number of versions of a test case that match version 1. If the design specification changes, it will become version 2. The module will have to be changed to match the specification, and this module will have a new version number. A new test case to match this specification

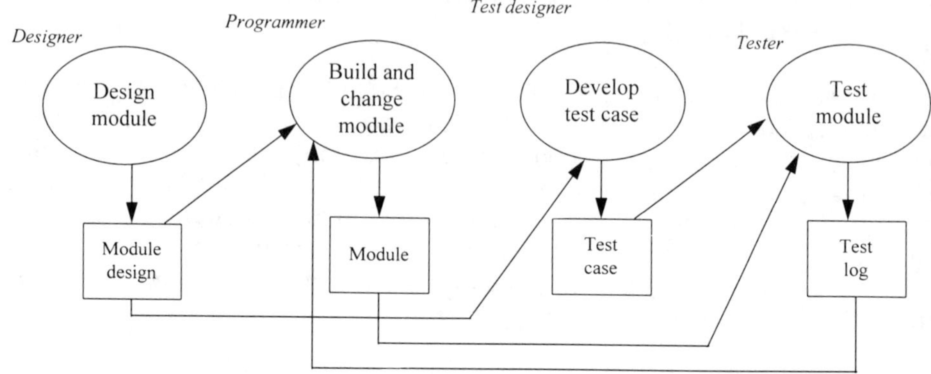

Figure 11.4 A sequential process.

version will also have to be created and it will have a new version number. What is important is that when a test is run that the test case and the module correspond to the same design specification version.

11.5.2 Joint Ownership of One Document

The process now becomes more complex, as a number of people can work on the same or related document. Now each person is given a copy, or version, of the document (in Figure 11.5, versions 1.1, 1.2, and 1.3). Each can suggest changes to their version. All team members must then agree which of their suggested changes will be permanently incorporated in the report, which in Figure 11.5 becomes version 1.4. Usually, such agreement on permanent change is reached through a discussion. Thus, although individual team members can make their changes asynchronously, decisions on a new permanent version usually require synchronous discussion and agreement.

In addition, it is possible to create customizations of some of the documents for specific needs (for example, creating an external version of an internal document by eliminating any commercial in confidence information).

11.5.3 Joint Ownership of Related Documents

The process here must coordinate joint activities on a number of related documents. This requires a combination of the processes in Figures 11.4 and 11.5. The combination is complicated by the fact that more than one version of one document can be made available to groups working on a related document. Thus a

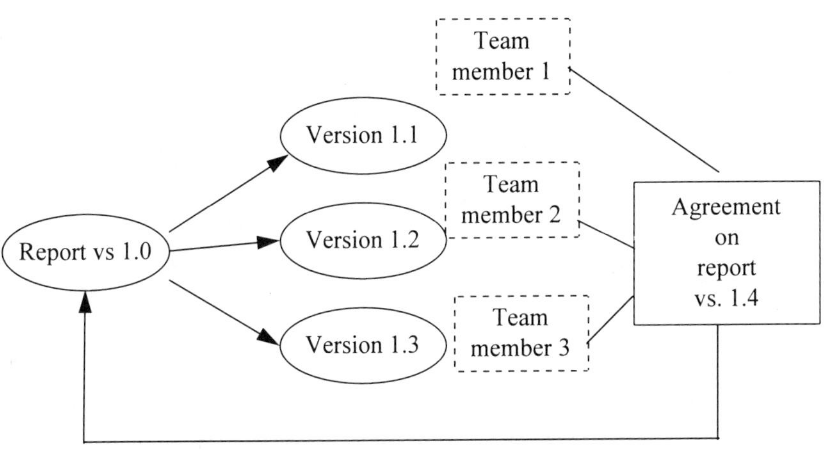

Figure 11.5 Process for parallel work.

group must not only agree to changes in their document, but also to agree with groups working on related documents. Figure 11.6 shows two such groups working on documents, document 1 (the budget section) and document 2 (the technical section). Versions of document 1 can be released for consideration by group 2, resulting in proposed versions of document 2. These can be fed back to group 1 when deciding on a final release version. Document configuration management must now support relationships between versions of document 1 and document 2. Furthermore, agreement on final versions of each document must consider suggestions made by the team preparing the other documents. It is also quite likely that representatives from each team may need to meet to agree on the final versions of both documents.

It is now also necessary to keep information about the relationship between versions of the documents-for example, in Figure 11.6, version 1.4 of document 1 led to the start of work on versions 1.4, 1.5, and 1.6 of document 2.

11.5.3.1 Case Study: Worldlink Consultants—Joint Document Production

Worldlink Consultants found that they are often developing proposals that require contribution from widely distributed experts. Usually, the managing consultant develops a brief following discussion with the client. Work to complete the proposal is divided into parts, and each part is assigned to a group of experts. The managing consultant is responsible for ensuring that all the parts developed by the groups are consistent with the original brief. Preparation of documents must often meet deadlines, and it is necessary to keep track and often expedite the work of the

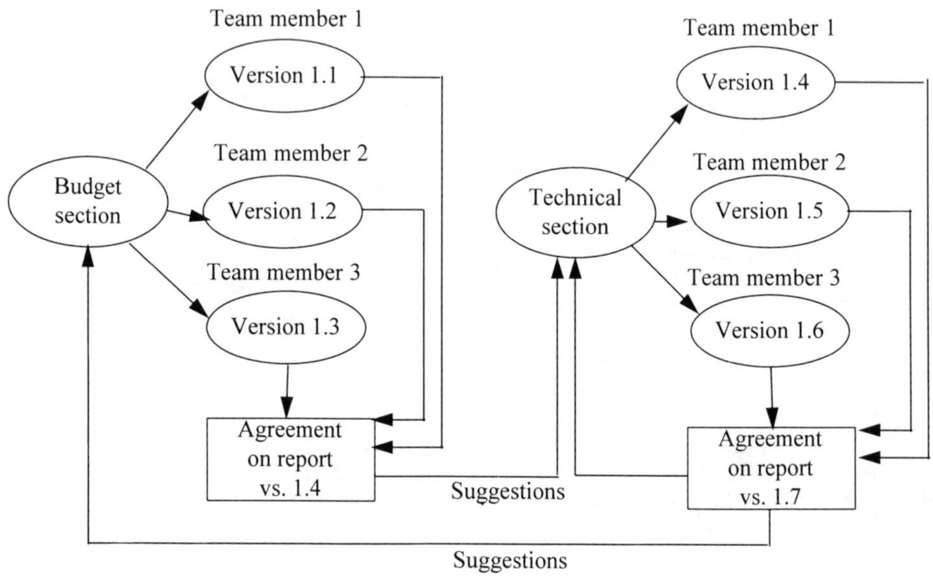

Figure 11.6 Joint development of related documents.

experts. It is becoming more difficult to keep track of proposals because of increased distribution of experts, the continuous changes made to documents, and the increased complexity of projects. There is thus increasing pressure to improve this process by somehow raising the general level of awareness of proposal status and notifying experts about their outstanding work.

The system developed by Workflow Enterprises and described in Chapter 10 looks attractive for this purpose. To do this, roles and responsibilities must first be more clearly defined and there should be a versioning policy. One approach being considered is to assign coordinators within the expert groups to facilitate this process. The coordinator will be responsible for sharing out the work within the group, ensuring proper reviews, and coordinating with the managing consultant to ensure document consistency. It is proposed to define three document states: draft (D), released (R), and issued (I). Drafts of proposal parts are prepared by members of a subgroup and reviewed within the group, with annotations made by reviewers. The comments are then consolidated by discussion between the subgroup members. While in the draft stage, each part remains within its subgroup without reference to other groups. Once the subgroup is happy with a draft, it will be released for consideration by the managing consultant through the coordinator.

Tasks in the workflow system must now be customized to this process. Tasks customized for coordinators will provide them with the ability to distribute documents to experts and reviewers. Reviewers can be nominated, with documents passed for review in the way described in Chapter 10. Progress is monitored by the coordinator and managing consultant, who will ensure that deadlines are met by detecting delays and expediting critical tasks. Also, general awareness about progress can be maintained. There is also the possibility of adding agents to expedite progress of tasks.

It is proposed that each coordinator keep a draft copy of their part on their workstation and allow members of their group to have copies. They can annotate changes to their copies, and the changes are consolidated by the coordinator following discussion with members of their group. Following agreement, a new draft version can be created, stored on the coordinators workstation, and also distributed to all members of the group. Once the group is happy with their part, it is released to the managing consultant. Versions will be numbered as (x.y.z) where x is incremented whenever work resulting from a client request arrives with the coordinator, y is incremented with each new change requested by the managing consultant, and z incremented whenever a new working version is created. In addition, the state of the document (D, R, or I) is also recorded. The managing consultant can require additional work by the group to ensure consistency with other parts. In this case, the released document again enters a draft stage (with y updated to indicate further work requested by the managing consultant). Once the managing consultant is happy with all the parts, the document is issued to the client. Following the clients response, additional work may be needed, requiring the issued document to again enter the draft stage (this time incrementing x in the numbering scheme to indicate that the further work was requested by the client). What remains now is to choose the document management system to be integrated with the workflow. Currently, documents are developed using Word files and the general consensus is that this will continue to be the case. Ways of integrating workflows to Word files maintained by the operating system are being investigated.

11.6 COMMERCIAL SYSTEMS FOR DOCUMENT MANAGEMENT

So far, this chapter has covered ways that can be used to keep track of documents. The next step is to look at services provided by commercial systems to implement them. Document management takes many forms in the commercial area. One is simply the storage of large volumes of documents, such as, for example, research reports. These are static systems with documents not being changed. Many dynamic document systems are *editorial systems* that allow stored documents to be updated. These may restrict access to people who can change them, often identifying different roles such as guest, for reading, reader, for creating views for others, author, who can change a document, and administrator who looks after the storage. Then there are the *versioning systems* that support cooperative development of documents. They provide facilities to keep track of successive versions of documents as they evolve.

Many of the systems use the client-server architecture, where a document stored on a server can be distributed to any number of clients. It is perhaps also fair to say that most commercial management systems concentrate on the structure of the documents and not so much on the process.

11.6.1 Versioning Systems

The usual version management system is a repository of documents that primarily allows users to *check out* a document from the system for their use and then *check in* the document when they are finished with it. The version number is updated at the time that the document is checked in. However, it is often the user's responsibility to keep track of why the new version was created or what document state it represents. Relationships between documents are not usually explicitly maintained, as for example, through hypertext links. Most systems assume that users will do this through the version numbering scheme. One way to do this is by uniformly changing the version number of each related document. Thus when one document changes to a new version, it is assumed that all related documents will also change to that new version number.

Backer and Busbach [1] define some desirable characteristics for versioning systems, and have implemented this in a system known as DocMan, which was developed as part the EuroCode project. One important characteristic is that versions once checked into a system should never be changed, with change only permitted to newly created versions, which can be distributed to a number of team members. They suggest that all versions of the same document be kept in a document folder. Their numbering scheme is based on two document states: draft and released. They define a two-digit numbering scheme, x.y, where x is the major part and defines a revision whereas y is the minor part and is incremented every time that a draft is released for updating. A revision may go through a number of drafts before it is released. The value of x is updated every time a draft is released.

One typical system here is Intersolv's PVCS. This provides a check-in, check-out facility where the version is updated whenever a document, which can be a file such as Word, is checked in. The document can be locked at the time it is checked out. The version numbering allows branching from a given version. Thus there is a numbering that takes the form x.y, but it is possible to create any number of branches from each version; these branches are labeled by four numbers, x.y.a.b, where x.y is the originating version. This allows customization of earlier revisions. A comment can be associated with each version to state the reason for creating this version. Another document management system is OpenDoc, which is supported by IBM and Apple. OpenDoc documents can be divided into a number of parts, and each part can be worked on independently. OpenDoc allows a number of drafts of a document part to be created, and different people can work on these drafts. The parts must later be manually reconciled. Commercial products such as PVCS or OpenDoc, of course, have many features that can best be found by consulting their relevant manuals.

11.7 SUMMARY

This chapter described some of the issues in joint document production. It identified major issues, in particular the maintenance of large document configurations, and keeping track of documents in that configuration. The chapter defined the important role of versions in configuration management and ways of integrating version control with a process that defines the flow of documents. Furthermore, it pointed out that an organization must have a versioning policy to effectively track documents.

11.8 DISCUSSION QUESTIONS

1. What is a document configuration?
2. What do you understand by the term *version*?
3. Why must configuration management support a process?
4. Why is it important to integrate process with version management?
5. What are the additional complexities when groups work on related documents?
6. Define some roles needed in asynchronous document management.

11.9 EXERCISES

1. Suppose a document is stored on the server in a client-server system, but a number of coauthors can change this document. Suggest a process that can be used to develop a document in this environment.
2. Define the type of document management services that would be needed in a virtual university.
3. Suggest a versioning numbering policy for document development by Worldlink Consultants.

References

[1] Backer, A. and U. Busbach, "DocMan: A Document Management System for Cooperative Support," *Proc. of the 29th. International Conference on System Sciences,* Hawaii, 1996, pp. 82–91.

Selected Bibliography

Babich, W,."*Software Configuration Management*" Reading, NJ: Addison-Wesley, 1986.

Buckley, F., *Configuration Management: Hardware, Software and Firmware,* IEEE Computer Press, 1993.

Greif, I., "Desktop Agents in Group Enabled Products," *Communications of the ACM,* Vol. 37, No. 7, July 1994, pp. 100–105.

Hawryszkiewycz, I. T, L. A. Maciaszek, and J. R. Getta, "Cooperation and Artifact Semantics for Asynchronous Distributed Cooperation," *Journal of Software and Systems,* Elsevier Science Inc., June 1996, pp. 179–188.

Intersolv, Inc., *Intersolv PVCS Version Manager,* Rockville, Maryland.

Mandviwalla, M., and S. D. Clark, "Collaborative Writing as a Process of Formalizing Group Memory," *Proc. of the 28th. Annual Hawaii International Conference on Systems Sciences,* Hawaii, 1995.

Neuwirth, C. M., et al., "Flexible Diff-ing in a Collaborative Writing System," *Proc. of the 1992 Conference on Computer Supported Collaborative Work, CSCW92,* Toronto, 1992, pp. 147–154.

Sharples, M., (Ed.), *Computer Supported Collaborative Writing,* Berlin, Germany: Springer Verlag, 1993.

Chapter 12

Workflows

LEARNING OBJECTIVES

- ❑ *Workflows*
- ❑ *Defining workflows*
- ❑ *Commercial workflow systems*

12.1 INTRODUCTION

Workflow systems have been attracting considerable interest because they allow a user to easily define and construct many of the predefined processes that are often found in networked business applications. They become the process tool that moves documents created in the enterprise. Workflow systems support processes that follow a predefined set of steps. The idea of a workflow is illustrated in Figure 12.1. Here, a requisition is made by a project engineer to get items from a supplier. The next step is for the requisition to be approved by the manager, followed by the order clerk making an order. The order is then sent to the supplier, and the items received from the supplier. Each of these steps is carried out by a person assigned to the step. The process usually results in document flows, and consequently the process is often defined in terms of document flows between persons assigned to the process steps. Each person is required to take some action based on the document and pass it on to the next person.

Writing workflow problems using conventional programming languages has been quite a lengthy and costly process. Programs must be written to store the documents, pass the documents from person to person, and generate any output reports. Workflow systems provide a way to easily construct workflow processes using a high-level definition language. Here, a process is defined as a process model

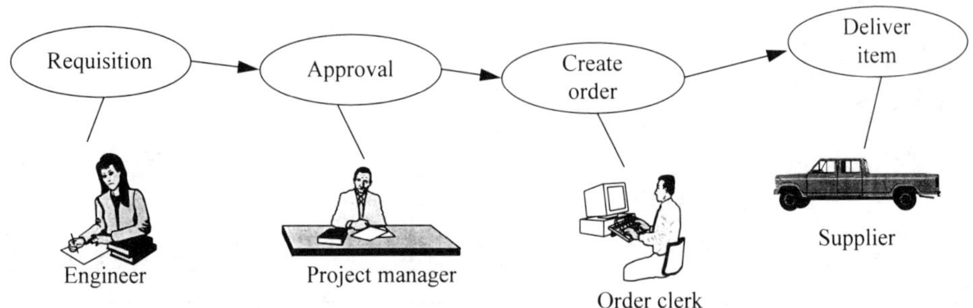

Figure 12.1 A workflow.

using a high-level definition method. The users define the document structures and process flows, often visually, and use the workflow generator to generate the system from this definition.

Most workflow systems, however, generate predefined processes and must often be accompanied by other services, such as e-mail, to give support to informal communication. Different workflow systems define and manage workflows in different ways, and steps are now in progress to develop some standards in this area. This chapter describes some workflow systems.

12.2 WORKFLOW SYSTEMS

The way that a workflow management system (WMS) fits into a platform is shown in Figure 12.2. On the one side, it provides the platform where applications can define the flow of work in the process. The user interface provides the language used to define the workflow. On the other side, the workflow system uses an information management system to keep track of artifacts and a communication system to distribute them. These two systems may themselves be services of the kinds described in earlier chapters. Thus, for example, LOTUS Notes includes these services as it can be used do create documents and has a communication system based on electronic mail. Any enterprise setting up a workflow system often has to choose all these components and integrate them into a working system.

In a network, the components shown in Figure 12.2 must be distributed across a number of workstations. One approach can be to place the WMS on one server, which becomes the workflow processor (as shown in Figure 12.3). This server stores the database and the process definition. The database can include stored forms as well as the transaction data such as, for example, accounts. Each person who is assigned a task in the process has a workstation. The workflow processor detects the process states and uses the stored process definition to send a message to

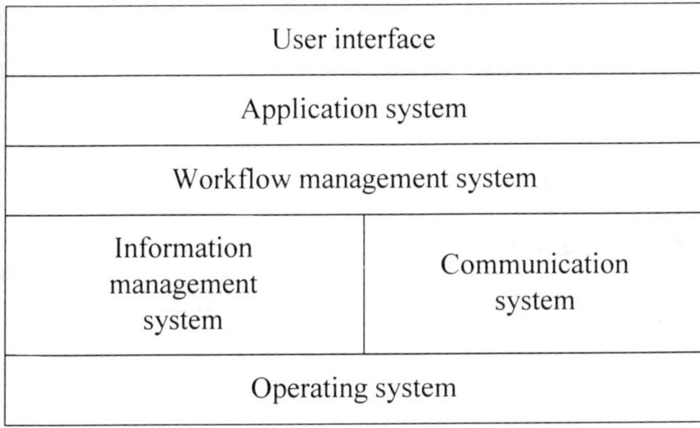

Figure 12.2 WFMS system.

the workstation whose role is to carry out the next process task. When that role completes its task, the role's workstation program sends a message with a form to the workflow processor, which then informs the next role of its task, and so on. The processor knows the location of roles and sends messages to those roles. Most workflows then use network services to move documents from one location to another.

It is also possible to distribute the different components. Thus in a client-server configuration, the database and even the workflow processing can be placed on client machines. The choice of configuration is often driven by performance issues. Workflow systems must be designed to ensure that a process is not held up unnecessarily and completes on time to the satisfaction of its users. Some systems also provide methods to define performance in terms of timings between different process steps and can monitor progress, reporting any problems with maintaining required performance.

In fact, performance is of such importance that a system known as Bonaparte has been produced with the sole aim of designing workflows that meet performance requirements. It is also unique in that it integrates workflows with the enterprise model. It supports designers by providing them with access to tasks and the organizational structure. The intention is to link Bonaparte to workflow generators. In this way, the process will first be adjusted to reach required performance standards and then used to generate a working system.

There are a number of important ways to describe workflow systems. These include the following:

- The way the workflow is defined;
- The way application programs are defined;
- The way users are supported;

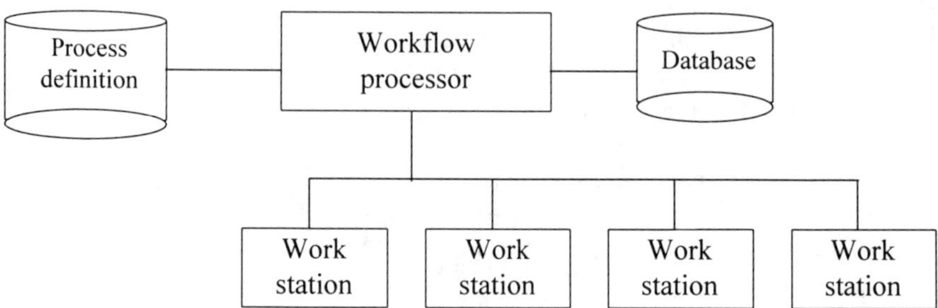

Figure 12.3 Workflow configuration.

- The way that process performance is measured and improved;
- How easy it is to change workflows.

There is now a large variety of workflow languages used to define workflows with little in common between them. Consequently, it is difficult to integrate systems built using different languages. As a consequence, there is now some emphasis on developing standards for describing systems. One of the first goals of such a standard is to develop a consistent terminology. There is also often a separation between the actual workflow and application development. The workflow defines sequencing of actions, whereas application development includes the design of data and processes that are included in the actions.

12.2.1 Terminology

One group-known as the Workflow Management Coalition, a loose group of workflow vendors and customers-is working towards defining a consistent terminology. Their goal is to develop a workflow reference model that can be used to describe any workflow system. At the time of this writing, their work is in its initial stages, concentrating initially on defining the components of the workflow process.

Some proposed terms include the following:

- The activity network that defines the sequence of activities that make up a process;
- Transition conditions that define movement conditions for moving from one activity to the next;
- A work item that defines an instance of an activity;
- An application tool that defines the process executed in an activity.

12.2.2 Defining Workflows

Workflows are defined either visually or by using a definition language. Many such languages use the idea of process state and define processes in terms of state transitions. They used formal tools such as, for example, Petri Nets to define such transitions. Workflow languages go beyond simply defining states and transitions, but must also include the role responsibilities and artifacts used at each step. It is now common for such definitions to be made visually. Most workflow definition systems define

- The document structures, usually as forms;
- The stages through document flows and the processes;
- The support given to users to maintain process flows;
- The documents used at each stage;
- Roles responsible for the transformation at each stage and persons taking on these roles.

Support to users is important. It concerns things like letting everyone know what is waiting for them, and reminding them of any outstanding work. Thus every day on arrival at their workstation, the users have a list waiting for them about their outstanding tasks and newly arrived work.

12.3 COMMERCIAL SYSTEMS

Workflow systems have been of interest to developers since the middle 1980s. Most of the early systems were built for specific uses or specific classes of use, but since 1990 there has been a prolific growth of systems and support tools for generating workflows.

12.3.1 Early Experimental Workflow Systems

The earliest systems in this area were developed in Europe as part of the ESPRIT program. One of the earliest examples is DOMINO.

DOMINO [1], developed at the German National Research Center (GMD), is one of the first workflow systems. It was developed primarily as a tool for specifying structured office procedures. It was written in the programming language C and used e-mail as the underlying messaging system. It supported two major concepts, forms and roles, and coordination was specified in a declarative way. It supported asynchronous coordination only, and members could be at different locations. It did not easily support dynamic process change.

Those with research interests can refer to a number of other early workflow systems. Prominand (ESPRIT Project, see [2]), for example, was designed to realize an electronic office environment supporting the concepts of roles (or office workers) and folders. Office workers are presented with an "electronic desk," where they can select a folder to work on. Folders correspond to documents flowing within offices. The folder consists of an envelope together with its contents. It was implemented on SUN workstations running UNIX with TCP/IP and SunView. Coordination was specified as steps through which an office document flows and, like DOMINO, supported asynchronous coordination with people at different locations. Folder contents can evolve by attaching notes and appendices.

12.3.2 Later Commercial Systems

Commercial workflow systems are now becoming available. Often these systems are integrated with other services to provide integrated platforms. LOTUS Notes, for example, is used with many workflow systems to provide an integrated document workflow system.

12.3.2.1 Action Workflow

One system that has attracted attention is known as action workflow, which has its origins in the work of Winograd's conversation systems that were described earlier in Chapter 3. *Action workflow* can be viewed to be made up of two parts: the process definition and application development. Processes in action workflow are based on the major steps in the action conversation. These steps are known as preparation, negotiation, performance, and acceptance, and form what is known as the *action workflow loop*. The notation used to describe such loops is shown in Figure 12.4.

Each loop starts with a customer making a request, called the preparation. There is then negotiation between the customer and performer on how to best satisfy the request. The performer then carries out any actions needed to carry out the request and the product or service is then accepted by the customer. The idea of workflow loops is to emphasize human interaction by formally defining the way that people make arrangements. Rather than simply modeling a flow as a document sent from one person to another, the goal of action workflow is to also include the processes of negotiation and acceptance that often accompany such flows. Action workflow also supports the basic loop with ways of handling exceptions and expediting information flows through reminding people of their requirements and monitoring process performance.

Workflow loops can be connected, so that a performer in one loop can request other services, also specified in terms of workflow loops. A business process can then be defined in terms of a number of connected workflow loops. Action workflow requires business processes to be defined in terms of such flow sequences. A set of connected lops is known as the *business process map*. The term

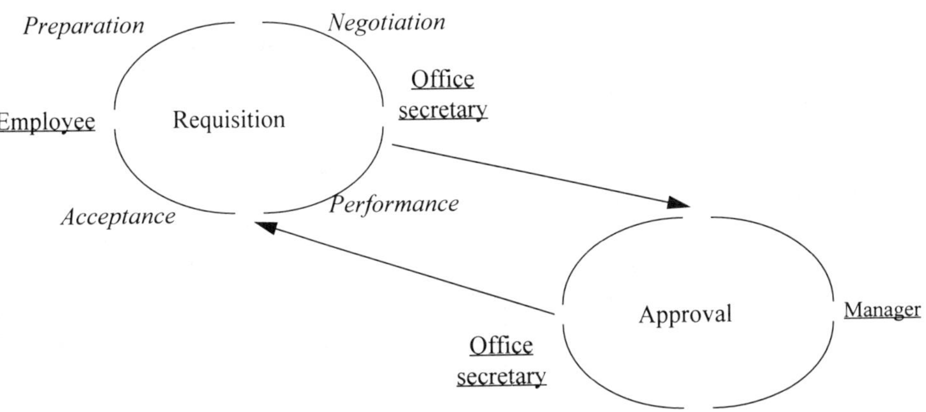

Figure 12.4 Action workflows.

primary loop is used to define a loop that activates another loop, which is known as the *secondary loop*. Activation of secondary loops can be made conditional.

In Figure 12.4, the first, or primary, loop defines an interaction between an employee making a request for leave through the department's secretary. There can be some negotiation on how long approval will take, or when a decision can be expected. The secondary loop is where the secretary takes the leave request to the manager for approval. Thus there are two interactions in the process. One is in the primary loop between the employee and the secretary asking for the request to be processed. The second is in the secondary loop between the secretary and the manager to obtain a decision on the leave application.

Action workflow can also specify the time for each of the actions and use this to monitor the process. Each point in a loop can also be defined as a script that defines the activity at that point in terms of operations carried out on data. An Interface has also been developed between action workflow and LOTUS Notes to enable LOTUS Notes documents to be passed using action workflow processes.

12.3.2.2 Other Products

There are a number of other products in this area; many workflows on top of e-mail as shown in Figure 12.5. They thus use e-mail to support structured workflows by connecting applications at the workstations through e-mail.

They can also be integrated with other products. For example, WorkMAN can be combined with LOTUS Notes to support the flow of documents between participants in a workgroup. Processes in this case are defined visually using a graphical interface. The designer uses a graphical interface to set up a workflow like that shown in Figure 12.1 for an approval process by:

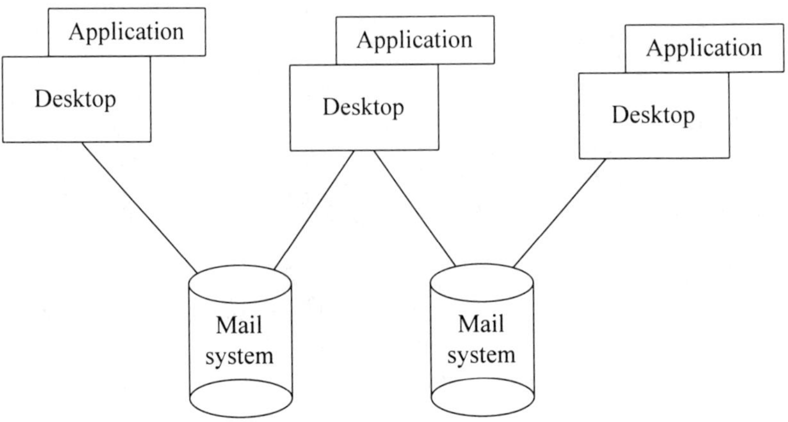

Figure 12.5 Workflows using e-mail.

- Defining the process roles as, for example, engineer, project manager, and order clerk;
- Creating links between the roles;
- Associating each link with a LOTUS Form (for example, a Request form) as the link between initiator and manager.

The terminology used in WorkMAN is to identify *stages* that represent points in the workflow process (for example, initiator, approver, and so on). The stages are linked through *links* (for example, request, approval, and issue). *Roles* define the function contributed by a user at each stage, and this function is accomplished by a *workflow task*. The links represent information flows between the stages. The flow of work is called a process.

Usually there is an initiator stage that starts a process. As each stage is completed, a document is sent to the next stage, initiating the participant at that stage.

Another commercial product here is ProcessWise, which was earlier marketed by ICL. It is a system designed to define business processes that may cross organizational boundaries. It is based on the following concepts:

- People;
- Roles;
- Systems;
- Services;
- Activities;
- Functions;
- Processes;

- Tasks;
- Documents;
- Objects;
- Products.

It primarily supports asynchronous operation participants at different locations. A graphical workbench is used to identify relationships between the people and their roles to activities, functions, and services. ProcessWise also includes facilities to monitor process performance.

12.3.3 Installing Systems

Workflow generators can be seen to be composed of two major components: the control component, which defines information flows; and the document component, which holds the information used in the flows. It is now common to integrate a workflow generator with a document-based system, which provides facilities for document generation. In such combinations, the workflow generator becomes part of the document platform. It is used to define flows whereas the information carried in the flows is defined using the document generator. Thus both WorkMAN and action workflow can be integrated into the LOTUS Notes system to enable workflows to be included in LOTUS Notes applications. Here, LOTUS Notes is used to define the forms, whereas the workflow generators are used to control document flows.

12.4 SUMMARY

This chapter has described methods for constructing cooperative systems that follow a deterministic workflow process. It outlined the principles behind workflow generators and described some existing systems. It stressed that workflow generators generally implement deterministic processes.

12.5 DISCUSSION QUESTIONS

1. What do you understand by a workflow?
2. Why do most workflow generators generate predefined processes?
3. What is the distinction between process definition and application development in workflow systems?
4. Why is process monitoring important in workflow systems?

12.6 EXERCISES

1. Try to model the process for software engineering using the workflow loops of action workflow.
2. Would a workflow system be useful in cases such as global planning, or in hospital systems?

References

[1] Kreifelts, T., et al., "Experiences with the DOMINO Office Procedure System" in Bannon, L., M. Robinson, and K. Schmidt, (Eds.), *Proc. of the European Conference on Computer Supported Collaborative Work, ECSW91*, Kluwer Publications, Doedrecht, 1991.

[2] Karbe, B., N. Ramsperger, and P. Weiss, "Support of Cooperative Work by Electronic Circulation Folders" *Proc. Conference on Office Information Systems*, Cambridge, MA, 1990.

Selected Bibliography

Abbott, K. R., and K. S. Sarin, "Experiences with Workflow Management: Issues for the Next Generation" in Furuta, R., and C. Neuwirth, (Eds.), *Proc. of the Conference on Computer Supported Cooperative Work, CSCW94*, ACM Press, 1994, pp. 113–120.

Action technologies, Inc., *Action Workflow Analyst User's Guide*, Action Technologies, Inc., Version 2.

Medina-Mora, R., et al., "The Action Workflow Approach to Workflow Management Technology" *Proc. CSCW92*, Toronto, 1992, pp. 281–288.

Rogers, S. B., *Novell's Groupwise 4: User Guide*, San Jose, CA: Novell Press, 1996.

Workman Product Description, Spring 1993, available from STARCOM Communications and Support.

Chapter 13

Meetings and Conferences

LEARNING OBJECTIVES

- *The meaning of meeting*
- *Reasons why people meet*
- *Electronic meeting rooms*
- *Task and process support for electronic meetings*
- *Videoconferencing*
- *Extended meetings*

13.1 INTRODUCTION

One of the most common things that people do is meet to discuss issues of importance to them. To most people, meetings take place face to face where people get together in a committee to discuss some issue and perhaps make a decision, or to have a discussion about a proposal, or come up with an idea. However, meeting is a generic term, and meetings can fall into any part of the space-time dimension discussed in Chapter 2. Computer communication is providing new ways to hold meetings in networked enterprises. One obvious way is to use videoconferencing that supports meetings at the same time but with people at different places. It is also possible to extend meetings over longer periods of time, sometimes known as extended meetings [1], which can be supported by discussion databases.

Apart from simply providing the ability to hold a discussion, computer communication support can facilitate meetings by providing easier access to documents during the course of the meeting, as well as other tools that facilitate

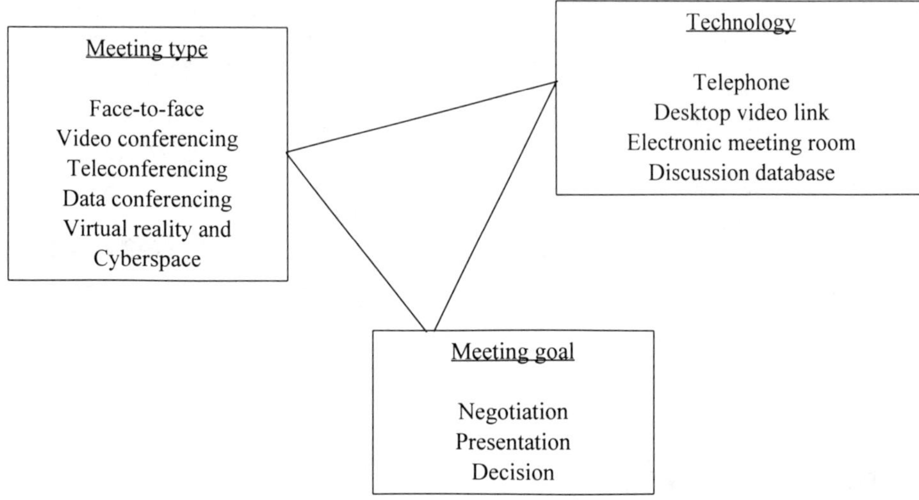

Figure 13.1 Meeting components.

meeting activities. This chapter describes some new ways of holding meetings made possible by the combination of computers and communications.

13.2 MEETINGS

The most generic definition of a meeting is a group of people getting together to reach some goal. People can meet for very many different reasons and their discussion can take place in many different ways. Typical reasons for holding meetings include are to:

- Make a decision;
- Discuss a topic to form some common opinion;
- Propose or evaluate an artifact design.

Meetings can be looked upon in the generic sense shown in Figure 13.1. There is the meeting type, a meeting goal, and the technology that supports the meeting. Computer communication technology has, of course, widened the possible kinds of meetings. Whereas most meetings of the past were of necessity face to face, now a greater variety is possible.

Computer communication technologies mean that meetings can now take any of the following forms:

- *Same time/same place,* where participants are supported electronically by things like electronic boards that display the current state of the discussion, keeping of minutes, and tracking of the agenda. Such meeting are often referred to as *electronic meeting rooms.*
- *Same time/different place,* whose goal primarily is to reduce travel by having people at different locations participate at the meeting, but providing them the same kind of support as with the same time/same place environment. These are often called *videoconferences.*
- *Different time/different place,* which is not yet widely used but introduces the new method of having an *extended meeting*

However, the three components shown in Figure 13.1 must be carefully chosen to fit together to realize an effective meeting. This choice is here described in the context of the meeting process.

13.2.1 Meeting Processes

Any meeting-irrespective of its location, the technology used, or location of its members-goes through the following steps:

- Arranging the meeting;
- Planning and preparing for the meeting;
- Conducting the meeting;
- Reporting on the meeting.

Computer communication can be used in any subset of these steps.

13.2.2 Arranging Meetings

Arranging a meeting usually means selecting the people for the meeting and contacting them to find a common meeting time. Most often this is done by someone contacting the people via the telephone, often a number of times before a mutually convenient time is found. The arrangement of meetings themselves is now often supported using computers. Firstly, for formal enterprise meetings, the computer can contain the lists of designated people who attend these meetings. Where a special meeting is needed, the computer can be used to select the most appropriate people given the meeting goal. Meeting times can also be electronically arranged using people's calendars and meeting schedules, which are made available online. The availability of such calendars on the network is a means of fostering collaboration in itself. Everyone is then aware of what everyone else is doing, thus making people feel like a team.

13.2.3 Planning and Preparing

The person responsible for the meeting can use the computer network to distribute any papers and information to participants, using e-mail. Meetings are most commonly characterized by their agenda, which defines the topics to be discussed.

13.2.4 Conducting Meetings

It is in the conducting of meeting that differences arise from face-to-face meetings. Now protocols are needed to define the way people contribute to the discussion. Important issues are the protocols used in the meeting, the way that information is managed, and decisions made.

Perhaps, after face-to-face meetings, the next most common meeting is a teleconferencing facility that uses telephones. The group has a speaker at the center of the table. The distant members speak over a phone. Anything spoken in the room is captured by the speaker and can be heard by the distant members. Anything spoken by the distant members is heard over the speaker. Even this simple extension requires a protocol. There must now be a conscious step to involve the distant speaker. The chairperson often indicates the times when these speakers are to speak and what is expected of them. Such meeting protocols are needed for most meetings.

13.2.4.1 Meeting Protocols

Protocols define the sequence in which people contribute to a meeting. In face-to-face meetings, protocols tend to be relatively informal. People contribute spontaneously to the discussion, or the chairperson maintains order by sequencing contributions to reduce simultaneous discussion getting out of control. Spontaneous contribution is not often possible in other kinds of meetings, and a formal protocol is needed. Any such protocol must ensure that everyone can contribute to the meeting and that everyone is aware of what is going on. The chairperson may raise an issue in some way, and then call for comments. This can be done by polling each person in sequence or by allowing them to make entries in some commonly accessible discussion database.

13.2.4.2 Managing Information

Again new opportunities arise in the way information is managed during meetings. In face-to-face meetings, most people have their set of papers and can make notes on them. Alternatively, there may be a blackboard or whiteboard where someone gathers comments and records them. Electronic meetings add to the variety of ways of conducting meetings. In electronic meetings, every participant has direct access to a computer and information can be displayed on that computer. It then becomes possible to change documents during the meeting, with the change immediately displayed to all participants. Thus, for example, a budget may be displayed to meeting participants and changed electronically during the meeting. Different alternatives can be discussed and one of these chosen during the meeting, with each alternative instantaneously displayed to all participants.

13.2.4.3 Making Decisions

Another dimension to meeting is the way in which decisions are made. In face-to-face meetings, it is often the case that a persuasive argument may gain the floor and steer through a decision while avoiding quantitative data, even if it is not fully supported. This becomes more difficult with electronic meetings where decision support tools can be easily used. Decision processes in such meetings can be supported by anonymous voting, easily displayed models, and decision calculations.

13.3 VIDEOCONFERENCING

Perhaps the simplest extension to a face to face meeting is a videoconference. Here, people are at different locations but they appear as video pictures on each other's computer screens. There are often technical limitations in the amount of movement that can be displayed in videoconferencing systems. Transmitting moving pictures requires wide bandwidth channels, which can be very expensive.

One of the most common forms of videoconferencing support is to simultaneously display an artifact for discussion at all site locations. More support is provided by letting participants make changes to the artifact, which can be simultaneously seen by all the meeting participants. Thus it becomes possible to discuss a design electronically, evaluating various options. Discussion about changes is made possible by a voice link that can be used simultaneously with the video link. Often this voice link can be a telephone.

The number of tools for task support in videoconferencing is steadily increasing, as are the functions that they support. Some, like Sketchpad, support the display drawings at a number of locations, allow them to be changed during discussion, and keep track of versions for later discussion. Effective videoconferencing, however, requires a seamless platform to enable people to easily move from one tool to another. Methods of constructing such platforms were described earlier in Chapter 9.

One important consideration in electronic meeting is whether they should simply follow the same process as face to face meetings, or whether they should encourage new meeting practices. Most electronic meeting support systems follow the face to face practices.

In a videoconferencing meeting, the participant is presented with a screen that contains the information shown in Figure 13.2. This shows images of the

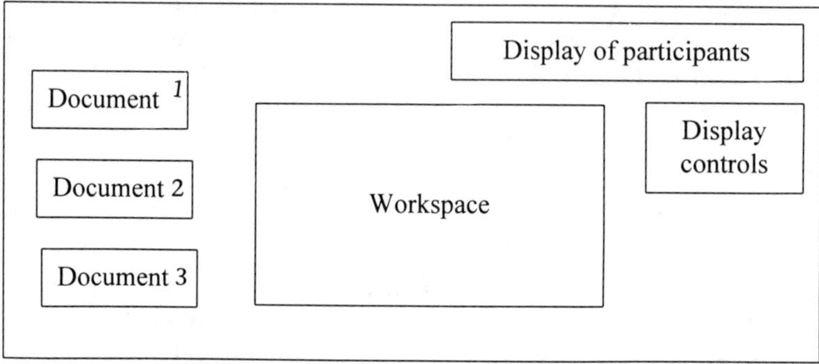

Figure 13.2 Electronic meetings.

other participants and a display control for the video and audio. A number of documents can be retrieved and simultaneously displayed at all terminals, with parts of the documents moved to the workspace and changed by joint agreement of the participants. These documents can include parts of the corporate database, meeting records, and any other relevant documentation. Such a screen could be set up using a toolkit such as GroupKit, described in Chapter 10. With large groups, a facilitator is often chosen to control the meeting protocol. Often, the protocol simply gives the floor to the participants in some sequence, and they can make their contribution when they have the floor.

13.4 ELECTRONIC MEETING ROOMS

Electronic meeting rooms are same time/same place face-to-face meetings that provide more meeting structure and support. Usually, participants are arranged around a table in the way shown in Figure 13.3. Each participant has a terminal in front of them to show any documents under discussion. The meeting participants can thus access information easily through workstations that are made available at each position, with any changes simultaneously displayed to all participants. The facilitator controls the meeting by following an agenda and displaying selected information on a large screen for discussion. Results of the discussion can be recorded by the facilitator and displayed on all screens for agreement. This allows proposed changes to be made very quickly, displayed, and quickly evaluated. The facilitator can use a number of tools, such as brainstorming or voting tools, to facilitate this process.

A number of electronic meeting room systems have been built to support same time/same place electronic meetings. These include the following.

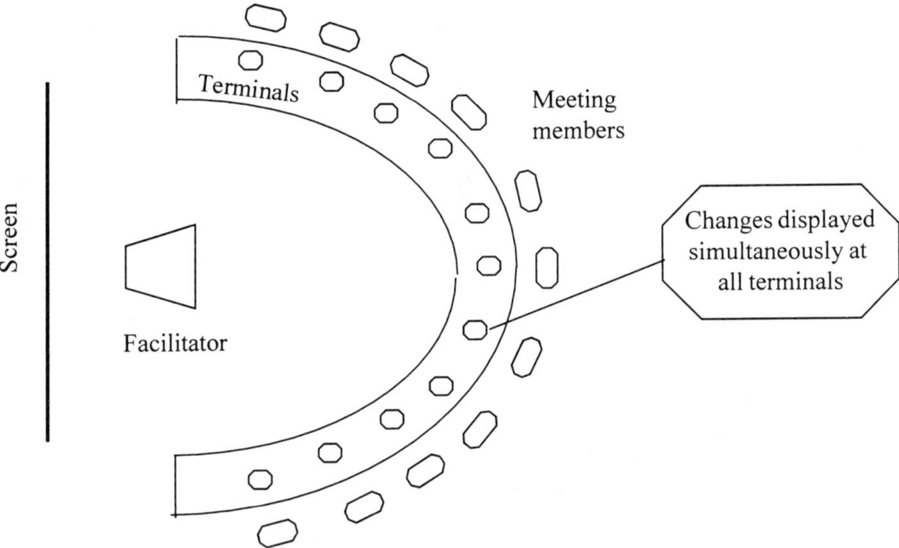

Figure 13.3 An electronic meeting room arrangement.

- The University of Arizona and called GroupSystems V [2].
- The Capture Lab at the University of Toronto [3].

Mantei, in her work on Capture Lab, has found that the arrangement of the meeting room itself is important in order to supplement the structured inputs with face to face discussion. It was also found important for the computer not to have a physically prominent place, and the usual arrangement is for the computer to be built into the meeting table. Another important consideration was the process used to give users access to a shared workspace, which in the case of Capture Lab is a screen in front of the table.

13.4.1 Providing Task Support in Electronic Meeting Rooms

The kind of services provided in meeting rooms is illustrated by GroupSystems V. It provides users with task support that includes such tools as electronic brainstorming, idea organization, voting, and retaining comments attached to ideas. Most electronic meeting rooms often contain a very large range of facilities. For example, some of the facilities in GroupSystems V are as follows.

- A brainstorming tool that allows participants to share ideas simultaneously and contribute anonymously to a particular question.

- A briefcase or filing system that participants can electronically bring to the meeting.
- A voting tool to enable participants to electronically vote on an issue with a variety of ways for balloting. These can include ranking, a Yes/No response or a 10-point scale.
- A dictionary for defining a consistent set of terms.
- A tool for organizing ideas by organizing comments and generating lists of the ideas together with their comments.
- A tool for categorizing comments made on a particular issue or idea. It can be used in conjunction with other tools, such as brainstorming, to generate lists of comments ordered in a specified way.
- A tool for ranking alternatives.
- A questionnaire that elicits participants' responses on various types of issues during a session.
- A policy formation tool that can be used to gather and edit information for a final statement by iteration through the participants.
- A group writer that is an editing tool, which allows members of a group to work simultaneously on a document. However, it only allows one participant at a time to work on a particular section, although other participants can view this work. Control over the section can flow from one participant to another.

GroupSystems V has been installed at a number of locations and there is considerable experience in its use.

13.4.2 Providing Process Support in Electronic Meeting Rooms

The electronic meeting must support the process by often introducing formal protocol rules for people to contribute to the meeting. Participants are *given the floor* in some way, and once they have the floor, they make a contribution to the discussion.

- Include a protocol for people to take turns at talking;
- Can provide private workspaces with better information support;
- Keep track of the conversation;
- Can focus meeting through provision of structured ways of discussion.

Protocols in electronic meeting rooms usually require some participants to take on well-defined roles. Mantei has described three approaches to meeting room protocols, namely,

- An *interactive* approach, where all participants have access to a keyboard when they wish to contribute to a meeting;

- A *rotating scribe* approach, where responsibility for making entries rotates between participants;
- A *designated scribe* approach, where one person makes all entries.

13.4.3 Some Outcomes

It is found that it is not possible to totally reproduce face to face by electronic meetings, and the human factor is now not so prominent. A study by McLeod [4] has indicated some differences between face to face and electronic meetings. These include the following:

- There is usually less personal identification with the outcome;
- There is usually less consensus on the outcome;
- The outcome is often one of higher quality then face to face meeting;
- It usually takes longer to come a conclusion;
- The meetings are more focused.

Some of these appear to conflict. For example, the decision is often one of higher quality reflects the idea of many people rather than those of only some. Consequently, no one seems to "own" the decision, leading to less personal identification, and sometimes less satisfaction, with it. Other supporting mechanisms here are as follows.

- Better premeeting support through collection of preliminary comments; meetings can then become more focused;
- Introduces more structure, thus reducing the possibilities of long meetings with no useful outcomes;
- Supports anonymity, thus bringing out more new ideas.

13.5 EXTENDED MEETINGS

Some people (Turoff, 1993) [1] have questioned the need to place time limits on meetings. Why should we not be able to contribute to a meeting at any time, especially when we have a good idea? Often, such ideas get lost over time, or because they simply do not fit into a formal meeting agenda, or because other people tend to dominate the meeting. Most of the tools described earlier in this chapter can be used to support extended meetings. However, what is needed is to have a more formal structure to keep track of meeting discussion. This structure must be adapted to the kind of meeting. Thus Figure 13.4 illustrates a meeting called to discuss an issue. A structured electronic discussion is maintained on this issue recording people's positions and arguments on the positions. A facilitator is often needed to maintain the discussion database and assist people to use it effectively.

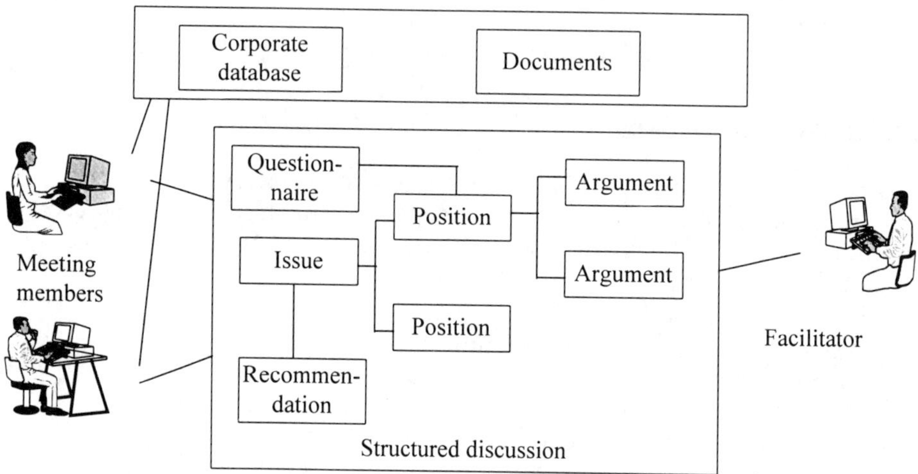

Figure 13.4 Support for extended meetings.

Additional facilities often needed for extended meetings, including summarizing comments, questionnaires, seeking votes, distributing recommendations, and so on, take place over an extended period of time. These again can be the responsibility of the facilitator.

13.5.1 Combining Extended With Face-to-Face

Many advantages are seen in combining extended with face-to-face meetings. The extended meeting can be used to collect information and consolidate people's positions prior to having a face to face meeting. Face-to-face meetings can then concentrate on debating the issues rather than simply airing and collecting information. This can often make face-to-face meetings more productive as well as reducing the number of such meetings. The sequence shown in Figure 13.5 can be effective in improving meeting productivity. Here, there is an initial face to face meeting to determine what is to be achieved. A discussion database is then constructed by the facilitator to match the meeting goal. Information is then collected in the discussion database, with face to face meetings called whenever conflict is to be resolved. Facilitators are quite important here as they guide the course of the meeting, deciding how to structure discussion and when to call face to face meetings. They also provide the support tools and document access needed to make these meetings effective.

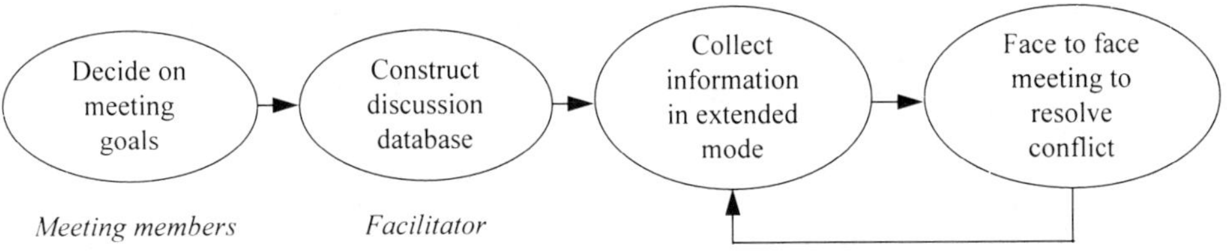

Figure 13.5 Mixing face to face and extended.

13.6 SUMMARY

This chapter described the different ways in which people get together to discuss some issues. It introduced a different meaning to the term *getting together*. Rather than restricting meetings to people physically present at the same location, it extended the idea of meeting to one of communicating to exchange information with participants physically distributed. The chapter then outlined the support needed for electronic meetings.

One of the most important aspects of conferences or electronic meetings, especially extended meetings, is to organize the discussion and arguments in a way that results in a successful meeting outcome. Such organization of arguments is often known as *discussion*, or *argumentation*, or *design rationale*, depending on the type of meeting. It is often the responsibility of the meeting facilitator.

13.7 DISCUSSION QUESTIONS

1. Why must protocols be more precisely defined for electronic meeting?
2. Are there any advantages to meeting participants being anonymous?
3. Do you think that electronic meeting systems directly model face-to-face meetings?
4. Name some task support tools useful in electronic meetings.
5. What do you understand by an extended meeting?
6. Do you think that the need to hold face-to-face meetings will disappear over time?

References

[1] Turoff, M., "Computer-Mediated Communication Requirements for Group Support," *Journal of Organizational Computing*, Vol. 1, 1991, pp. 85–113.

[2] Nunamaker, J. F., et al., "Electronic Meeting Systems to Support Group Work," *Communications of the ACM*, Vol. 34, No. 7, 1991, pp. 39–61.

[3] Mantei, M., "Observation of Executives using a Computer Supported Meeting Environment," in *Decision Support Systems*, Vol. 5, Elsevier Publishers, B.V, 1989.

[4] McLeod, P. L., "An Assessment of the Experimental Literature on Electronic Support of Group Work: Results of Meta-Analysis," *Human-Computer Interaction*, Vol. 7, 1992, pp. 257–280.

Selected Bibliography

Dennis, A., et al., "Task and Time Decomposition in Electronic Brainstorming," *Proc. of the 29th. Annual Hawaii Conference on System Sciences*, Hawaii, 1996, pp. 51–58.

Lewe, H., "Computer Support and Facilitated Structure in Meeting - An Empirical Comparison of their Impact," *Proc. of the 29th. Annual Hawaii Conference on System Sciences*, Hawaii, 1996, pp. 24–33.

Part C

The Design Process

Chapter 14

Building Cooperative Systems

LEARNING OBJECTIVES

- ❑ *Alignment*
- ❑ *Evolution of CSCW systems*
- ❑ *Strategic CSCW design*
- ❑ *CSCW design methodologies*
- ❑ *Soft systems methodologies*
- ❑ *Platforms for evolution*

14.1 INTRODUCTION

One theme that this book introduced is that shown in Figure 1.6. This was to emphasize that the design of networking systems must begin with an understanding of how people work and then continue by choosing services to support this work. So far the book has described some ways in which we can describe work practices, particularly in terms of group culture and structure, and given a number of examples of such practices. The book has also described a large number of services. It now remains to define the design process itself.

This chapter begins to describe design. Readers should note that in the past many cooperative systems have evolved rather than been strategically introduced. This has been the case with organizations as well as cooperating individuals from different organizations. With individuals, this usually begins by using networks to improve personal communication across distances. In organizations, a typical scenario is where an operating unit becomes familiar with some group support software, tries it, finds it offers some advantages, and begins to use it. Systems, in this case, are usually developed quite informally. Development starts with a brief description of what is needed and then uses group support software to meet this need. Other people see benefits derived from the use of this software and begin to apply it in their work.

However, this approach is slowly changing as enterprises begin to realize the advantages that can be gained through better communications support. For this reason, some organizations have considered an alternate approach of introducing an intranet with a platform of services to improve enterprise work practices. Such platforms were outlined earlier in Chapter 7, and some commercial examples were given in later chapters. The platform in this case should be flexible and allow users to change the way they use it as they learn more about it. Such platforms must not only provide the technical tools needed by people in their work, but also support evolution to new social structures using the new technology.

This chapter describes these two alternative approaches and then outlines methods that can be used in design.

14.2 DESIGN TO ACHIEVE ALIGNMENT

Design of collaborative systems differs significantly from the way that the transaction-oriented systems are designed. One way to explain this difference is in terms of alignment of technology towards some organizational need. A number of alignments have been identified and are shown in Figure 14.1. One alignment is from technology to task, to simplify and make the task more effective. Technology to task alignment is probably the simplest kind of alignment and characterizes transaction system design. Here, the task can be precisely defined in functional and deterministic ways. The inputs and outputs can be defined and technology to use the inputs and produce the outputs chosen. There are often indirect effects on the culture, usually through the redesign of jobs.

Another alignment is technology to the culture, which requires technology to support the ways in which people work together, formally and informally, in the collaborative environment. E-mail is a good example here. We can introduce e-mail as an added feature to support the informal communication processes between users. Users are free to choose to use it, and can adapt its use to their particular situation. Thus, in some cases they may use it for communication on a person to person basis, or they may decide to use it to collect comments, or to distribute information to selected groups. Social needs are thus satisfied as the technology is adapted to the way the group wants to work.

Network design for enterprises usually emphasizes alignment of technology to the culture and the organizational structure. It differs in that it places great emphasis on the way people work in detail, and how they interact whether working on the same artifact or in the process of making a decision. Network design thus

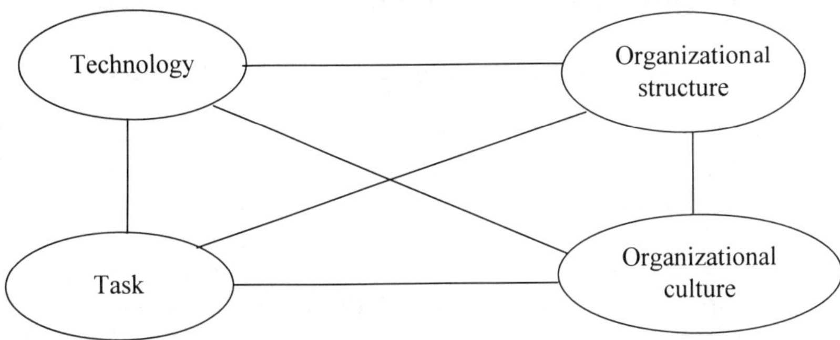

Figure 14.1 Technology alignment.

emphasizes people and not tasks, as is the case in most transaction-based design. Analysis must thus begin with people and what they do.

New systems must then support the new information flows expected in the organization, while at the same time distributing information and supporting communication consistently with the culture. Thus, in a hierarchical structure, systems must support flow between hierarchical levels. This may include collection of information needed in the development of plans, approvals for activities, or assignment of tasks. At the same time, informal flows between individuals carrying out related tasks are also valuable to expedite common projects. Such lateral flows will only be supported if they are consistent with the culture. In open groups, on the other hand, information flows may be quite different. Now, tools for collecting ideas, discussion, and joint decisionmaking will be more important, and everyone must be aware of what everyone else is doing.

14.2.1 Design for Growth

There is also another difference in designing cooperative systems. Whereas transaction-based systems are relatively stable once put into use, cooperative systems continually evolve over time-new technology becomes available and the social culture changes, as does the nature of work. However, as a rule, people often do not want sudden changes to their work practices, but want to gradually learn to use technology better as they learn about it. Thus it is often necessary to begin with a system that simply supports existing work practices, but ensure that any supporting systems can naturally evolve gradually with changing social factors, tasks, and technology. As a result, research [1] in this area has shown that the design should produce a system that has little impact on existing practices but provides a new basis for growth. The designer is thus faced with a complex environment. Not only must they have initial minimal impact, but also they must provide a platform for growth.

14.3 TWO BROAD ALTERNATIVES FOR NETWORKING ENTERPRISES

There are two alternative ways for implementing cooperative systems in organizations, namely:

- The *evolutionary* (or bottom-up way): Here, increasingly complex functionality is introduced as organizations become familiar with new technologies. This bottom up introduction can happen by default as operational units adopt new techniques. Usually it proceeds through the connectivity levels described in Chapter 2, beginning with emphasis on information exchange. The idea here is that the social structure changes gradually as people learn to use the system.
- The *strategic* approach: Here, there is conscious re-engineering of the enterprise towards cooperative systems. The enterprise makes a conscious decision to improve its communication and coordination processes by enterprise-wide changes. Strategies are introduced early and define the goals to be achieved by cooperative networks, the responsibilities for individual units, and the technical strategy, especially the structure of internal intranets and their links to public networks. Here, the learning is more intense with cooperative systems introduced with a well-defined goal, related to the organization's critical mission.

14.3.1 The Evolutionary Approach

The evolutionary approach is based on one of the principles in Chapter 3-the introduction of groupware systems should have minimal initial impact on the group, but should allow the group to introduce subsequent changes as they see ways to improve their work practices with the software. Such change, however, depends on the group culture and may itself affect that culture. Thus, for example, highly structured organizations may not support changes without approval, thus affecting both the rate of change and its nature, which usually has to mirror the structure. Open groups are likely to be more experimental and introduce more radical and experimental changes more quickly.

A number of stages can be identified for the evolutionary approach, almost like Nolan's stages for the whole organization with gradual movement to the higher stages. The stages are described in the following sections.

14.3.1.1 Stage 1—Information Services

Here, users get access to sophisticated information services, and generally use these services in information retrieval. However, provision of such service should go beyond simply providing users with access to volumes of information. Having such access can simply mean that users spend a large amount of time searching through this information to find items of value in their work. Such access must also include assistance for simplifying the search by notifying users about changes to items of interest to them. Alternatively, software agents can be developed to assist information searches.

14.3.1.2 Stage 2—Networking for Information Exchange

Services are now extended to enable users to connect to each other and exchange information using computers. The kind of information services provided may include distribution of information, discussion databases, bulletin boards, and so on. Information may include informal messages or transfer of documents, and can often be met through the provision of e-mail services. Again, where networking goes beyond the simple exchange of messages to asynchronous collaboration, a notification scheme may be needed to maintain awareness.

14.3.1.3 Stage 3—Provision of Generic Collaborative Services

This stage improves communication by providing additional generic applications such as teleconferencing. Usually, the networking introduced in stage 2 is extended by mail-enabling desktop applications such as word processors for document transfers. Such enabled applications follow closely some business process and their goal is to support tasks in that process. Activities in this stage include

- Selecting the artifacts to be made available in the workgroup;
- Defining the collaboration roles;
- Defining a function to be applied to the artifact and roles responsible for its function;
- Defining the workgroups;
- Supporting discussion and design.

The design often results in a number of databases and documents accessible to workgroup members. The process of using them is to some extent ad hoc—usually a change to one database results in a notification being sent to one of the members.

14.3.1.4 Stage 4—Application Development

Usually, stage 3 introduces the typical collaborative generic services in the organization (for example, collaborative document preparation). However, it does not extend those services to business applications. This occurs in stage 4, which begins to introduce application development, often through the redesign of existing applications or the addition of new databases using tools such as LOTUS Notes. It will typically follow steps such as those described below.

Step 1

The first step is used to develop or redesign typical applications, usually those that simplify day to day operations and may be

- Customer tracking;
- Distribution of information, such as catalogs;
- Scheduling of business activities;
- Inventory tracking;
- Provision of assistance;
- Executive information systems including strategic plans;
- Tracking cases in legal systems;

Redesign here includes development of a database that stores the information as a set of artifacts needed by the application. It includes methods of presenting the information at the interface in ways that enable them to act on the presented information in natural ways.

Step 2

Step 2 is used to extend applications by formally defining the processes to support a number of team members. To do this, it is necessary to:

- Define how group members are to distribute information about artifacts;
- Provide supporting communication applications such as electronic conferencing or electronic mail.

The result is usually a closer integration of the applications. It is also usual to superimpose collaboration technologies onto existing systems here by, for example, providing teleconferencing facilities or supporting meetings to discuss actions that involve a number of artifacts or applications.

Step 3

Step 3 is used to define the communication protocols in more detail. Here, detailed and formal workflows are defined where applicable. As well as workflows, this may include notifying team members of events in the system and maintaining schedules such as, for example, to-do lists for team members.

14.3.1.5 Stage 5—Organization-Wide Systems

The final stage is to provide a platform for organization-wide use. This platform must both integrate a number of collaboration services with each other and use them to integrate the applications developed in stage 4. The services must be aligned with the organization's mission using workflow support. People in the organization can select the services for their particular task and can use new services whenever their task changes. The enterprise-wide platforms will provide services for workgroup formation and support distant work. Provision of organization-wide systems may require an application development policy to set priorities and assign responsibility for the applications. The policy will also define how workgroups should be formed. Choices will have to be made on how to form people into groups and the communication support to be provided to each group.

14.3.2 Strategic Enterprise Introduction

The strategic approach usually follows a decision made at board or executive level of an organization to introduce information technology to support teamwork in achieving the organization's mission. The way the strategy can be formulated is shown in Fugure 14.2.

- The goals to be achieved by the cooperative systems in the activities found in business processes, distinguishing between business to business networking, consumer networking, and support for internal processes;
- The technical strategy in terms of how the network services needed to improve information exchange, interpersonal relationships, and work process support in the activites, and the degree of integration with corporate databases;
- The responsibilities of individual units in implementing parts of the network and integrating it with their operations;
- The division of responsibility for the network between individual enterprise units and the enterprise itself, with support to be provided by the enterprise to the units.

One major difference from the evolutionary approach is that units no longer pursue their own plans, which can often lead to systems that cannot be easily integrated into corporate-wide systems. Instead, broad strategies are set and enterprise-wide standards set early to reduce later integration problems. Unit goals are clearly defined, and units themselves are concerned with defining projects within the enterprise-wide strategy. There is, however, a certain element of risk involved in that sufficient flexibility must exist in the plan to cater for rapidly changing technology.

The strategic approach must emphasize the objectives of cooperative systems early in terms of the three major elements: information exchange, interpersonal relationships, and work practices and their relationship to business goals. An approach that only stresses the technical strategy without carefully analyzing its purpose and making this purpose widely known within the enterprise should be avoided. Often, the appearance of such systems may be totally differently interpreted [2] by the organization's personnel from its original intention. Such misinterpretation may often lead to casual nonproductive use, or, alternatively, people

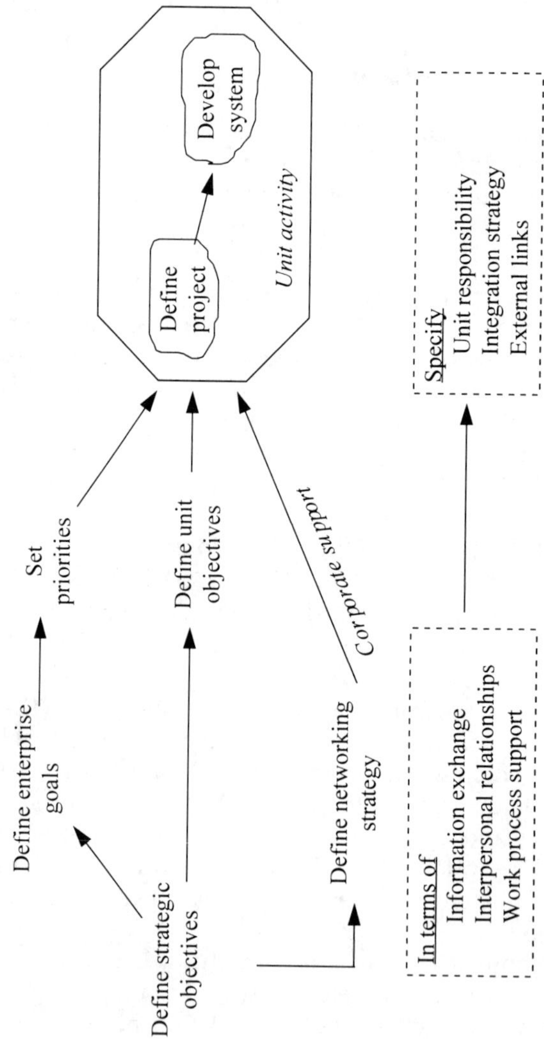

Figure 14.2 Elements of strategic planning.

may just avoid the effort to use the new technology. Consequently, there must be some emphasis placed on developing common terms of language that can be used to discuss collaboration between users and providers in meaningful and unambiguous terms. Part of the design process described in later chapters will be to define such a language.

14.3.2.1 Technical Strategy

The major elements of technical strategy usually include

- The distribution of systems between public networks and internal intranets;
- The intranet strategy; in particular, will it be an enterprise-wide responsibility or will units develop their own intranets that can be integrated as needed;
- A statement about the level of corporate support and its nature, whether simply financial, or including core technology support, or providing integration services;
- The method of integration between enterprise units and with existing legacy systems;
- The core technologies to be used.

The trend now is to introduce cooperative systems within enterprises through an intranet. The major choice here is how to distribute responsibilities between units and the core technologies. Some possibilities were described earlier in Chapters 9 and 10. Then the kind of services to be provided must be carefully considered. Simply providing elementary level 1 services such as e-mail does not mean that it will be used effectively in meeting the organization's mission. On the other hand, all the organization's personnel must be informed of the business reasons for introducing an Intranet. They must also be trained in its use. This book thus suggests a systematic approach to the strategic introduction of intranets, whether at unit or enterprise-wide level. This is to analyze the kinds of services using the organization's mission, to define precise unit projects as shown in Figure 14.2. These services should then be provided on a platform based on the core technologies defined by the technical strategy. Furthermore such services should be produced gradually and users educated on how to use them in their everyday work. Ways of carrying out such analysis are suggested in the following chapters.

14.3.2.2 Prototyping

It should also be mentioned that the strategic approach itself is implemented through prototyping rather than direct introduction. Prototyping follows the iterative cycle shown in Figure 14.3. Implementation begins by installing an initial system. People are trained in using the system, perhaps through a workshop, and

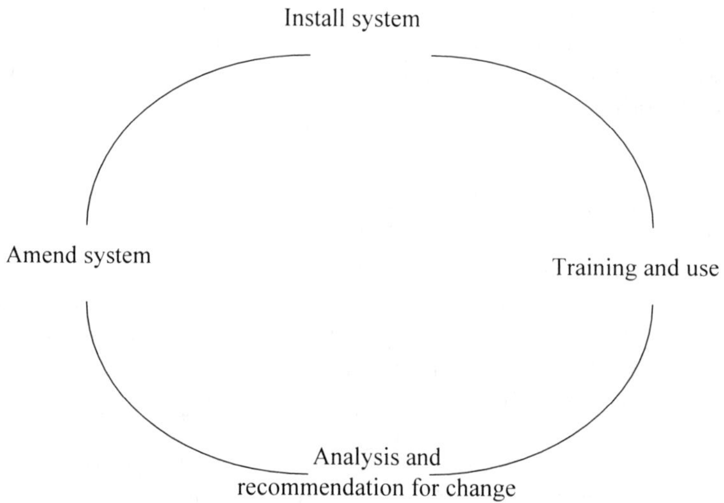

Figure 14.3 Prototyping.

experience is gained with it in the field. Changes are then suggested and made, and the cycle repeated. Such a prototyping approach is needed because of the difficulty of precisely specifying the requirements of supporting systems given the detailed nature of interpersonal relationships. The broad goal can be defined in a relatively straightforward manner, but the detail nature of work calls for gradual introduction and experimentation to match the work practices.

Case Study: Global Planning Support

A strategic approach to developing a communications network must consider the needs of the communities that will use this network. Development of a strategic approach to support global planning is initially concentrating on gathering the requirements from these large distributed communities. It is not feasible to hold frequent meetings to gather such requirements. An alternative to using electronic means to collect information is thus being developed. The idea is to simulate an extended meeting by creating an issues agenda with issues recorded against the agenda and comments on issues from all potential users recorded. This is achieved by using a WWW site. Annual workshops are than held to develop the strategic plan using the gathered comments.

Information required for the strategic plan is gathered in the context of the organizational strategic issues, in this case improvements to health status, to ensure that information technology services contribute to this goal. Thus

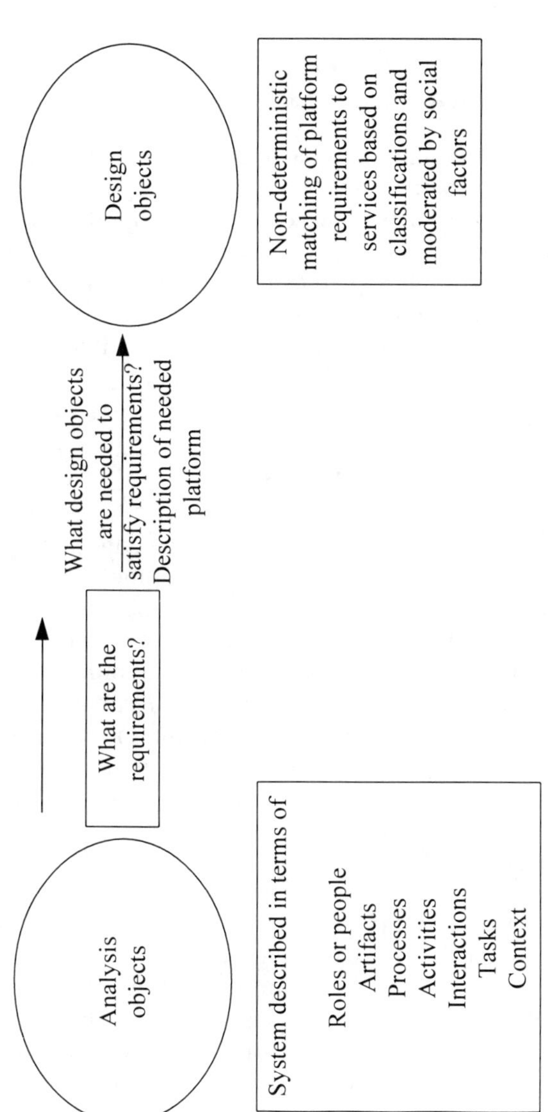

Figure 14.4 A view of design.

technology issues are raised in the context of health issues to maximize such contribution.

14.3.3 Comparing the Approaches

To some extent the evolutionary approach is less risky as systems are introduced gradually as people learn about them. The evolution begins with aligning technology to tasks, with social alignment proceeding gradually. However, the second approach if properly managed, will give an organization additional competitive advantage much faster by providing an intranet with enterprise wide applications. The question here is whether we can go directly to stage 5. To do this we need some methodology that can identify social as well as technical needs, and what is more to integrate the two.

14.4 METHODS FOR DEVELOPING SYSTEMS

Cooperative systems development commences once enterprise priorities are defined. Perhaps the most important part of this development, especially where level 3 support is to be provided, is to understand work in greater detail. It becomes important to identify existing work practices, suggest ways of improving them, and then provide a platform of services that supports the improved practices. Going into the details of work practices can, however, often result in loss of direction and a random approach to design. For this reason, it is important to be able to eventually introduce a *systematic approach* into the design process. This systematic approach is introduced here, but will be described in detail in the following three chapters. We can begin to describe the systematic approach using the simple idea shown again Figure 14.4. This idea comes from the design of transaction-based systems, where systems analysis defines the objects in our work system. Designers then convert these objects to a design made up of design objects.

A view of design like that shown in Figure 14.4 has proven very useful in designing transaction-based systems. For example, entities and relationships are the analysis objects in database analysis. The design objects can be files or databases. Designers are often provided with well-defined mappings to convert the analysis objects to design objects; that is, to convert an entity relationship diagram to a set of files. The important lessons that come from transaction based design are as follows.

- Unambiguous representations for work and design objects;
- A sequence of steps followed in the design;
- A design method and techniques used when going from step to step;

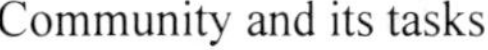

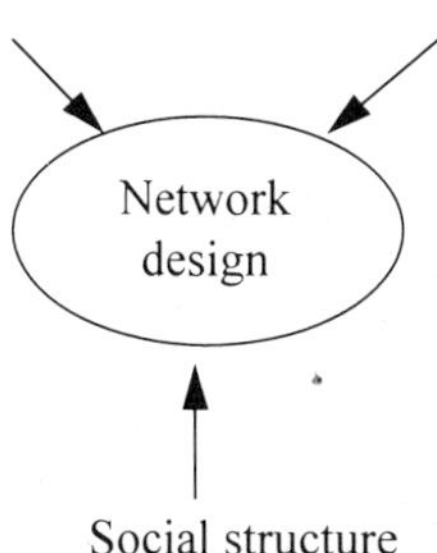

Figure 14.5 Major influences on design.

- Tools to support these methods and techniques.

However, such lessons must be modified to suit the nature of collaborative processes, which are different from transaction systems. In transaction-based systems, activities can be precisely specified and conversions usually deterministic. Most transaction systems have a fixed set of transaction types that are used to predefine a sequence of steps. This is not usually the case with CSCW systems, where the same input can result in different outcomes. Here, work processes themselves are not precise and can change from day to day. Furthermore, transaction-based systems concentrate on tasks rather than people, again making a significant difference in early analysis.

What is needed to design a cooperative system is a systematic method for choosing the best services for a given set of requirements. This choice must be based on a language or set of terms that can be used to rationalize service selection considering technical factors but moderated by social factors, which usually call for rapid changes to requirements as social factors change. Thus design must include another important feature-using social processes as a moderator in choosing design objects and supporting flexibility on the platform. It must also include representations that can meaningfully represent the complex interactions that characterize interpersonal communications.

The design method itself must include the three important influences shown in Figure 14.5, namely to identify

- The community to be supported and the important tasks in the community and interactions between community members,
- The social culture of the community, including its social structures, which also will impact on the processes to be provided by the system;
- The available services and how to use them.

Furthermore, as all of these components change over time, designers must ensure that user applications can naturally evolve gradually with changing social factors, tasks, and technology. A systematic approach to design needs techniques to identify these factors and choose the best services for them. Users should then be able to combine these services in ways that match their work practices.

14.4.1 Soft Systems Methodologies

There has been considerable work, especially at the University of Lancaster [3], on soft systems methodologies. Soft systems design is premised on the idea that prescriptive and structured approaches are not useful for designing systems that are not well-defined. The emphasis here must be on developing an understanding of the systems and their central issues. The methodology stresses seven stages, namely,

- *Stage 1:* Identify the problem situation.
- *Stage 2:* Express the problem situation.
- *Stage 3:* Define the relevant systems, identifying the major concepts that can be used to describe the system.
- *Stage 4:* Develop a conceptual model that uses these concepts to describe the relationships between the system components.
- *Stage 5:* Verify the conceptual model against the problem situation in stage 2.
- *Stage 6:* Define feasible desirable changes.
- *Stage 7:* Take action to improve the problem situation and repeat steps if needed.

Details of these steps can be found in the literature [4]. What is important here is that the soft systems approach stresses the importance of getting a better understanding of systems. The first two steps are concerned with understanding and expressing the problem. Soft systems methodologies, at least in their initial stages, do not require detailed specifications of how systems work, but require a broad view of system operation. A proposed tool for this purpose is the rich picture, and a number of such pictures have appeared in earlier chapters.

The next two stages call for a more precise definition of the system. There is not a prescriptive way of going towards this definition, but tools such as the rich picture assist in searching for a better structure. The main goal is to develop what is known as root definitions, or concepts, that can be used to generalize the model, and then create a conceptual model that expresses relationships between the concepts. The next two chapters will describe some concepts that can be used for this purpose.

14.4.2 Representations That Include the Social Process

One important observation as far as group support is concerned is that there is no known way of analyzing their processes. The usual way of interviewing people is not satisfactory as it is difficult to describe in an off-site interview how a highly dynamic group process works. Try, for example, to describe the meeting you went to last using some formal model.

Even soft system methodologies may not be easily adapted here as they more easily express static relationships rather than dynamics, which must usually be expressed in natural language attached to rich picture components. Rich pictures are also complemented with scenarios that illustrate the most common activities in the system. These representations are described in detail in the next two chapters.

14.4.3 Combining With a Set of Steps

Perhaps the strongest suggestion from transaction system design is to define a problem-solving process that provides a systematic approach to design. Although we cannot adopt the same process that is used in transaction-based systems, we can abstract the general problem-solving steps needed in design. The following four steps have been found extremely useful in practice:

- *Physical analysis*: to develop an understanding of the system, through interviews, observations, rich pictures, and other tools. This phase determines how the system works now.
- *Logical analysis*: to describe the system in unambiguous terms meaningful to all people in the process. The goal is to determine what is being done.
- *Logical design*: to develop a set of logical requirements by specifying what is to be done and the kinds of service objects needed to support teamwork.
- *Physical design*: to propose how the logical requirements will be satisfied by choosing the services needed to realize the collaboration.

It is then important to provide the tools needed to support this process. Rich pictures are often suitable for physical analysis. When logical analysis begins, it becomes more important to use precise constructs and language to describe the collaboration process and its requirements. This will create a conceptual model that includes concepts or a classification scheme that describes system components, but not necessarily their detailed operation. This step can be made more effective by having a set of concepts that are generally applicable to CSCW systems such as, for example, the interactions described in Chapter 3. The next step is to specify the services to implement the system. Again, the classification schemes described in Chapter 7 become useful, as we can specify the repositories, communication services, and processes to realize the requirements. The final step is to select the services to meet the specification. Design can then be seen as a mapping from the language constructs to the platform services to provide an easy to use platform of services.

The mappings used to go between the stages are not deterministic, as is the case in structured systems. Design is thus based more on guidelines rather than on deterministic conversions. The resulting methodologies will have substantial differences from transaction-based design. These differences are described in the next few sections.

14.5 TESTING THE SYSTEMS

A chapter on system development would not be complete without reference to how systems are to be tested. Testing is needed to assure that a system will work without error once it is put into practice. Transaction-based systems are often tested by initiating a transaction and seeing if it results in changes defined for the transaction. Usually, transactions do not require any specific responses. Testing of cooperative systems introduces another dimension into this process. It is that testing must include people at more than one end interchanging information in nondeterministic ways. Testing thus becomes much more difficult because it requires a number of persons to be involved in the test, with a large number of communication patterns between them.

Testing must distinguish between technical and social acceptance. One way to test CSCW systems for technical correctness is by using typical scenarios, which are often identified during analysis. This is often referred to as system testing. Roles identified in these scenarios must now be taken by people involved in the test, and they should then follow the scenario steps. Social acceptance usually requires evaluation in real situations. Thus, for example, for Workflow Enterprises, scenarios for unit testing and integration testing were defined and a dummy run of documents made for each scenario. This should find technical faults in the system. When system tests are completed, it becomes necessary to carry out a trial to develop a software system to evaluate its social acceptance.

14.6 SUMMARY

This chapter described approaches to designing systems to support networked enterprises. It described alignment as the fundamental issue in design. It then made a special distinction between evolutionary growth and strategic development, suggesting that the latter is more difficult to achieve and requires a methodology. The need to define a strategy that defines enterprise goals and technical strategies was stressed, with importance placed on services to go beyond e-mail and include the broad range of services for information exchange, interpersonal relationships, and work process support.

The chapter made the important observation that network design must provide for systems to evolve as they are used. This contrasts with transaction system design, where little evolution is usually expected. The chapter then described what is meant by a methodology; in particular, the need for a representation, a set of steps, and a process to go from step to step. An important consideration is to *find a model* that can be used in the analysis to describe requirements in unambiguous terms. Such a model must be general, but also include concepts that closely correspond to the user's mental model of collaborative work. All of these aspects will be described in the next chapters.

14.7 DISCUSSION QUESTIONS

1. Why is the alignment of task to technology easier than other alignments?
2. Describe alignments that occur in each of the evolutionary stages.
3. What do you understand by a strategic introduction of CSCW systems?
4. What special steps are needed to ensure successful strategic introduction?
5. Identify the generic problem-solving phases.
6. What should be the goal of a CSCW design?
7. Do you think rich pictures can improve the understanding CSCW systems?
8. What would be a useful set of concepts for describing collaborative systems?
9. Why is it more difficult to test group support systems?

14.8 EXERCISE

Which of the two approaches, evolutionary or strategic, would you suggest for Worldlink Consultants and global planning support.

References

[1] Applegate, L. M., "Technology Support for Cooperative Work: A Framework for Studying Introduction and Assimilation in Organizations," *Journal of Organizational Computing*, Vol. 1, 1991, pp. 11–39.

[2] Orlikowski, W. J., and D. C. Gash, "Technological Frames: Making Sense of Information Technology in Organizations," *ACM Transactions on Information Systems*, Vol. 12, No. 2, April 1994, pp. 174–207.

[3] Checkland, P. B., and J. Scholes, *Soft Systems Methodology in Action* Chichester, New York, NY: John Wiley and Sons, 1990.

[4] Patching, D., *Practical Soft Systems Analysis*, London, UK: 1990.

Selected Bibliography

Olson, G.M., and J.S. Olson, "User-Centered Design of Collaborative Technology," *Journal of Organizational Computing,* I, 1991, pp. 61–83.

Chapter 15

Analyzing Work Practices

LEARNING OBJECTIVES

- ❑ *Analyzing work*
- ❑ *User scenarios*
- ❑ *Ethnography*
- ❑ *Modeling methods*
- ❑ *Rich pictures*

15.1 INTRODUCTION

The previous chapter called for a systematic approach to design. This approach is to start with a representation of how the current system works and then continue to provide support for this system. Finding out how people work in groups is often one of the most difficult parts of design. As such, analysis must go into the detailed ways of how people actually exchange information, the rules they use in discussion, and so on. These are things that one is often not aware of when doing them- how many people can, for example, describe the process that they followed in a meeting? A recent September 1995 issue of the *ACM Communications* [1] was almost entirely devoted to this topic- defining it as making work visible.

Detailed work practice analysis is further complicated by the fact that the detailed practices are often not predefined or documented. The analysis must place greater emphasis on social processes and not just look at how tasks are carried out. Social processes tend to be less structured and must portray the actual interactions of people, their feelings and opinions, the reasons why they do certain things, and how they interpret situations in their environment. The more important aspects of social models are

- Cognitive issues that concern ways that people understand and reason about systems and are often concerned with designing interfaces that improve such understanding;
- Organizational issues and people's relationships within organization structures, and how such relationships affect people's actions;
- Behavioral issues within different group environments and how they effect people's contributions to the group.

Such descriptions must also include the concrete objects that one finds in any system; in particular, the artifacts used by people and the people themselves. It is important in collaborative systems to combine these two views in a way that will lead to improved collaboration.

This chapter will describe some methods of analyzing and representing collaborative systems. It continues the call of Chapter 14 for a good system representation and for a "language" composed of well-defined terms to describe CSCW. It also outlines some representations that are have been proposed earlier by others for describing systems.

15.2 FINDING OUT ABOUT SYSTEMS

Analysis of work practices must capture the essence of what is being done, why certain actions are taken, and how people communicate. It must also go further and not only define the interactions between people, but also show how these interactions fit together into processes. The term *communities of practice* often appears. It denotes groups of people that work in some common context and somehow interchange their ideas or information.

15.2.1 Ethnography

Ethnography is not a new field, but a new approach to analyzing computer system requirements. It concentrates on observing systems without in any way affecting how they work. This approach can be important in describing social systems, as it is important in this description not to impose the analyst's viewpoints onto the people in the system. In this way, it becomes possible to actually observe what people do and the reasons for their doing so. The goal is not to superimpose the outside view onto a system, and consequently identify the wrong problems, but to actually see what goes on from the insider's viewpoint. The main characteristics of ethnographic studies are that:

- Analysts observe or possibly even participate in such activities;
- Any interviews are conducted *in situ*, possibly as informal discussion rather than a formal interview;
- There is emphasis on examining the interaction between users;
- There is emphasis on tracing communication links;
- There is detailed analysis of artifacts.

The emphasis here is on observing system activities (and perhaps analysts even actively participating in these activities), and on observing the system from within rather than from the outside. There are a number of advantages in the ethnographic approach. The users are not disturbed in their activities and information is gathered directly, and not from an informal description obtained through interviews. The two most important ways of carrying out ethnographic analysis are through participation or observation.

15.2.1.1 Analysis by Participation

Ethnographic studies, because of their emphasis on interaction, are particularly important in studying the way that groups of people work. Their goal is to study the dynamic social situations that occur in such environments. It is usual here to identify communities or workers and analyze their interactions. One way to gather information is by actually participating in group activities. The analyst becomes a member of a team, perhaps in an indirect capacity assisting other team members.

15.2.1.2 Analysis by Observation

The goal here is to observe what people do in an unobtrusive way. The best way to do this is by video recording. It is important in video recording to ensure that the presence of the video itself does not alter behavior while at the same time collecting sufficient in-depth information to make useful observations. This requires considerable skill both in the placement of video recording equipment and the setting of the video camera itself. Once a video is complete, analysis commences. Analysis is usually carried out in conjunction with participants in the work setting. The result of the review session is a script of user activities and their interaction with each other.

15.2.1.3 Analysis by Interviewing

Sometimes, especially with large groups, it may not be possible to observe the whole group or participate in all its activities. In that case, some interviewing may be needed.

However, there are some other characteristics of interviewing that can lead to incorrect assumptions about a system. One is the danger that because the interview is usually carried out outside the working environment, the interviewee may distort replies to questions. There is the possibility of exaggeration or emphasizing the less important aspects of activities. There is also the danger that the interviewer may have preconceived ideas about the system and its needs, and try to impose these ideas into the study or interpret replies in terms of these preconceived notions.

A further disadvantage of data gathering based on individual interviews is that it assumes that an individual's work is relatively routine in nature and does not change over time. This, however, may not be the case in some team situations where tasks are dynamically and spontaneously distributed between users. We thus have a situation where there is a dynamic, or time-varying, division of labor with the rules for this division themselves changing. Interviewees often find it difficult to describe these situations in exact terms, thus leading to possible misunderstandings about the system. Ethnography aims to overcome some of these problems.

15.2.1.4 Some Analysis Techniques

Ethnography itself includes a number of techniques. The most important of these [2] are

- *Analyzing people's roles:* This identifies how a particular person feels about their work and the kinds of problems that they encounter. The goal is to gather information on the practices followed by people carrying out a particular task.
- *Analyzing interaction:* Interaction analysis defines how users work together in groups. A simple diagram may be produced initially, showing the various system roles and their relationships. Then scripts may be developed to describe such relationships in more detail. This often leads to the identification of communities of practice that are spontaneously formed around a particular task.
- *Analyzing location:* A study is made of what happens at a particular place over a period of time. Often, the study produces a set of snapshots of activities during that period.
- *Analyzing artifacts:* The emphasis on artifact analysis is to determine how it fits into the flow of work rather than on artifact structure in its own right. Important aspects are
    - How an artifact flows through the system;
    - Use of artifacts by teams rather than individuals;
    - Considering the artifact as a work space in its own right.

- *Analyzing tasks:* The analyst studies the processes within a system and the roles of individual users. Emphasis is on the information needed by the user and what the user does with the information and where it is obtained.

Although ethnography has been mainly applied in the study of detailed interactions within the dynamic group environment, we can learn from it for more general application in analysis. For example, we may identify the interactions at a broad rather than detailed level, and then use these interactions as a guideline in our further analysis. Furthermore, many ethnographic analyst's identify or center their observation on specific objects, which then become parts of the objective model. People, roles, locations, and artifacts can all be such objects.

15.2.2 Outputs From Ethnographic Analysis

Outputs from ethnographic analysis are usually scripts that describe occurrences of processes in the observed environment. Another term used to describe work practices is *scenarios*, which are examples of processes followed by people carrying out their work. A large number of scenarios must often be analyzed to find those that are most typical of the system. Scenarios are often descriptive in nature and are often gathered by interviewing and observation. There is no standard way of documenting scenarios, with the method used left to the observer.

15.2.3 Quantitative Measures

The outputs from ethnographic analysis are usually descriptions, and often do not include quantitative data.

There have also been a large number of experiments in CSCW work, and each of them have used their own documentation techniques to describe group activities. The chosen documentation technique usually stresses the particular collaboration property that is being studied. Figure 15.1 shows one used by Olson and Olson [3]. The particular study concerned the proportion of time that people spend on particular interaction within an activity. The proportion is illustrated as a part of the circle. The study also looked at how people move between different activities. Such movements are illustrated by the directed arrows.

The thickness of the transition line indicates the frequency of the transition. The activity itself is divided up into a number of actions that indicate what actually goes on in the activity. Another variant is for the circles that represent the activities to be of different sizes. In this case, the size is proportional to the total time of the activity. Other measures that have been used in similar studies are the time people spend in synchronous and asynchronous activities or time spent waiting for other people to complete related work.

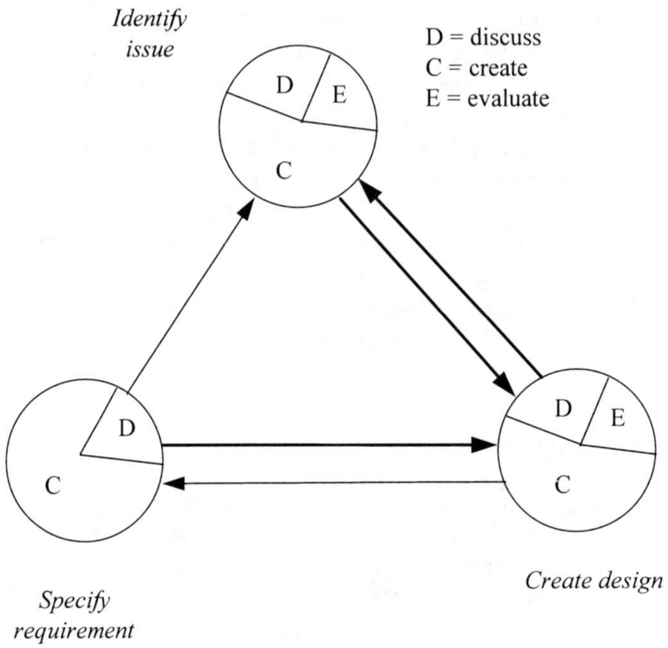

Figure 15.1 Modeling processes.

The interesting research here is how to choose the activities. Are they to be specific to a user application, or an instance of a generic service such as voting, brainstorming, and so on? The activities can contain artifacts, and actions can then be assigned to specific roles. We then have the bridge from the activities to the things; that is, roles and artifacts.

15.3 REPRESENTING THE FINDINGS

Scripts and scenarios are one way to describe the results of findings. The next step is to somehow represent these findings in ways useful for later design. Whereas scripts tend to describe particular occurrences, what we need are more general system representations. Such representations must improve people's understanding about the system being modeled. They should reduce rather than add to ambiguities and enable meaningful discussion between a variety of users. Representations in these early stages are by nature close to physical and must somehow convey the essence of the work without letting detail result in unnecessary distraction, but yet letting the reader be aware if this detail.

15.3.1 Some Contemporary Models

An important modeling requirement is to have a good representation method. The representation method describes the system in terms of unambiguous modeling concepts. It should represent the concrete system objects as well as the behavior of these objects. A good representation should be visual and represent the entire system in one diagram. It should also be possible to select and elaborate particular model components in detail when needed.

15.3.1.1 Object-Oriented Methodologies

Earlier, Chapter 7 introduced object modeling as a way of representing the service objects. Here, concrete system elements are represented by objects and their behavior by methods. Each object in a class has a number of methods, where each method describes the effect on the object by a given input. Object-oriented models encapsulate data with procedures. Thus, Figure 15.2 could be seen as a high-level representation of the coauthoring system. Here, there are four objects. Each object is made up of three parts: the name, for example, BUDGET; the attributes, in this case TEXT; and methods that can be used to operate on the attributes. The REPORT is composed of the three parts, TECH, BUDGET, and LEGAL, and includes methods to integrate the parts and check for consistency. It is, however, difficult to include people's interactions in this model. All we can do is show the methods that people can use.

The encapsulation of methods with data in the object model makes the object-oriented model easier to change. New functions that apply to an artifact can thus be easily added to the object that represents the artifact.

15.3.1.2 Dataflow Models

Structured systems analysis centers around major modeling tools; in particular, dataflow diagrams and entity relationship models. An example of such a model for coauthoring is given in Figure 15.3. Here, the data component is described by an entity relationship diagram that shows the major data components. Flow of information is shown by the dataflow diagram.

Because many flows in cooperative systems are not predefined, the data flow diagram can become cumbersome and confusing, as it would contain a large number of flows in order to cater for all possible situations.

15.3.2 A Comparison of Modeling Methods

The modeling methods described above have been traditionally used to design transaction-based systems and concentrate on aligning technology to tasks and

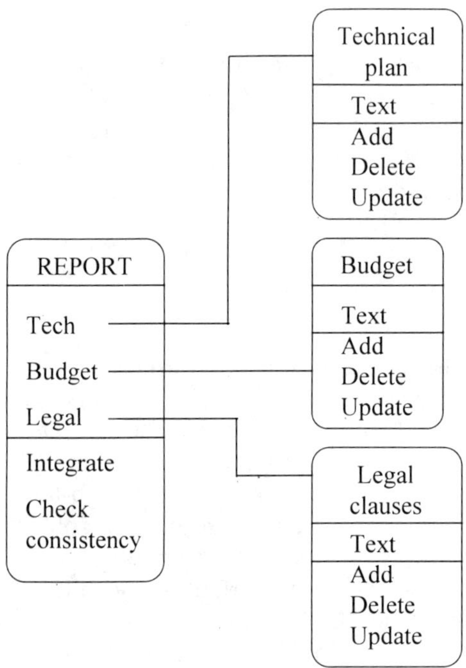

Figure 15.2 An object-oriented model.

with predefined processes. The general comments about applicability of such tools for group system design are

- The existing transaction-based models initially place little emphasis on people and how they interact. The actions to be followed by people are usually defined after the system is defined. This is contrary to most CSCW systems, where people interaction is the important issue.
- Existing modeling methods are often unsuitable for modeling collaborative systems. For example, it is difficult to model a group meeting using dataflow diagrams because of the large number of data flows and their relatively informal and nondeterministic nature. Similarly document states, so important in joint responsibility, cannot easily be represented by most data modeling tools.
- Object-oriented methods have some advantages because they can better represent autonomous agents. However, it is still not clear how more dynamic groups like meetings can be modeled using the object-oriented approach.

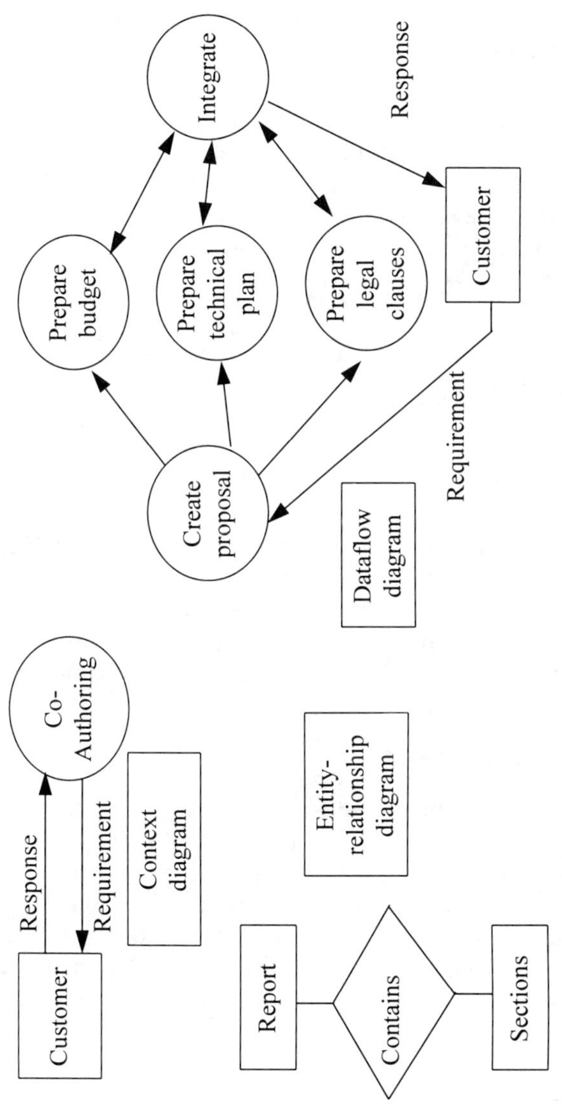

Figure 15.3 A system model.

- Data flow diagrams are not very useful because they assume a predefined environment. This is not the kind of environment found in most cooperative systems, as it is often not possible to predict what will happen in group situations. They are dynamic, and what happens next usually depends on some previous output. The number of outputs is so large and unpredictable that they cannot be prespecified.

All these methods, however, are relatively prescriptive and assume activities that can be predefined. They do not easily capture the dynamics of work, where interactions can change from moment to moment depending on some, perhaps unpredictable, outcome. They also have little emphasis on people. For this reason, we turn to the more open techniques found in soft systems methodologies, particularly the rich picture.

15.4 RICH PICTURES

The usefulness of rich pictures in analyzing cooperative systems was stressed in Chapter 14, and rich pictures were also used in earlier chapters. Rich pictures emphasize the people and their roles in the systems, and also their personal interrelationships. These relationships can be expressed using the classification described in Chapter 3. We have used the idea of rich pictures earlier in this book to describe cooperative systems. Often the representation was open, where constructs were used freely to describe systems. However, once we begin a more systematic approach, a standard way of drawing rich pictures becomes attractive.

15.4.1 Standard Constructs

Any standard should stress the main ideas to be captured in the rich picture. In cooperative systems, these are often interactions between people , artifacts, and roles in the initial analysis. The standard construct can then take the form shown in Figure 15.4. Here, the interaction appears as a cloud that describes the kind of interaction. The roles and artifacts are then linked to the cloud. The roles appear as icons, whereas artifacts are represented as rectangles. A role's perceived task is also shown-for example, the artist sees her task as commenting on the artifact.

15.4.1.1 Case Study: Specialist Designer

Figure 15.5 is an example system using a rich picture. Here, the specialist designer has now expanded to produce designs and artifacts for general sale as well as for specific clients who may ask for special designs. The services of a market researcher have been engaged to get ideas of the market potential of some proposed designs.

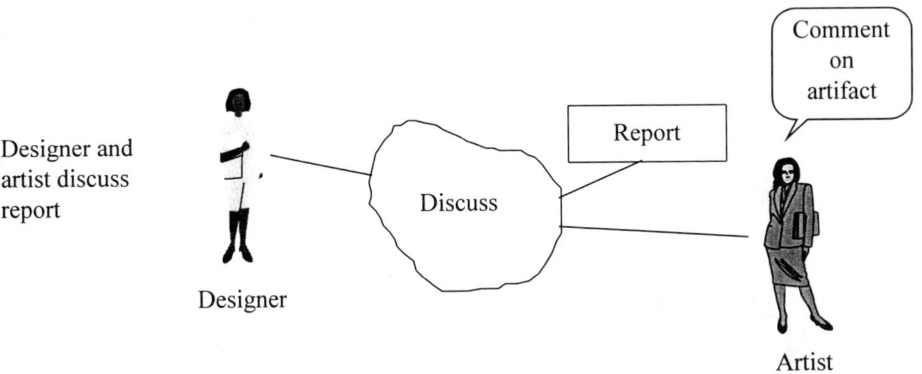

Figure 15.4 Stressing roles, artifacts, and interactions.

The designer must now also make arrangements to have the artifacts produced and distributed.

As a result, the scope of work is now much larger than when only special clients were looked after, and some of the activities described may need further elaboration. Perhaps what is important now is the ability to customize designs to customer needs. Customers can now discuss their requirements with the assistant, perhaps examining alternative designs. It then becomes important to keep records of alternatives considered in earlier designs and during customer discussions so that ideas can be quickly raised and illustrated when a new need arises.

Production of designs intended for volume production are outsourced to contractors for production. Arranging production involves negotiation of a contract as well as some planning of work. So does arranging transfers to distributors, where third parties such as transport or delivery firms may be involved. Such detailed expansion may be needed here to describe the individual relationships and describe interaction in more detail using the classification given in Chapter 3.

Above all, the rich picture emphasizes people. It is important that the rich picture identify relationships between people in terms of interactions between them. It should also bring out the major artifacts that are used in the interactions. But its most important element is its emphasis on people and the relationships between them. Also, remember the rich picture is often a sketch and may not become a formal document. Some ways of keeping more formal records of analysis are described in the next chapter.

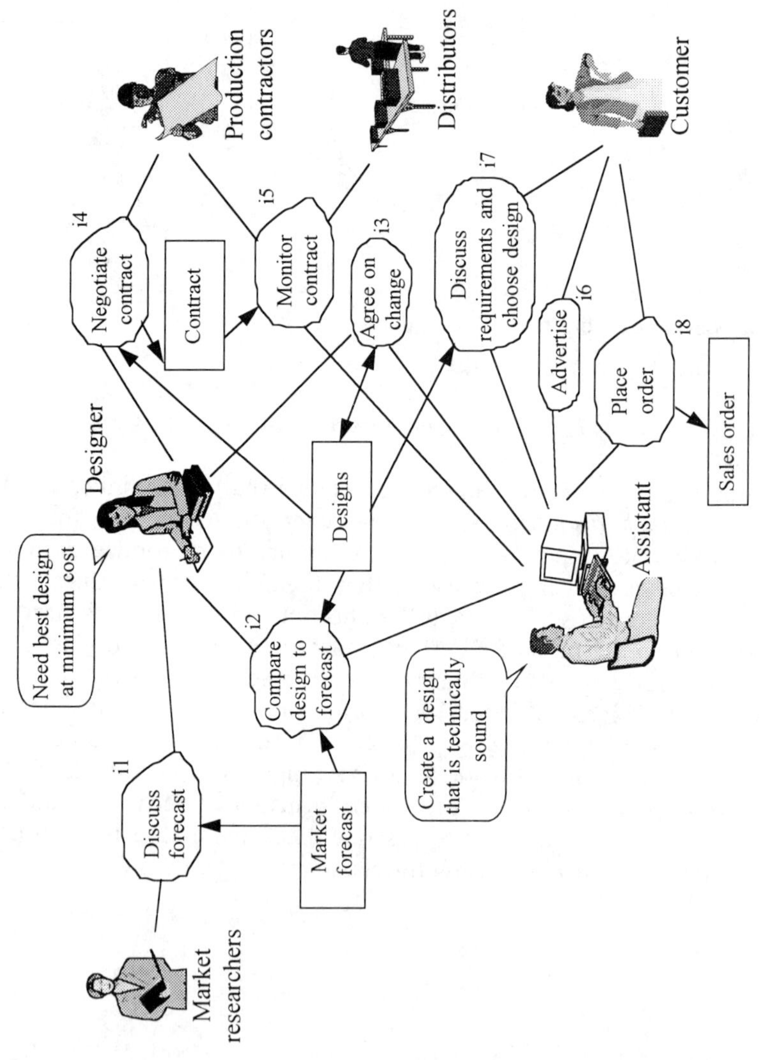

Figure 15.5 The specialist designer.

15.4.2 Extending With User Scenarios

The rich picture is a relatively static representation and does not describe the dynamics of a system. The dynamics are not easily expressible in their entirety because of their indeterminate nature and hence representations such as the dataflow diagrams are not useful here. Instead, an alternative is to define typical user scenarios as illustrations of behavior in a system. A typical scenario is often an informal script.

15.4.2.1 Case Study: A User Scenario for the Specialist Designer

The designer often needs to clarify an issue with the producer. This may either be some unclear part of the specification or a change requested by a client. In either case, the designer usually prepares notes on the current contract with the producer and calls the producer on the phone. The phone discussion may go on for several minutes, with notes taken at both ends. Often a fax is sent to explain some of the detail and discussion resumed sometime later. Once the issue is resolved, the contract may need to be amended. This again leads to the transmission of suggested amendments. These are discussed over the phone until some agreement is reached. This user scenario is initially defined as a script, but later elaborated in more formal terms as a process. Ways of doing this will be described in the next chapter.

15.5 SUMMARY

This chapter described approaches to developing a model of cooperative systems. It defined the major modeling requirements, especially the need to integrate social, objective, and subjective components in any model. The chapter then described some existing modeling methods and how they can be used to model cooperative systems. The models described in this chapter indicated the kind of concepts that can be used in analysis of group support systems although most of them identified concrete objects and the relationships between them. The interesting question here is how to include social factors into the model. Obviously, we cannot simply add social properties as objects, as the relationship between the two classes of objects is not deterministic. It seems that the best way may be to include social descriptions in the scenario or as a separate description of the context in which interactions take place. The next chapter will describe an approach for modeling and representing CSCW systems based on the ideas described in this chapter.

15.6 DISCUSSION QUESTIONS

1. What are the three main dimensions for CSCW modeling?
2. What is primarily represented by social modeling methods?
3. Why is it important to go into the details of work practices in analysis?
4. Why are transaction-based methods unsuitable for CSCW design?
5. What are communities of practice?
6. Do you think that ethnography is a good approach to modeling social viewpoint?
7. Why is it so important to consider social factors in CSCW systems?
8. What is the main purpose of user scenarios?

References

[1] Suchman, L., (Guest Ed.),"Representations of Work," *ACM Communications,* Vol. 38, No. 9, Sept. 1995.

[2] Jordan, B., "Ethnographic Workplace Studies and CSCW," *Proc. of 12th. Interdisciplinary Workshop on Informatics and Psychology,* Shaerding, Austria, North-Holland, June 1993.

[3] Olson, J. S., and G. M. Olson, "Groupwork Close Up. A Comparison of the Group design Process With and Without a Simple Group Editor," *ACM Transactions on Information Systems,* Vol. 11, No. 4, 1993, pp. 321–348.

Selected Bibliography

Hammersley, M., and P. Atkinson, *Ethnography: Principles in Practice,* London, UK: Routledge, 1983.

Gooch, G., *Object Oriented Design with Applications,* (2nd Ed.), Redwood City, CA: Benjamin-Cummings, 1991.

Hawryszkiewycz, I. T., *Introduction to Systems Analysis and Design,* (3rd Ed.), Sydney, Australia: Prentice-Hall, 1991.

Chapter 16

Conceptual Representation

LEARNING OBJECTIVES

- ❑ *Modeling concepts*
- ❑ *Interactions and activities*
- ❑ *Describing processes*
- ❑ *Describing interactions*
- ❑ *Logical modeling*
- ❑ *Detailed process specifications*

16.1 INTRODUCTION

Chapter 14 introduced a systematic way to design CSCW systems by proposing a general cycle of physical analysis, logical analysis, logical design, and physical design. Chapter 15 described the initial analysis that creates a rich picture and a set of scenarios. The next step is to concentrate on the logical aspects and describe the physical interactions in logical terms using generic concepts. This chapter defines a set of concepts or terms that can be used in a modeling "language" to logically describe cooperative systems. The language is not a programming language, but a set of concepts or terms used to describe in a precise and unambiguous way. Its goal is to avoid situations where people have different interpretations of the same system, as often happens when ad-hoc terms are used to describe and solve a problem. It also provides a way to keep track of objects found during analysis to ensure that they are not forgotten during design. To be useful, the terms or concepts must be

- *Simple but meaningful* and correspond to actual system objects to reduce the cognitive gap described in Chapter 6. Thus people should in some way be represented by concepts that are directly related to people, artifacts by symbols that correspond to artifacts, and so on.
- *Limited in number* to make it easier to choose the best symbol to represent a system object, thus avoiding the confusion that can result if there are too many concepts,.
- *Easily recognizable* so that a person on seeing a symbol quickly associates a meaning with the symbol. If this is not the case, and some interpretation is required, then ambiguities can arise leading to confusion about a model.
- *Applicable to collaboration* including terms that describe how people interact in collaborating.
- *Widely applicable and sufficiently abstract* to apply to a large variety of problem domains.

A further requirement is that these concepts be in some sense complete so that they can describe all collaborative systems. This is often difficult to prove when the modeled systems themselves have no formal definition. Some factors that can be used to judge the completeness of a modeling method require that the modeling method, for a particular class of system, must:

- Have a theoretical underpinning. This is only possible where the system being modeled is formally defined. For example, criteria for complete relational languages are only possible because the relational model can be formally defined.
- Be able to represent existing alternative models. In this case, there may already be existing models for a given class of systems. A proposed model should be able to model the type of systems modeled by the existing model and also include additional components.
- Be empirical in the sense that its application to a large number of practical systems can be illustrated.

16.2 LOGICAL ANALYSIS

The concepts described in this chapter emphasize people collaborating on tasks that involve a variety of artifacts. They include concepts to describe the static and dynamic objects of collaboration. The description begins with the static components and then describes modeling of dynamics. The way that such descriptions are used in the overall design cycle are then described in Chapter 17.

16.2.1 Static Components

The most important modeling component is the context within which collaboration takes place. This context sets the entire work situation [1] and includes cultural factors, missions, goals, and context components such as roles, tasks, and artifacts. People interact in the context in the variety of ways made possible by the services in the context.

The book thus proposes that the first logical modeling step is to represent the context (or at least its most important components) in a single diagram (the rich picture) that allows designers to see or visualize the complete picture, including the different people and roles, the artifacts, and the major activities. This will give the designer the total picture in which collaboration will take place and the kinds of tasks that people will have. The major concepts used initially are

- *Contexts:* Contexts describe the situation in terms of tasks, roles, artifacts, processes, and relationships between people and their current status. They also include components that describe the organization's cultural factors.
- *Artifacts:* Artifacts represent the organization's information base. These can include files as well as reports, designs, videos, and so on. The artifact can include data values as well as methods to operate on these values, and constraints and rules about the values.
- *Cultural factors:* Cultural factors including things like the usual way of decisionmaking, important mission criteria, policies, and so on. They can include standards or system rules that may be used in the decisionmaking.
- *Roles and actors:* Roles have been described earlier in Chapter 3 and represent people's responsibilities within a system. An actor is assigned to each role. Actors are often selected using organizational rules. Responsibility changes can thus be made by only changing an organizational rule. A person can thus have many roles and there is nothing to prevent a person from transferring knowledge (informally) between roles, as often happens in organizations.
- *Tasks:* Tasks define what has to be done. They define what each role must accomplish or, alternatively, what is to be expected of the role. This may be something like complete a design to specification or test a program.

Analysis centers around finding system objects and recording them in these terms.

16.2.1.1 Recording the Model

An organized or systematic design process must have an organized way of keeping track of information discovered about a system. This can be simply lists of objects found in analysis and stored in a data dictionary. Alternatively, it can be a dia-

Roles	Artifacts	Tasks	Culture
Designer	Market surveys	Evaluate market	Current design practice
Assistant	Design drawing	Produce design drawing	Market preferences
Contractor	Evaluation criteria	Change design drawing	
Distributor	Production contract	Check design drawing	
Customer	Sales summaries	Select producer	**Activities**
	Distribution contract	Create contract	Creating a design
	Customer requirement	Select distributor	Arranging production
		Evaluate customer need	Making sales

Figure 16.1 Defining static elements in a data dictionary.

grammatic representation. The next few sections describe possible ways of documenting the information, perhaps not in a sequence that it would be developed, but in way that introduces concepts in a gradual way.

Case Study: Identifying the Design Context for the Specialist Designer

Chapter 15 described a rich picture developed to describe the design context. The next stage is to identify all the objects discovered in this analysis and present them in some organized way. The best way is to create a data dictionary that simply lists all the concepts. Figure 16.1 illustrates this very simple idea—a listing of the objects in the context. Some such objects are listed in Figure 16.1, but this is not a complete list for this case. For example, the number of tasks can be quite large, and again Figure 16.1 only shows a subset of tasks. Usually such lists grow gradually as designers build up their understanding of the system. One value in maintaining such a list is to ensure that all the objects discovered in the analysis are considered in design.

16.2.2 Dynamic Components

The context also includes dynamic components. Dynamics describe how work is organized, and in particular how people work together (or interact) to accomplish their tasks. In this book, dynamics are described by activities and interactions. Activities are *recognized* in an enterprise as *producing well-defined outputs*. The output is usually a new or changed artifact and is produced using information provided by other artifacts within the context. Activities are made up of a number of interactions, which describe how people collaborate to produce the outputs from the inputs. Interactions and activities are defined as follows:

Interactions: Interactions define what people actually do when they collaborate. The interaction thus includes a number of roles, a number of artifacts, and a script that defines the actions carried out by the roles during the interaction. The kind of interaction described in Chapter 3 is also included.

Activities: Each activity is within a context and produces some organizationally recognized artifact. It can include a number of interactions needed to create that artifact.

16.2.2.1 Distinguishing Between Tasks, Interactions, and Activities

It is often difficult to make precise distinctions between tasks, interactions, and activities. Firstly, it is important to remember that a task defines what is to be done, whereas interactions and activities define how people work together to accomplish their tasks. Thus a task such as "make design drawing" defines what must be done; that is, a design must be produced. The interaction then describes how the roles work together when they made a design drawing. There may be interactions when one role asks another to review a design, where they brainstorm ideas or when they discuss alternatives. The task itself may be broken up into many subtasks. There can also be an organizationally recognized activity called "create design" that defines how the design is created. Often it is convenient to start by defining the major activities and showing their relationship.

Another example occurs in business process support, described in Chapter 5. Here, there may be an activity that results in a letter of intent to form a network. This may involve tasks like "find a partner." To do this may require interactions like "discuss interest in following opportunities" with potential partners.

Case Study: Activities in Specialist Design

Figure 16.2 illustrates a way to describe activities. It shows how work in specialist design could be broken down into three activities: creating designs, making sales, and arranging production. How the system is broken down into activities is often at the discretion of the system designer. However, one guideline is to choose activities that satisfy one major criterion; that is, the production of a recognized output-in our case, a design, a contract, or a sale.

Each activity may have many people interacting with each other. Each such interaction usually carries out a subtask, usually requiring a finite amount of time. Thus a meeting or a message interchange is an interaction. Furthermore, an interaction is characterized by a fixed set of roles, an artifact, and a task. The roles in the interaction (of which there may be one) carry out a task on an artifact.

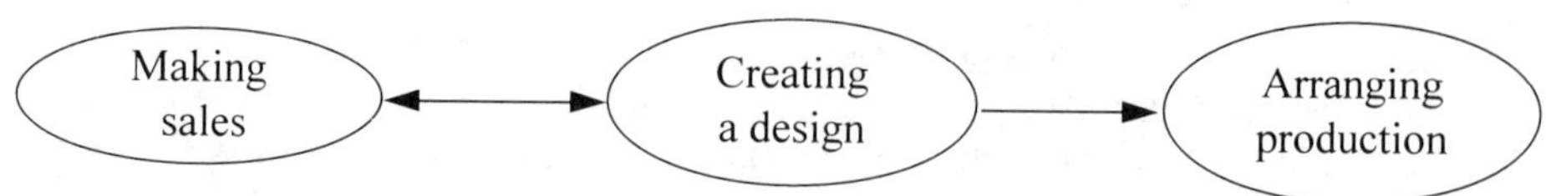

Figure 16.2 Showing activities.

16.2.2.2 Modeling Interactions

Interactions are the central elements of cooperative system design. They represent the personal relationships that must be supported by collaborative services. It is also important that the design identify the logical rather than physical interactions; that is, what people do rather than how they do it. These interactions will be used later as guidelines for choosing services to support the interactions. It is, however, often difficult to directly provide a logical set of interactions. It is easier to see the physical processes and thus directly represent them. Going to a logical design requires thorough examination of scripts or an in-depth look at the interactions. One can then go through a script and look at each interaction, identifying its kind by the classification described in Chapter 3. The kind of interaction indicates the service needed to support it.

16.2.2.3 Recording Interactions

Again we have found that the best way to keep track of interactions is by listing them, although they are grouped by activities. Such a list will show the roles that participate in each interaction and the artifacts that they use.

Recording Activities and Interactions for the Specialist Designer

Figure 16.3 shows how interactions are listed and described. Usually there can be a separate table for each activity, but in our case we have included interactions from a number of activities for illustrative purposes. Apart from roles and artifacts, the interaction also describes the interaction kind and the time/space relationship of the roles. The one important component is a number assigned to each interaction (i1 through to i8). This is later used to make sure that each interaction is considered during design. It can also correspond to numbering in the rich picture in Figure 15.5.

Interactions are named using a logical emphasis, as for example, "discuss forecast" rather then "market analysis." or "discuss change" rather than "change drawing." The TIME/SPACE column is there to give an idea of the time/space relationship of the interacting role (with ST meaning "same time" and DT meaning "different time").

Listings like that shown in Figure 16.3 tend to be developed gradually with comments indicating social processes or scenarios that cannot be directly represented. In a large system, there can be one such list for every major activity. The

Activity: Creating a Design

Interaction		Roles	Artifacts	Time/space	Type of interaction	Current implementation
Name	Number					
Discuss forecast	I1	Designer Mar. Res.	Forecast	ST/DP	Discuss	Phone Face to face
Compare to forecast	I2	Designer Assistant	Designs Forecast	SP/ST	Discuss	Face to face
Agree on change	I3	Designer Assistant	Designs	SP/ST	Discuss	"
Negotiate contract	I4	Designer Contractor	Designs Contract	ST/DT/DP	Negotiate	Phone, Fax etc.
Monitor contract	I5	Assistant Contractor	Contract	DT/DP	Clarify	Phone Letter
Advertise	I6	Customer Assistant	Designs	DT/DP	Broadcast	Mailout
Choose Design	I7	Customer Assistant	Designs	ST/DP	Choose	Phone Face to face
Place order	I8	Customer Assistant	Order	DT/DP	Request	Letter

ST = Same time, SP = Same place, DT = Different time, DP = Different place.

Figure 16.3 An example of interactions.

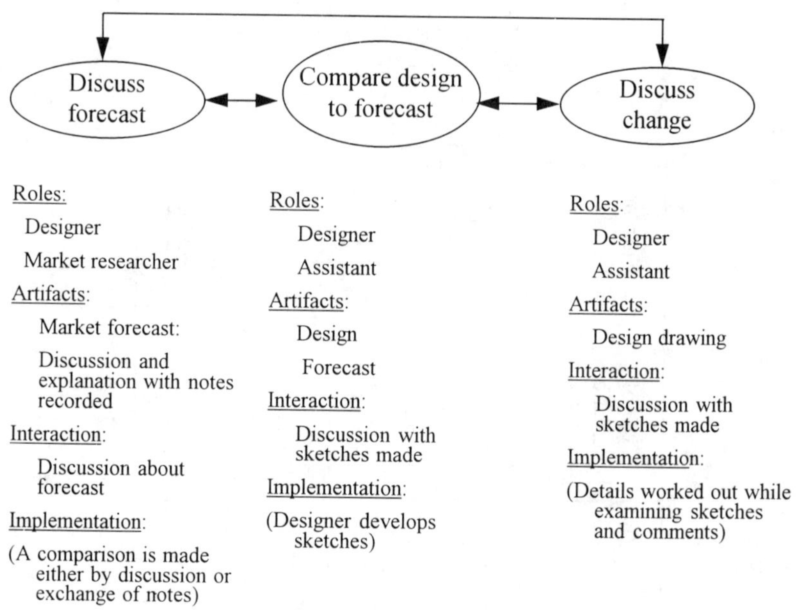

Figure 16.4 Interactions in creating a design.

most important is that they define the context (including social contacts) and artifacts, and indicate broad activities.

16.2.2.4 Recording Interactions Diagrammatically

An alternative representation is one that shows the interactions diagrammatically as processes within an activity.

Case Study: Describing Interactions for the Specialist Designer

The representation in Figure 16.4 describes some of the interactions in the activity "creating a design."

Interactions are again described by their roles and artifacts as well as by short informal scripts and how they are currently implemented. The location of roles may also be important as to whether the interaction is synchronous or asynchronous. For example, in comparing to market, the interaction is either a discussion or an exchange of notes. The description will later be used to map the interaction to a service using the classification scheme described in Chapter 7. The service will probably be one of sharing views or interchanging messages.

It is important to remember that all activities take place in a context. It is possible for roles within the activities to create new activities, or even to define their own context. It should be noted that the creation of new objects is the responsibility of roles. Activities themselves cannot create new objects. Thus, semantically, the context often determines what occurs in activities, whereas roles in an activity can change the context. Activities themselves can dynamically change (by adding or deleting interactions or transitions). Such new interactions can arise spontaneously during an activity and cannot be always predefined. Interactions tend to be more fixed in their structure.

16.2.2.5 Defining Processes

Processes are defined by the sequencing of interactions and activities. Processes can be specified diagrammatically or as state transitions. Just showing transitions diagrammatically as lines between interactions is usually sufficient, as in most processes, transitions are usually dynamically determined as the process proceeds.

Processes apply both to activities as well as interactions. Although it is possible to represent processes by formal rules, the nature of collaboration makes this too restrictive. In our model, we represent processes simply as possible transitions, and leave it to the user to choose the most appropriate transition.

Case Study: Defining Processes for the Specialist Designer

Figure 16.2 already showed a representation of processes in terms of transition diagrams. The same method can be used to describe how interactions fall into processes. Figure 16.5 shows an activity process described in terms of transitions between interactions, shown by arrows. The double arrows indicate the possibility of iteration. The completion of each activity becomes an event, which initiates the next interaction. What Figure 16.5 says is that after we "create a design," we may negotiate it with a designer, discuss further design changes, or further consult market analysis.

Such processes are often rough sketches that help us to consolidate scenarios into a larger picture. The transitions are important as they provide the guidelines for integrating services. Figure 16.5, for example, indicates that it should be possible to easily move from discussions on design changes to those with market researchers and contractors while displaying any information created in design in the discussions.

16.2.3 Decomposing

In large systems, it is often necessary to subdivide the activities into smaller activities and describe each activity by interactions in the way shown in Figures 16.3 and 16.4.

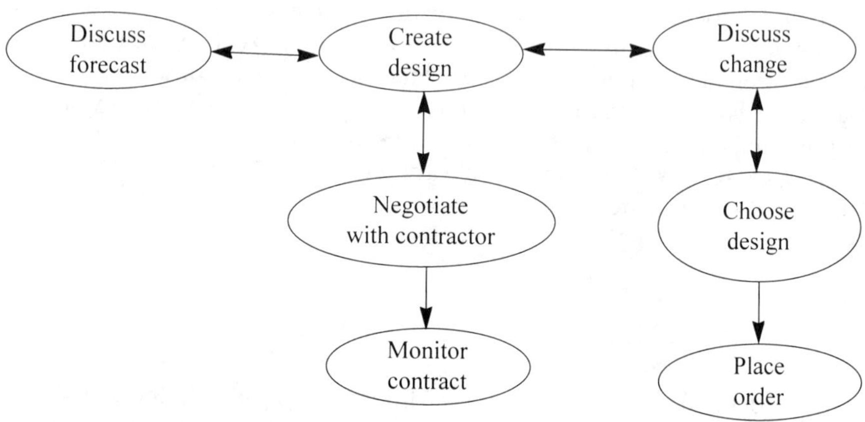

Figure 16.5 Modeling activities.

16.3 GOING INTO DETAIL

So far, the diagrams have only shown high-level objects and defined processes in terms of transitions. Usually, relatively little formality is required in practice when analyzing collaboration because of its dynamic nature, and the way that processes have been defined is usually sufficient to choose a platform. However, it is possible, if one wishes, to become formal using ideas of state transition as described in [2].

16.3.1 Modeling Processes in Detail

An activity is modeled as a collection of roles involved in a number of interactions, with each interaction having a well-defined task. The activity also includes a number of artifacts from the environment. So far, the model proposed here showed possible changes between activities in the process without going into the rules or reasons for such changes. To describe such rules introduces formality into process definition. One question is how much formality is needed when defining processes. This depends on the problems. Where workflows are to be specified, the transition rules between states should be specified. However, specification of such rules may not be useful in non-predefined processes, simply because all possible rules cannot be prespecified. Often such rules are created as the process is executed.

To model a process in detail requires process states to be identified. Rules can then be specified in terms of these process states.

16.3.1.1 Case Study: Defining Patient States in the Midvale Community Hospital

The critical process in the health system goes through four steps. The completion of each of these steps can be seen as reaching a state. Thus there are states such as

- Patient admitted;
- Patient discharged;
- Symptoms analyzed;
- Test completed;
- Patient discharged.

Rules between states can then be specified as formal rules such as, for example,

- If state = (patient admitted) then set state = (begin examination);
- If state = (test completed) then set state = (continue examination).

However, examination of the whole process indicates that not everything can be specified in terms of states. An examination can have many interactions that happen almost spontaneously. For example, the doctor talks to a patient, then consults a specialist, and following this may order a test. It would not be very productive to define states such as (initiate patient discussion), or (specialist request), or (look at chart) as the rules for moving between these states are totally nondeterministic.

Consequently, there is a balance to be reached here. Usually, the high-level activities that follow some known sequence can be modeled formally and defined as workflows. Lower level interactions, especially those involving task groups, should be modeled as informal transition diagrams.

16.3.2 Constructing Objects Adding Local Features

One implementation of such detailed structures could be to use an object-oriented approach. Each concept can become an object that includes local features such as

- *Properties:* values stored in the object;
- *Methods:* programs stored in the object;
- *States*: states allowed for the object;
- *Rules*: object rules.

These objects can enter states, which in turn set trigger states in other objects. The basic semantic here is that a trigger state value initiates an object method. The basic rules for objects takes the form

if state = x then new state = y.

Here an object state, *x*, causes the object to go to a new state, *y*:

s1 : <method *x*> —> s2;

Here a state ,s1, initiates the execution of a method, method *x*. The object is set to a new state, s2, when the method completes.

if state = *z* then initiate (port, port state).

Here, a state in an object sets a port to a new port state. This is often used in activities to describe the activity process.

One possible class of objects here can be coordination objects. These objects defines the standard process rules, such as conversations, with roles as object parameters. We can select one of the *coordination objects* and instantiate its role parameters to create the process.

16.4 SUMMARY

This chapter stressed the importance of having a common set of terms for describing CSCW systems. It defined eight concepts for this purpose: contexts, cultural factors, roles, tasks, artifacts, activities, interactions, and processes. It then showed how they can be used in modeling. The chapter concentrated on high-level models, as these are usually sufficient for design. However, it also described how more detailed models can be constructed.

16.5 DISCUSSION QUESTIONS

1. What are the important cultural factors at your place of work?
2. What is the difference between activities and interactions?
3. What do you understand by the term *task*?
4. What is the difference between actor and role?
5. Describe some common artifacts.
6. What do you understand by *transition diagrams*?
7. Why is it important to identify logical activities?
8. When are detailed process specifications needed?

16.6 EXERCISES

1. Develop a logical representation for Worldlink Consultants.
2. Develop a logical representation for the virtual university, identifying its major logical activities.
3. Develop a logical representation for film production.

References

[1] Suchman, L., *Plans and Situated Action: The Problem of Human-Machine Communication*, Cambridge, MA: Cambridge University Press, 1987.

[2] Hawryszkiewycz, I. T., "A Generalized Semantic Model for CSCW Systems," *Proc. of the 5th International Conference on Database and Expert Systems*, Athens, Greece, Springer-Verlag, Sept. 1994, pp. 93–102.

Selected Bibliography

Divitni, M., et al., "A Multi-agent approach to the design of Coordination Mechanisms," *Working Papers in Cognitive Science*, Roskide University, 1995.

Gintell, J. W., M. B. Hoed, and R. F. McKenney, "Lessons learned by Building and Using Scrutiny, a Collaborative Software Inspection System," *Proc. of the Seventh International Workshop on Computer Aided Software Engineering*, IEEE Press, 1995.

Hawryszkiewycz, I. T., "A Design Method for Choosing Services for Large Distributed Teams," *Proc. Second International Conference on Design of Cooperative Systems, COOP'96*, Juan-Les-Pins, France, INRIA, June 1996, pp. 515–533.

Chapter 17

Designing the System

LEARNING OBJECTIVES

- ❑ *Design process*
- ❑ *Specifying requirements*
- ❑ *Identifying user communities*
- ❑ *Choosing services*
- ❑ *Evaluating services*

17.1 INTRODUCTION

Chapter 14 described two ways for developing collaborative systems for networked enterprises. One was by gradual evolution, starting with information access and evolving to integrated platforms. The other was a strategic approach, usually resulting in the development of an intranet. The evolutionary approach is best suited for small groups that are looking for new ways to conduct their business or their groups within larger enterprises. The strategic approach is best suited for enterprises that want to improve critical enterprise-wide processes that affect many groups.

The goal of both approaches is to develop a platform of services to improve communication within communities and enterprises. Earlier chapters suggested that the services should be gradually integrated into work processes. This calls for a systematic design method, especially for the strategic approach. Development should be managed as a project that starts by identifying user needs and finishes by producing a service platform that satisfies those needs. This chapter will describe such a systematic approach. It describes a design process that begins by using the

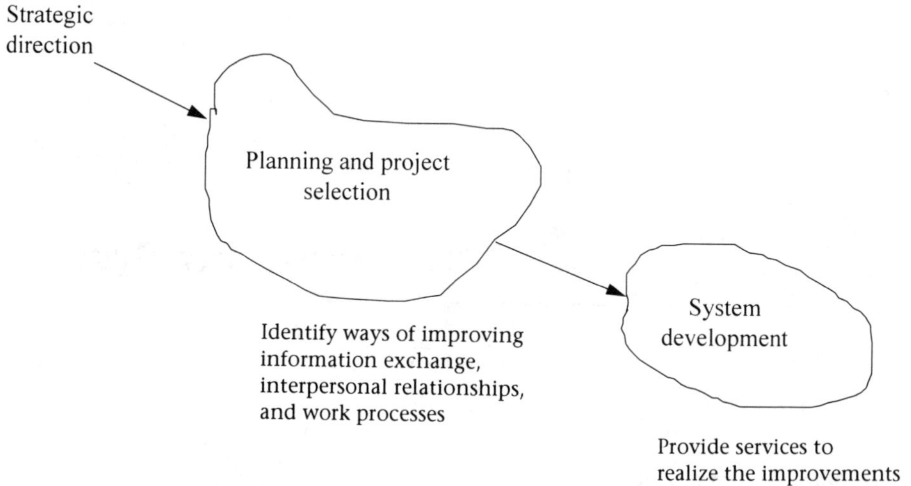

Figure 17.1 Development stages.

analysis methods described in the previous two chapters to identify user needs. It then continues by showing how to choose the network services to meet those needs. It describes both the choice and justification of such services and their integration into a platform.

17.2 THE MAJOR DEVELOPMENT STAGES

There are no widely accepted methodologies for developing cooperative systems. However, the two major stages shown in Figure 17.1 are almost natural. Here, the first stage is to develop a proposal that defines a new way of carrying out business, concentrating on improving ways for information exchange, interpersonal communication, and work process support needs. The second stage is to provide the services to realize the improvements. This project-oriented approach is applicable to both the evolutionary and strategic approaches, although in the strategic approach design is constrained by a defined strategic direction identified in the way described in Chapter 14.

The first stage addresses a problem in the broadest terms. Often, this follows on from the strategic direction-for example, the strategy for the *specialist designer* may be to increase sales through widening contact with potential clients. Possibilities are then better advertising and better ways of managing design documents to make them easily available in discussions with customers, producers, and within the office. System development addresses the technical, social, and managerial issues of realizing the broad goal. Technical issues concentrate on choosing services that will realize the goal. Usually, services are selected from a number of *alternatives*.

Social issues concern the effect and impact of the selected network services on existing processes and relationships, requiring designers to ensure that the services support and enrich existing social relationships. The managerial issues require designers to address the critical business problems rather than technical issues and to do this in a cost-effective manner. Designers must thus select any services using a defined set of *criteria* and *evaluation arguments*. Furthermore, these criteria and evaluation arguments must be expressed in CSCW terms.

17.2.1 Identifying Projects

The first stage concentrates on identifying ways that collaboration can improve critical business processes, concentrating on physical and logical analysis. The steps here are to:

- Identify situations of maximum benefit.
- Define strategic objectives.
- Identify system (organization)-wide benefits.
- Select and define project.
- Construct implementation plan.

Stage one identifies the major objects and processes in the environment and concludes with a report that identifies the potential of improving these processes. Process improvements are defined in terms closer to CSCW rather than business needs, stressing critical factors such as awareness, group memory, and so on. The process steps are broadly identified in Figure 17.2.

They begin with broad requirements and eventually reduce them to objectives and what must be done to achieve them. These are described in a project proposal in terms of new businesses processes, stressing support for communication, and the effect on these processes on work practices showing ways to make them socially acceptable and enriching the current work environment. It will show the impact of new processes in terms of typical scenarios. Priorities for services are set here, and then detailed system development commences. Where there are a number of roles, critical roles are identified for priority support. Effects on the workplace will be described in ways such as:

- Ensure quick availability of the latest documents within the software development process.
- Provide easy access to the latest proposals within planning, with the ability to easily contribute or comment on developments.
- Reports will now be distributed electronically.
- Provide an enterprise directory of people's expertise.

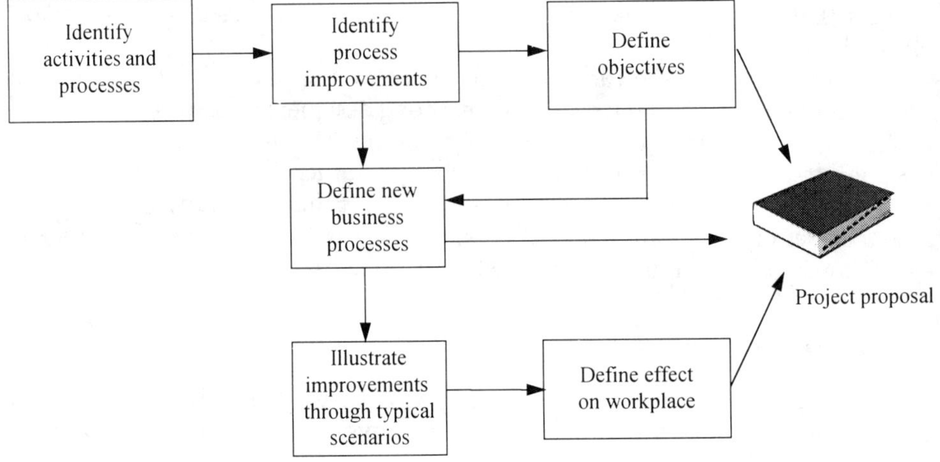

Figure 17.2 Project selection.

- Provide a system to find the best expert for a client problem.
- Provide a document management system to keep track of the best designs for a specified customer need.

For example, the *specialist designer* may propose a major objective to "focus on communication with customers, in particular, to make it easier to quickly adapt to changing customer needs and produce customized designs." The proposal may be to concentrate on improving information exchange with customers and to improve market analysis with a view of becoming better aware of changing trends.

17.3 SYSTEM DEVELOPMENT

System development is described in Figure 17.3. The three phases identified here are

- Phase 1: Developing an understanding using the methods described for physical and logical design in the previous two chapters.
- Phase 2: Defining logical requirements.
- Phase 3: Selecting services and a platform.

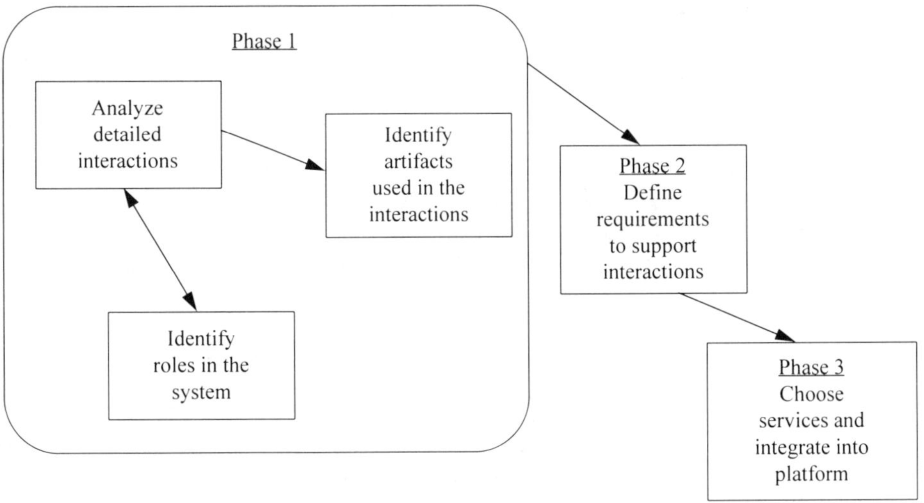

Figure 17.3 System design.

17.4 PHASE 1—DEVELOPING AN UNDERSTANDING

Phase 1 begins by identifying the objects in the system. Figure 17.4 suggests a possible way to do this. It is to:

1. Begin by identifying the major activities, which are best identified as groups of interactions carried out to produce a well-defined artifact. A rich picture can be drawn by showing the activities.
2. Define roles in each activity and enter them into a rich picture.
3. Expand each activity into a rich picture. Define the task of each role and artifacts used by the role, while at the same time looking at the interactions between the roles, including both formal and informal interactions. The processes followed by the interactions may also be drawn at this stage to see how they fit into the activity.

The process can be quite iterative, with, for example, new roles emerging as interactions and tasks are defined.

Once the analyst is happy with the rich picture, a more detailed description of interactions is developed using the logical terms described in Chapter 16, supplemented with notes showing any detailed concerns and social issues. Each interaction must be examined in turn to identify its roles and artifacts and describing personal relationships within the interaction using scripts. It produces a listing of

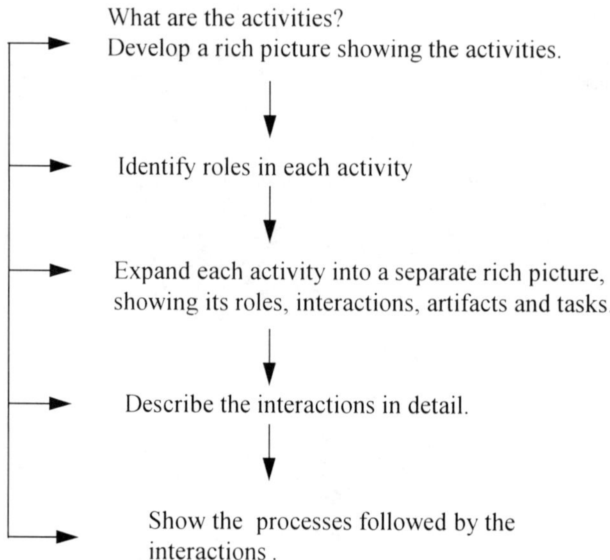
What are the activities?
Develop a rich picture showing the activities.
Identify roles in each activity
Expand each activity into a separate rich picture, showing its roles, interactions, artifacts and tasks.
Describe the interactions in detail.
Show the processes followed by the interactions .

Figure 17.4 Phase 1—analysis.

these objects showing the relationships between them using diagrammatic forms like those in Chapter 16, when necessary.

During the analysis, it is also necessary to identify any problems within the context. These problems must be identified in terms of collaborative key factors, such as awareness or group memory (defined in Chapter 4). The strategic direction must always be the driving force in identifying the problems, identifying ways in which cooperative support can improve the business process. Such improvement is expressed in terms of the key factors (for example, reducing lead times in producing some item or improving quality). These problems are then summarized at the conclusion of the phases and used to present a management report that identifies potential benefits of supporting the processes with collaborative technologies. Thus, in summary, the phase concludes with:

- Identifying deficiencies in the key process or mission in terms of collaboration factors;
- Showing potential improvements to processes;
- Producing a report that describes the findings.

It is these problems that become the driving factors in phase 2.

17.4.1 Some Phase 1 Guidelines

One important factor here is to produce a logical picture of interactions to use later in design. The interaction kinds described in Chapter 3 will be useful here to

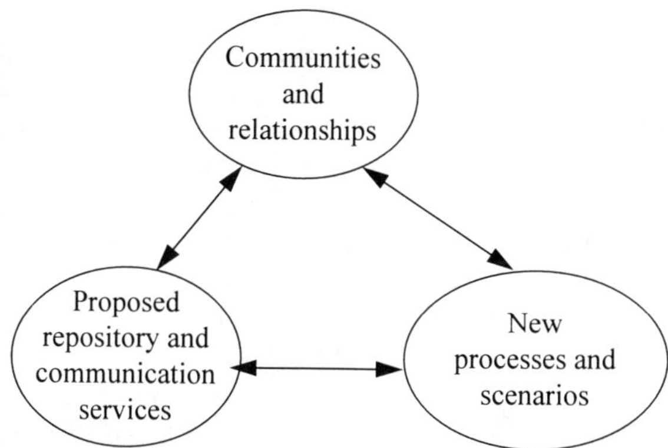

Figure 17.5 Designing the new system.

distinguish between those interactions that can be expressed simply as information exchanges and those, such as negotiation, which require extended personal interactions. These will then provide the guidelines for choosing the most appropriate network services.

17.5 PHASE 2—SPECIFYING NETWORK SERVICES

The goal of this phase is to define the repositories, communication services, and processes to be supported in the new system. It has the main activities shown in Figure 17.5.

17.5.1 Re-examining the Group Relationships

Design begins by specifically identifying the community to be supported. Current group relationships are examined and ways of improving them suggested. It is here that the issues discussed in Chapter 3 become important. The designer must look at the kind of group structure to be supported and its culture. This will impact information flows, as these must be consistent with the culture. Thus hierarchical systems will require higher level management to monitor activities in their area of responsibility, thus suggesting workflow systems with monitoring capabilities. More open cultures, on the other hand, will need services that enable each team member to be continuously aware of the activities of the other members, while making it easy for management to obtain status reports or information quickly on an as-needed basis. Often, the kind of community to be supported indicates the kind of group structure and identifies the roles that must be specifically defined. For example, a peer group community will suggest easy distribution of information with perhaps a new facilitator role nominated to manage this distribution. Clerical groups, on the other hand, are more likely to require workflow systems, requiring a manager or coordinator role.

The kinds of communication should also be identified here, making a distinction between the three kinds described in Chapter 1—information exchange, personal communication, and workflow process. The types of groups also give an indication here of the kind of communication support needed. Committees or task groups may concentrate on interpersonal relationships, and will suggest a need to support synchronous interaction that can be supplemented with asynchronous discussion databases. Clerical groups are more likely to emphasize information exchange and workflow support.

A new rich picture for the proposed system is often drawn here showing the new roles and interactions and a list of interactions made. This list will be the same as that in Figure 16.3 but will now show any new interaction proposed for the new system. One extremely useful technique here is to also define new scenarios together with any new roles (such as facilitator) needed in these scenarios to support the workload. These scenarios will also identify any new workflows within the system. These scenarios must identify any impact on the existing social relationships and business processes, and show how these changes improve the process. They will also identify the kind of interactions that will occur in the new system, and one of the final steps here can be to begin to add any new interactions (and change existing ones if necessary).

17.5.2 Defining Repository Requirements

Now, more detailed design work begins by identifying information requirements in more detail. The ways of communication provide some guidelines for designers, as shown in Figure 17.6. First of all, information exchange usually indicates the document structures and databases exchanged. In addition, existing artifacts also provide a good indication here, as much of the information to be used in cooperating processes already exists. Personal communication requirements identify repositories needed to support interpersonal exchanges. These will include needs for new discussion databases and bulletin boards, as well as setting newsgroups, if necessary. These kinds of repositories were described in Chapters 8 and 10. In contrast, work process support will require monitoring databases that keep people aware of process status. These are often included as part of a workflow support system, such as, for example, LOTUS Notes.

The interactions play a useful role in identifying repository needs, as they specify what information is needed by roles in the interactions and the methods used to process the information. Any support services, such as spreadsheets, needed to process the information will also be identified here.

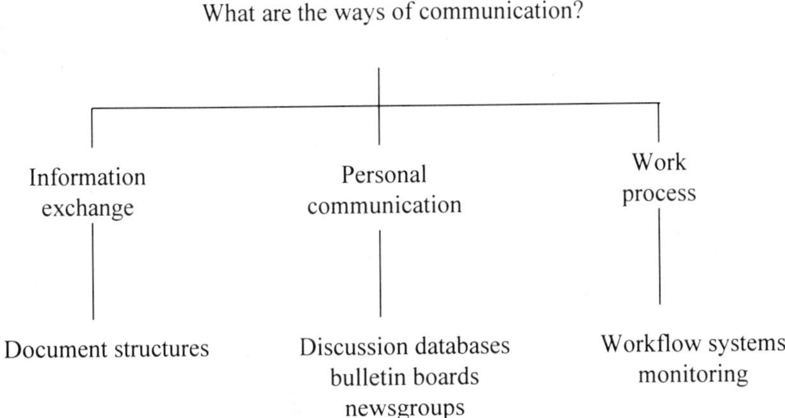

Figure 17.6 Choosing repositories.

17.5.3 Defining Communication Requirements

Once the roles and repositories are known, designers can begin to choose locations at which the repositories will be kept and how the different people will get access to them. This, in the first instance, requires the time and space relationships of the roles to be identified. Once this is done, it becomes possible to determine where communication links are needed and, in particular, identify where synchronous communication is desirable. Then the ways that repositories will be accessed can be specified.

Once the links for information flows are identified, it becomes necessary to look more specifically at some of the more technical issues. Information volumes must be checked and workloads evaluated.

17.5.4 Verification

Verification is used at this step to ensure that all interactions identified in phase 1 are considered in design. Verification can also be assisted by tools that clearly describe the selection process. There is some advantage here in producing a diagram that identifies these services, using the methods described in Chapter 7. We illustrate such a diagram with our example.

17.5.4.1 Case Study: A Logical Requirements Diagram for the Specialist Designer

Two major ways of communication can be identified for the specialist designer. One is the information exchange that takes place between the designer and the market researchers. This often has to be supplemented with some personal inter-

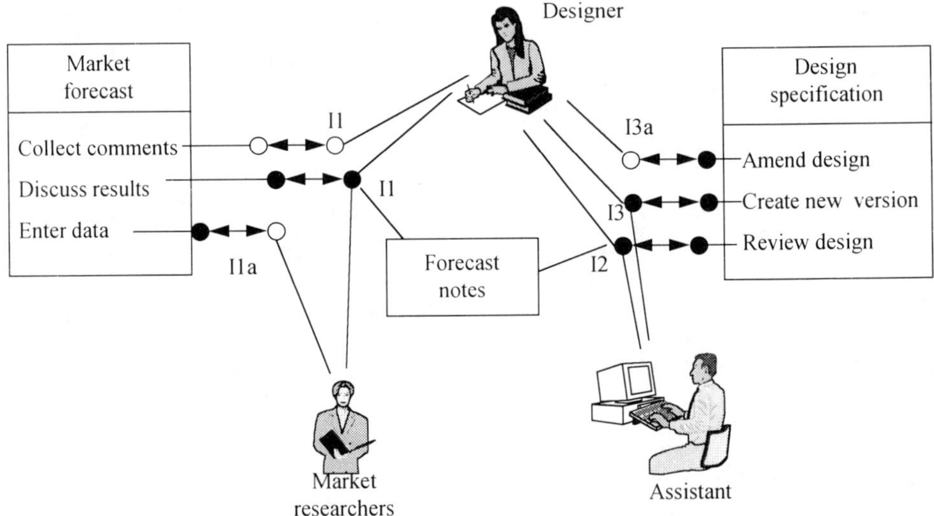

Figure 17.7 A logical representation.

changes. The relationship between the designer and the assistant is predominantly one of personal exchanges while examining design documents. The main repositories that are needed are then quite clear. They were identified in the analysis of Figures 16.1 and 16.2: a market forecast and design specifications are the central information repositories. The question then is whether to provide repositories to support personal exchanges. This was seen as unnecessary. The discussion between the researcher and market forecaster is often quite informal, and telephone conversations with notes kept may be sufficient. These notes may be kept on a small document database. As far as the design interactions are concerned, a better approach will be to keep track of any discussion recorded against design documents, using document versioning system. These relationships can now be illustrated diagrammatically as shown in Figure 17.7.

The diagram in Figure 17.7 shows the major roles and artifacts. Communication services are then added and shown by the links between the roles and artifacts, specifying the interactions supported by the links. Thus, for example, interaction i1 is supported by an asynchronous link that allows the designer to access the market forecast and by another link that allows the designer to discuss it with the market researcher. Interaction i3 requires support for synchronous interaction while discussing design documents. Two additional interactions, I1a and I3a, have been added to allow the market researchers to make entries to the market forecast and for the designer to change designs.

Figure 17.7 only shows some of the interactions—those concerned with design and the first three interactions in Figure 16.3. It is important to note here that all identified interactions will eventually be included in the logical design.

Extension to the expanded needs for the specialist designer will require attention to be paid to the workflow support system. This will allow the assistant to monitor the progress of work arranged with production companies and track the distribution of products. Thus work by the production contractor could become an external task that is monitored by the assistant. This, of course, requires the contractor to have access to a matching workflow system.

A new repository has also been introduced in Figure 7.7 to store notes about forecasts. It is used during discussions in interactions i1 and i2. This enables the notes to be used to record ideas and record how alternative designs can satisfy market needs.

17.6 PHASE 3—SELECTING COMMERCIAL NETWORK SERVICES AND PLATFORMS

Once logical requirements are defined, selection of specific commercial network services can begin and their integration into platforms defined. The aim here is to propose the best groupware services for a system and ways of integrating them into a platform. These may be any of the network services described in earlier chapters, and include

- E-mail: This can be used to broadcast or collect materials;
- Bulletin boards: For supporting interpersonal relationships;
- Newsgroups;
- Discussion databases;
- Document systems: For keeping track of documents;
- Videoconference link: This can be used to broadcast or in meeting situations;
- Sketch pad: To draw diagrams that synchronously appear at more than one location;
- Calendar systems: For maintaining appointments.

There is of course a variety of other services provided by networking systems.

The process can become quite ad hoc unless some systematic approach is used to keep track of potential solutions, given the large number of available services. The systematic approach is needed to ensure that selection focuses on process improvements identified for the project. It must also ensure that design concentrates on choosing those services that have the potential to improve the key collaborative factors, rather than those that are technologically interesting. Figure 17.8 suggests a set of steps that can be followed in the selection.

The first step is to list possible network services against the interactions. Then, the most useful services are selected as candidates for justification, applying

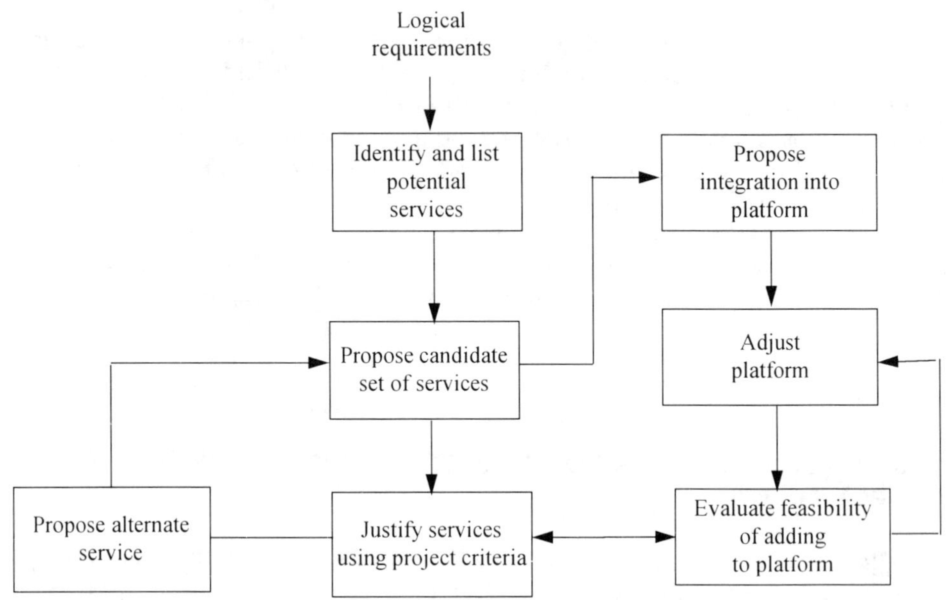

Figure 17.8 Physical design.

broad criteria. Thus, for example, some may be rejected because they do not conform with current enterprise standards or are just obviously too expensive. Once a candidate set is identified, each service is evaluated as an individual service to see its impact on the process and how it helps to achieve the original objective. At the same time, a possible platform can be proposed and a second evaluation made to see whether the initial evaluation changes once additional factors that come up in integrating the service are identified. If not, acceptable further adjustments can be made to services using the initial candidate list.

17.6.1 The Selection Process

The selection process to some extent already commenced during requirements specification. This identified where bulletin boards or discussion databases are needed, or what information should be stored in documents.

The process of selection then becomes one almost of exhaustion. It is to look at the potentially endless list of tools and match them to each interaction. For example, where a bulletin board is needed; whether it should be on the WWW, using one of the available systems; whether it should be specially designed or LOTUS Notes used instead. Because of the potential size of this task, it is often useful to identify the most critical interactions, derived from strategic requirements, and start with them. Similarly, it is useful to identify potential core technologies here and start with services provided or easily integrated with these technologies. Some options may then be discarded because they do not fit with this platform.

Again, one of the best ways to start selection is by examining interactions. The framework that distinguishes between ways of communication can again be useful here. Thus, where there is a need for information exchange, the kinds of services described in Chapters 8, 9, and 10 may be useful. The WWW, for example, can be a candidate for the wide distribution of general documents, whereas a document management system may be considered necessary where there is joint work on developing documents.

A systematic way to record potential services is proposed here and illustrated in the following example.

17.6.1.1 Case Study: Selecting Services for the Specialist Designer

Selection of services for the designer can commence once a set of specifications are available. These specifications were shown in Figure 17.7. The designer now begins to build in a table like that shown in Figure 17.9.

The first column in the table corresponds to either an interaction or an activity in analysis and, together with the numbers in the second column, can be used in verification. The top horizontal row identifies the services, whereas the way that a service is used is described by the references (labeled as xi) in the intersection of a row and column. Thus, for example, the following entries would be made

- x1: The WWW is used to get easy and quick access to market analysis results.
- x3, x5: The documentation system is proposed for keeping track of changes and alternative designs through automatic versioning.
- x7: Discussions between designer and assistant can be supported by videoconferencing with access to design documents, and the ability to make notes.
- x9: Videoconferencing expedites discussion with contractors (but only if they have this facility).
- x13: A workflow system based on LOTUS Notes is proposed to monitor contractors and distributors by using documents to describe agreed tasks and minitoring them using views.

The table is built up gradually as ideas develop and is useful, in fact, in assisting people to identify possible services. Each identified interaction and activity must appear in this column. The table also identifies the level of service integration needed. Checks are made to make sure that all interactions are included in the selection.

Interaction / Service		WWW	E-mail	Docs	Bulletin board	Video conferencing or face to face	W/F	Integration requirements
Name	No.							
Discuss forecast	I1	X1	X2					Need to view forecast in discussion
Design to forecast	I2			X3		X4		Access to alternative designs needed
Discuss change	I3			X5	X6	X7		Ability to change or make notes
Negotiate contract	I4			X8		X9		Access to latest contract needed
Monitor contract	I5		X10	X11	X12		X13	Ability to keep track of documents
Advertise	I6	X14	X15					
Chose design	I7			X16		X17		Ability to examine various designs
Place order	I8	X18			X19			Consider direct entry to order file

Figure 17.9 Identifying services.

17.6.2 Evaluating Services

Structured rationale can be used to choose the most appropriate of the candidate services to satisfy the interaction needs. Support tools are proposed to structure the design rationale. Possibilities here include the following:

- A chart of process gains and losses and defining of the design criteria in terms of gains that can be made to the process. Generic guidelines for specifying such gains and losses have been given earlier in Chapter 4, but must be converted to the concrete terms of the application.
- Matching the services to process gains to select the most appropriate tools. *A design summary table* is proposed here to include arguments for selecting particular tools. These arguments are made by expressing how each tool meets the criteria using evaluation criteria.

17.6.2.1 Evaluation Criteria

An approach that is useful here is that of evaluating the effect of each service against the process. The process losses gains described earlier in Chapter 4 and illustrated in Figure 4.10 can be used as possible criteria in evaluating services. One guideline that comes from the work of Nunamaker and others [1] is to focus on processes and tasks and see how any services:

- *Improve the process structure*: This defines changes to the steps to be followed by a process and how they can be improved. It can include adding new roles like a facilitator to smooth workflows or simply removing unnecessary steps.
- *Provide better process support*: This provides better support for a process. Examples may be quicker transfer of information or the ability to communicate better across distance. It may also include better scheduling of process tasks by using an appointments system. Improvements to awareness of the process may also come under this heading.
- *Provide support for task structure*: This is to help structure the information in a task. Examples are the IBIS structure for arguments in decision support, keeping track of documents, and so on.
- *Provide better task support*: This actually makes it easier to carry out a task (for example, better access to information, recording of results, and so on).

17.6.2.2 Justifying Services for the Specialist Designer

The table in Figure 17.10 justifies the choice of services shown in the table in Figure 17.9. Thus each interaction is examined in turn to choose the best service to

Activity: Design

Service		Gains				Losses	
Name	Ref	Information Access	Improved Awareness	Group Memory	Improved Communication	Cost	Distraction
WWW	I1	Anytime access	Through advertising				
	I6, i7,i8	Facilitates advertising			Customers can view designs at any time		
Doc. Mgmt.	I2, i3			Simplifies tracking design records			
	I4, i5			Simplifies task of keeping contracts			
	I7			Easy access to alternate designs	Designs quickly made available to customer		
W/F.	I5		Easier to track production				High cost of development
Video Conf.	I2, i3					Not needed as work is at same location	
	I4					Most contractors don't use it	
	I7, i8				Easier to describe tricky issues	High cost	

Figure 17.10 An evaluation table.

support that interaction including the improvements that can be made to support personal relationships.

In Figure 7.10, the interactions are grouped for each service. The table entry then specifies the benefits or losses of each service for the interaction using selected evaluation criteria. The actual choice has then look back at the project's original objectives. Thus, videoconferencing may not get high priority because of its high cost and limited application, especially in customer contacts. It may be used for discussions with the contractor, but only if the contractor is willing to install such a system.

Each service is evaluated against the interactions for its contribution to process gains and losses. The evaluation table assists the evaluation. Here, rows correspond to the components in the model together with identified issues. Next to each of these are the alternative tools chosen for the evaluation. The columns describe the evaluation structure. The process selects each interaction and looks at the alternative technique for implementing the component. It evaluates whether the service improves the process structure or support, and task structure and support.

The arguments in the table must be *directly related* to the process gains and losses, with a tool justified only if it contributes substantially to the process. The arguments must also satisfy specific evaluations; for example, the service *improves* awareness of latest customer status or it results in *better coordination* between client and salesperson.

Process evaluations must also consider the general social effect of computer-supported collaboration. Important question are effects on the level of satisfaction and consequently contribution to the group. The acceptance of such work in organizations is also important. Possible outcomes to measure are

- Direct measures such as travel time and time to do a task;
- Improvements to output quality;
- Reduction in internal tasks such as making copies, sending faxes, and other minor jobs;
- Reduction in the number of meetings;
- The degree of satisfaction.

In general, the expectation is that task support and structure would apply to the interaction, whereas process support and structure would apply to the activity. The table, of course, only indicates the suitability of particular tools. Decisions to adopt a particular tool must still be made. These will be made on cost grounds and the preferences of intended users.

17.6.2.3 Identifying Risks

As well as identifying advantages, potential risks must also be noted if a particular service fails to get acceptance. In our case, the risk may be that spending time in choosing and learning to use the system may detract the designer from her work and lose business. Easy learning and quick adoption will minimize this risk. Alternatively, this risk may be minimized by introducing any system gradually, aiming at one interaction at a time.

17.6.3 Designing Platforms

After services are identified, the next step is to integrate them into a platform. These must be combined into a platform that supports the processes in the enterprise. The following steps are usually followed:

- Step 1: Identify the critical services.
- Step 2: Choose the core technology for these services.
- Step 3: Identify services to be added to the core technology.
- Step 4: Evaluate the feasibility and return to step 2, if needed.

Two main goals are seen to be important for the *specialist designer*. One is to determine the impact on external contacts, in particular, advertising and ordering, and the other is document management to keep track of designs. It could well be said that the recommendation may be to begin to use the WWW for communication, as it services many of the interactions. The WWW would be of greater value if it were integrated with a document management system to keep track of designs and thus better realize the objective of quickly getting access to information and making it available to customers. The WWW also gives the opportunity to advertise to customers and also to get indications of customer interest. It can also be integrated with design documents, providing a seamless transition from design activities to customer activities.

Support for monitoring external contractors may be delayed as it is not likely that most will have workflow support. However, it may be worthwhile to consider the idea of using e-mail or even forms on the WWW to support contractor input. This will probably rise in priority as the number of contractors grows. Eventual implementation will require careful choice of the document management system and its specific integration with the WWW, possibly needing the development of CGI modules. It will also be necessary to provide a customized interface, often a multimedia interface, that allows users to select any of the services and also to move from one service to another. A possible extension then is to add forms for use by external contractors to log progress of their work. This may require the development of a CGI program both to gather the inputs and integrate the inputs with the internal databases that store records about contracted work.

17.7 SUMMARY

This chapter described a way of designing CSCW systems using design methodologies. It described a methodology that begins by modeling the systems and describing its collaboration requirements in terms of critical factors relevant to the mission. The next phase is to identify the platform services that will satisfy these requirements. The platform is then developed in an evolutionary way by gradually adding the identified services.

17.8 DISCUSSION QUESTIONS

1. Why is it necessary to identify problems with respect to the mission?
2. Why should these problems drive the design process?
3. What are the advantages of using a systematic approach in design?
4. Why is it necessary to redesign group relationships?
5. Describe some repository services that are used solely to support interpersonal relationships.
6. What do you understand by process task and structure?
7. Do you find the design table in Figure 17.9 useful for rationalizing the choice of tools?

17.9 EXERCISES

1. Identify some process gains and losses for the virtual university and suggest the services needed to obtain the gains. What are the risks in providing such services?
2. Design a set of WWW pages to allow the specialist designer to keep track of contractors.

References

[1] Nunamaker, J. F., et al., "Electronic Meeting Systems to Support Group Work," *Communications of the ACM*, Vol. 34, No. 7, 1991, pp. 40–61.

Selected Bibliography

Hawryszkiewycz, I. T., "A Design Method for Choosing Services for Large Distributed Teams," *Proc. Second International Conference on Design of Cooperative Systems, COOP'96*, Juan-Les-Pins, France, INRIA, June, 1996, pp. 515–533.

Chapter 18

Future Networking Trends

LEARNING OBJECTIVES

- ❑ *Evolution or strategic introduction*
- ❑ *Seamless platforms*
- ❑ *Agents supporting platforms*
- ❑ *Interface requirements*

18.1 INTRODUCTION

That enterprise networking will become a dominant method of work is almost inescapable. The growth and need is there—the ever-shrinking globe will make it inevitable. There is little doubt that the use of computer communications will grow, and probably grow very quickly both between and within organizations. It will grow in the interchange of information, in interpersonal communication, and in supporting work processes. There will at the same time be greater pressure to provide technology platforms that match the variety of enterprise needs and to make the platforms adaptable to changes in these needs.

People in enterprises will be continuously calling for better and better communication services to support them. This in turn will call for ways of combining the many technologies into ever more powerful networks to provide these services. In most of the earlier chapters, the book made suggestions as to the kinds of services that will evolve and how work in enterprises will change. It is this relationship between organizational structure and technology that will become important, as they both drive each other. The intranet will most probably become the dominant direction where enterprises wish to improve internal processes. At the same time, as links between enterprises will no doubt increase, so will the use public networks; in particular, the Internet. Ultimately, this will become a combination of private intranet and public Internet that will become common, where internal operations are supported by the private network whereas the external interface is through the public network.

The definition of the term intranet will in itself probably evolve. Initially, the definition to most people means Internet technologies provided for exclusive use by members of one enterprise. But what if the Internet technologies continue to both diversify and integrate with others? The definition will thus go beyond limiting itself to Internet technologies to include any groupings of technologies collected into a network to support the networked enterprise. The rest of this chapter uses this wider definition of the intranet.

18.2 THE MAJOR PARAMETERS

This chapter sees three major parameters playing a major role in defining this evolution. These are objective, level of connectivity, and kind of communication:

- The *objective* defines the enterprise role played by the communication network, in particular whether the goal is to use communication for internal, consumer, or business networking.
- *Level of connectivity* is the support given to these groups as defined in Chapter 2, ranging from basic communication for information exchange, through generic collaborative services, to a seamless level that provides work-specific contexts.
- *The kinds of communication* cover the variety described in Chapter 7, including services to exchange information, to support interpersonal relationships, and to support specific workflows (including integration with corporate legacy systems).

Figure 18.1 summarizes some of the macro trends that are now appearing. The major directions that are now becoming evident are

- Internal support, usually starting with basic connectivity by providing e-mail, which is gradually being extended to provide support for more specific workflows. This results in the growth of intranets within organizations, usually based on platforms from one vendor, such as, for example, LOTUS Notes.
- A growing range of consumer services based on the Internet or other public networks, or other private networks, can be expected to grow to gain access to an increasing number of consumers.

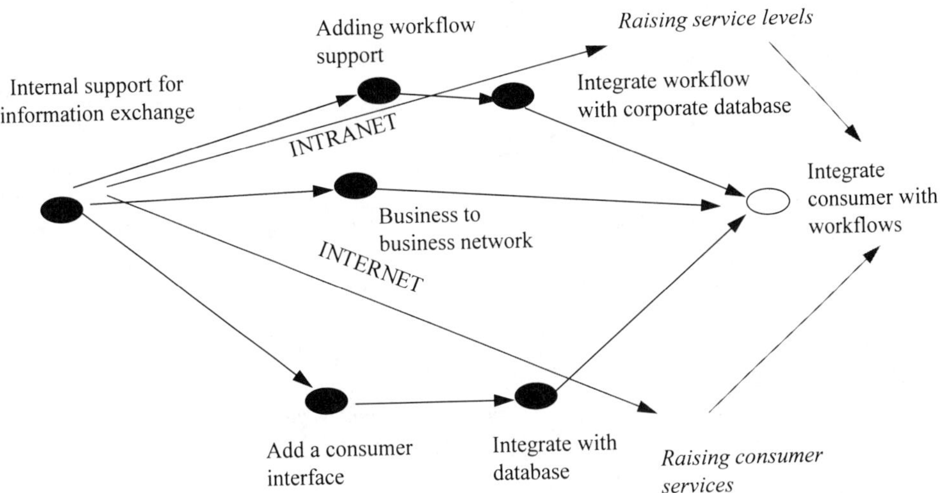

Figure 18.1 Trend scenario.

- A growth of business to business communication, starting with support for interpersonal communication but extending to business processes as business alliances are formed.

Figure 18.1 also shows a convergence of these three directions towards greater integration of the services with the corporate database, both with the internal and the consumer support services, leading eventually to an integrated network.

The networks will provide increasing support to teleworkers and external contractors, mainly connected to the intranet (but ways of doing this are still evolving).

There is also a growth of business networking arrangements. These are more likely to be based on the Internet initially, but again evolving to an intranet, perhaps provided by an external provider. Just like with internal operations, these are more likely to initially concentrate on interpersonal relationships and then move into more task-specific areas, finally being integrated with internal business processes.

The trend described above will itself have an impact on the development of supporting technologies with emphasis on:

- Gradual evolution with growth of value-added communication services, most often based on a public network, as they are needed, but not stressing integration in any way.

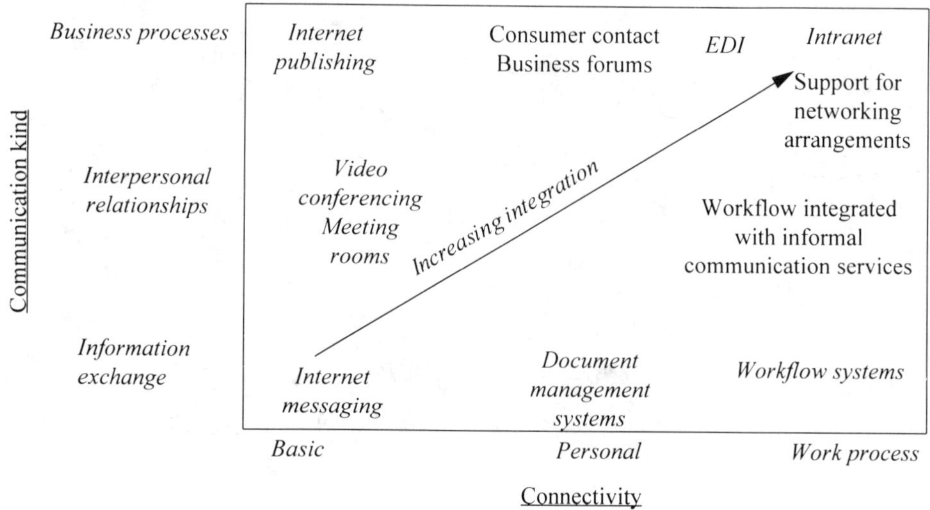

Figure 18.2 Growth of value-added services.

- The development of an intranet. initially providing level 1 services but then gradually evolving to higher levels of support, based on a single technology platform.

Such trends will require enterprises to develop appropriate technical strategies.

18.3 TECHNICAL STRATEGIES

Networking requirements can perhaps be best split into the two levels (first outlined in Chapter 7): networking services and collaborative services. Another way to look at this is as the network infrastructure to connect the community of users, and the value added services to support collaboration, including workflow, document, and conferencing services, without necessarily integrating them.

The networking services are gradually reaching a point where it is possible to connect most vendor systems through the increasing use of standards, the best known of which is TCP/IP. It is, however, the next level of service that will exhibit the greatest volatility in the next few years. This can perhaps be illustrated by looking at the relationship communication types and the level of connectivity that they will need. This is illustrated in Figure 18.2.

Here, at level 1, there are now a large number of candidate services, with Internet services for information interchange, videoconferencing for interpersonal relationships, and the use of WWW publishing for business processes. Level 2 calls for more sophisticated personal services, such as joint document management, the development of interactive bulletin boards, or discussion forums such as those described in Chapter 11. Then, towards level 3, these services are increasingly integrated. Such integration raises the question of standards to make it possible to easily connect value-added services into seamless platforms with such level 3 interfaces.

18.3.1 Choosing the Platform

Will there be standards that will simply allow us to plug any two services together? Such standards will be essential to make it possible for networks to be quickly set up and to have platforms constructed from a number of services. It is hard to envisage widely adopted standards within the next three to five years, given the rapid evolution of the technologies. The field itself is not yet clearly defined enough to even start to define the standards.

In the absence of such standards, two directions have been identified here in Chapter 7—using middleware or basing integration on a core technology. For intranets, one can perhaps envisage a most likely strategy where a core technology is adopted and augmented with middleware to give it an ever-growing possibility of linking to a variety of products. This is certainly the trend with the two main contenders for core technologies at the time of writing, those being the WWW and LOTUS Notes, which provide complementary services. Apart from there being an interface between the two, there is an ever-growing list of value-added products to link them to other services. Thus products to link the WWW to database products are now growing exponentially. So are services that can be added to LOTUS Notes.

Will the emphasis be on the Internet or will there be a growth of intranets? Perhaps the safe prediction here is that most organizations will prefer the Internet for consumer networking, at least in the very early stages. Internally, the evolution will be towards intranets. What technical platform will dominate the latter? This is a very interesting question, with major discussion centering on the LOTUS Notes versus WWW technologies. It is now apparent that what will happen is a gradual merging of these platforms to integrate their complementary services, perhaps together with other platforms.

18.3.2 Networking Platform Requirements

Successful introduction of the intranet will require platforms that are seamless and support gradual growth. Such platforms must introduce services in an evolutionary manner to allow their users to gradually learn ways of using the new technologies while at the same time building up their services in an integrated manner. Such platforms will have to emphasize interfaces to support the ever-growing number of users who are not computer specialists. They will also see a growing introduction of software agents to assist these users.

One important question is whether platforms will have to be customized to different role responsibilities within the enterprise. Furthermore, will the roles themselves have to be able to customize platforms as work patterns change? The latter approach will have to be favored, as it will become easier for enterprise workers to change the platform themselves as their work evolves, rather than expecting networking experts to do so.

A likely development is to integrate communication services into operating systems and provide them as a standard feature of desktop systems on personal computers. In that way, the operating system will take on many of the features of the WWW, perhaps initially through applets moving between the two. But this means that the WWW will no longer be predominant and may in effect no longer be needed, as the operating system will gradually take on many of the features of the WWW and their integration will make them look like one system.

18.3.3 Interfaces

The third important development is the interfaces developed to support groups. A generalized interface will usually begin with a user on signing on to obtain information about the latest changes to the context, accessing any artifacts relevant to the change, and then taking action depending on the state of the context. The services at the interface are those provided by the generic services platform together with support for combining these services into higher level applications.

The second important interface issue is how to maintain awareness and interpersonal relationships across distances. This will require considerable research into both presentation of other people's activities as well as determining the balance of synchronous and asynchronous communications needed to maintain effective working relationships.

Careful development of the intranet to align it with enterprise needs will provide the interfaces through which technology is made available to users. Furthermore, services provided by the intranet will change the way that work will be carried out in enterprises.

18.3.4 The Growing Importance of Security

In addition, networking with consumers through the Internet will continue to make security a prominent feature of development. This will place heavy emphasis on the balance of networking between public networks, especially the Internet, and private intranets. In the short term, it is expected that security concerns will accelerate the trend to greater emphasis on intranets and secure providers, with the Internet primarily used for consumer access (although gateways between the two will be a major feature). Support for new work practices (especially teleworking) will probably play an increasing role in networks, further supporting the need to integrate the Internet—with its wide access to such workers—to internal enterprise intranets.

18.3.5 Software Agents

Much of the work of supporting platforms will be delegated to software agents. There have been suggestions that such knowledge about using platform services be included in software agents that can adapt platforms should user responsibilities, locations, or tasks change. This will continue to be a growing research area.

18.4 DEVELOPMENT STRATEGIES

That enterprises will change in the future is almost inevitable. What is important here is how computer communications will be part of this change. Ultimately, one may see people in enterprises using secure networks to easily establish connections to discuss issues related to their work. One important question here is whether such networks will gradually infiltrate networked enterprises or whether strategic decisions will be made to provide cooperative services throughout them. To date, most such systems were introduced gradually almost in an experimental manner. The promise of intranets based on the WWW tend to favor a more strategic approach once the advantages of networking become clearly identified. Enterprises that wish to provide support for teleworkers or mobile workers will need to make strategic decisions on how to provide such support. The trend to electronic commerce will also tend to favor a more strategic approach.

Given considerable investment in existing systems, the trend to networking at the enterprise level will only eventuate through the re-engineering of existing systems. One can thus imagine a scenario where the following occurs:

- Organizations learn about workgroup computing through new applications;
- Use this knowledge to re-engineer existing systems.

18.4.1 Re-engineering

Business is now going through a relatively important phase of re-engineering business processes in consumer-led rather than function-led culture. The need here is both to increase integration of existing legacy systems and to make systems more adaptable and open to adapting to changing consumer requirements. Workgroup computing can be expected to have a significant role here, as such adaptability across functions will require considerable collaboration between functional units.

The question comes up whether such re-engineering requires rewrites of systems or whether there is a way of using existing legacy systems.

There is an approach of integrating legacy systems into cooperative paradigms through gates to objects in an integrative interface. Thus a metaphor such as a client-server interface can be placed on top (or front-end) legacy systems. This calls for changes where the legacy system must conform to the gate specifications of the integrating metaphor. The degree of change depends on the functionality to be provided at the interface. For example, the ability to move information between legacy systems may require major changes, whereas simply making data available at the interface may require only minor changes. In the latter case, integration may be through a new application.

18.4.2 Meeting Social Needs

So far, this chapter has concentrated on technical developments. However, throughout the book has emphasized that successful systems must match technical and social needs. From this perspective, one can suggest that successful systems will result only where development strategies are aimed at meeting some of the principles identified in Chapter 3. An approach of building communities of practice will no doubt prove beneficial. It is often worthwhile to start with a small but committed community and, once adopted, extend growth to wider communities.

The ultimate result may perhaps be the *virtual organization*. Here, we have an organization that is formed for a short period of time to meet some goal. It may include representatives from physical organizations whose only mission is to provide the services needed by rapidly changing virtual structures. Such organizations or enterprises will arise as support for interpersonal relationships increases and is further integrated with business processes.

18.5 WHO WILL PROVIDE THE SERVICES?

Provision of services can also be expected to evolve. So far, this chapter has assumed that intranets will be provided by the organization's personnel, whereas private networks will have their own staff. But who will service the new communities of loosely collected entities that need support—the sports associations, industry groups, and so on? New provider models are expected to emerge. So far there has been a growth of Internet providers that provide the network connection, but not the customized services needed by the community. One possible model for such support is shown in Figure 18.3. It includes the providers of basic technical services, the Internet and its services, LOTUS Notes, or some yet new core platform. The user community is an identified group that is part of a loose enterprise or organization relying on funding from individual community members. There is then a facilitation group that identifies the community needs, obtains licenses from the provider for the use of a subset of services, and markets these services to the com-

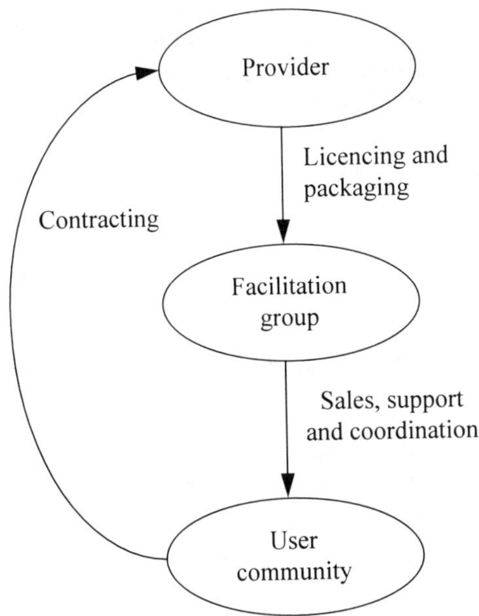

Figure 18.3 Supporting disparate communities.

munity. Contracts and fees collected by the provider are then used as a basis for royalty payments to the facilitation group.

Within the framework of this chapter, the facilitator group will be the group that will raise the level of service through levels 2 and 3 to support the community work practices. In that respect, it will need to educate the community in networking possibilities to enable it to fully exploit its capability. This task is expected to be long and gradual, as learning to use the technologies will need to compete with the day to day activities of community users. This model can also be applicable within enterprises that set up a facilitator group to set up service platforms for particular groups within the enterprise, also requiring approaches that facilitate gradual learning and integration of services into everyday work.

18.6 SUMMARY

This chapter described networked enterprises developing in ways where they are integrated more and more closely into the everyday communication processes of people in the enterprise. They go beyond simple information exchange, and more and more into seamless platforms that can be easily adapted to changing work practices. The chapter sees the evolution of platforms to support such integration.

Platforms will most probably be based on core technologies and supplemented by middleware products. Communities will grow around these networks, and legacy systems will need to be redeveloped towards cooperative work, based on ever-growing communities of practice. Finally, the chapter commented on who will develop these systems. Again, the likelihood is that development itself will be networked, including service providers, marketers, and those who customize systems to enterprise needs.

18.7 DISCUSSION QUESTIONS

1. Why is evolutionary development a more likely scenario for the introduction of cooperative systems?
2. Will work process supports dominate the intranet?
3. What trends would you envisage for interpersonal communication?
4. What will be the major requirements of interfaces in the future?
5. Why is integration with the consumer likely to become a high priority?

Glossary

Activity	A number of interactions that combine to achieve an enterprise outcome
Actor	A person or program that undertakes a role
Alliance	A joining of two entities to carry out a common goal
Artifact	An object that stores information
Asynchronous	Contributing to a conversation at different times
Awareness	Knowledge about the state of things
Brainstorming	Spontaneous development of ideas
Business process	The way an organization carries out its business
Chaos theory	A theory to describe behavior in unpredictable environments
Concept	A term with a specific meaning used in a language
Connectivity level	Kinds of communication supported
Concurrent engineering	A way of speeding up production by carrying out closely related tasks at the same time
Context	The work situation including roles, artifacts, and policies at their current status
Conversation	Interchange of information between participants
CSCW	Computer-supported cooperative work
	Expected way of doing things
Empowerment	The placing of responsibility (usually on a group)
Electronic mail (e-mail)	Exchange of messages using computer
Electronic meeting room	A special room that supports meeting with computer services
Facilitator	Role that provides services to enable a group to work together
Generic service	A service that can be adapted to a particular class of need

Goal	Defines what must be achieved
Group	A number of people with the same interests
Group memory	A history of actions and reasons for taking the actions
Innovation	Finding better ways for achieving goals
Interaction	A group of people working together on a task
Interface	Display that allows a user to interact with the computer system
Lean organization	An organization concentrating on its main business
Language	A way of speaking about a domain
Legacy system	Existing systems built using earlier technology
Meeting	A conversation between people in direct contact
Mission	A broad statement of organizational aims
Mobile worker	A worker who moves from one location to another
Notification schemes	Ways of keeping people informed about system changes
Platform	A set of services made available at the computer interface
Private workspace	A workspace that can be seen by only one person
Process	The way that things are done
Process structure	The set of steps used in a process
Process support	Methods used to facilitate the process
Repository	A collection of artifacts with methods to access and transform them
Role	A responsibility within an enterprise
Seamless	Connections made transparently of operating system
Service	A way of communicating within supporting groups using computer technology
Shared workspace	A workspace that can be seen by many people
Situation	The current state of the context
Synchronous	Contributing to a conversation at the same time
Task	What must be done
Task structure	The way that a task is carried out
Task support	Methods used to support the way a task is carried out
Team	A number of people working to the same goal
Teleconferencing	A meeting conducted via telephones
Telework	Work carried out at a remote location
Tool	An object that can be used to transform an artifact
User	A person using computer facilities
Videoconferencing	A meeting conducted via a video link
Virtual	A logical view of the world that may not physically exist
Visualization	Providing a natural view on a computer screen
Workflow	A set of steps used to carry out an activity
Workspace	Where users record the results of their work
Work practice	Ways in which people carry out a task

About the Author

Igor Hawryszkiewycz is Professor of Computing Science at the University of Technology, Sydney. Prior to that he was at the University of Canberra and received a number of visiting appointments in the United States, Germany, and the United Kingdom. He completed bachelor and masters degrees in electrical engineering at the University of Adelaide and a Ph.D. in Computer Science at MIT. His early research was the design of databases, which resulted in numerous publications and one of the earliest textbooks on database design in 1980. Since then he has published two more books on database and information systems design. Since 1987, his work has focused on cooperative work, especially on methods for designing systems that support distributed groups. He has applied this methodology to a number of real-life problems, resulting in a number of publications on its application to real problems. His work has always emphasized application of research results to industry and he has consulted to a number of government departments and industry, as well as developing successful public courses.

Index